APPLICATIONS

OF Electrical Construction

THIRD EDITION

APPLICATIONS
OF Electrical
Construction

THIRD EDITION

Robert K. Clidero
Kenneth H. Sharpe

Irwin Publishing
Toronto, Canada

First Edition published 1975
SI Metric Second Edition published 1979
Third Edition published 1991

Edited by Kate Revington
Designed by Jack Steiner Graphic Design
Typesetting and illustrations
   by Trigraph Inc.
Cover photograph by Birgitte Nielsen

Canadian Cataloguing in Publication Data

Clidero, Robert K.
   Applications of electrical construction

3rd. ed.
Includes index.
ISBN 0-7725-1719-3

1. Electric engineering.   I. Sharpe, Kenneth H.   II. Title.

TK452.C55 1991   621.3   C91-093548-3

ISBN-13: 978-0-7725-1719-7
ISBN-10: 0-7725-1719-3

6 7 8 9   09 08 07 06

Printed and bound in Canada

Published by
Nelson
1120 Birchmount Road
Toronto, Ontario M1K 5G4
1-800-668-0671
www.nelson.com

Contents

Preface

As classroom teachers, we are aware of the need for student textual materials that explain the non-mathematical theory behind electrical devices and equipment.

Drawing on our experience as journeymen electricians and high school teachers, we have aimed, in this text, to explain as simply as possible modern electrical products and their applications in electrical circuits. We have also tried to downplay the use of complex technical terms and to emphasize the reasons for electrical installation practices. This *Third Edition* has been given extensive review not only by us but by the manufacturers of illustrated products. Always, our intent has been to provide readers with the most recent and up-to-date changes in product design.

Many chapters have been lengthened by the insertion of new product and technology information. New illustrations provide a clear understanding of recent changes in both technology and wiring codes. Also, a new chapter (Chapter 22) provides readers with a fuller understanding of the specialized tools used in the electrical industry.

Since the electrical industry has not yet adopted the SI (metric) system of measurement, the commonly used imperial measure tables appear with metric equivalents throughout the text. The text accurately reflects an industry in transition.

It is our hope that high school students, apprentices, electricians, and the general public will find this text a useful tool for understanding how electrical theory is translated into practical terms.

We would like to thank all those persons and organizations whose co-operation made this text possible.

Robert K. Clidero
Kenneth H. Sharpe

1

The Three Wire Distribution System

Electrical power is supplied to the home by the local hydro utility. Because there are so many appliances available today, two voltages are needed. Lighting and receptacles for such *small appliances* as radios, toasters, teakettles, frying pans, and electric drills require a *120 V* supply. *Large appliances,* such as electric stoves, clothes driers, some air conditioners, and electric heaters, operate on *240 V.*

On the North American continent, a frequency of 60 Hz (cycles) is the standard. The *frequency* of an alternating current system is the number of pulses of current that pass along the conductors in one second. On a 60 Hz system, there are 120 pulses per second. One positive and one negative pulse make up one *cycle*, or *hertz*. (See Fig. 1.1) Residential voltages operate on a *single-phase* system. Both single-phase voltages come from a transformer that has a single primary winding.

Obtaining Two Voltages From Three Wires

When two dry cell batteries of 1.5 V each are connected in series, 3 V are obtained. (See Fig. 1.2) By connecting a wire midway between the two cells, a 3 wire system providing two voltages is created. The distribution transformer

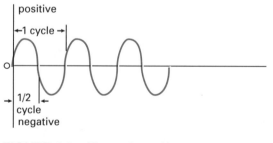

FIGURE 1.1 Three alternating current pulses

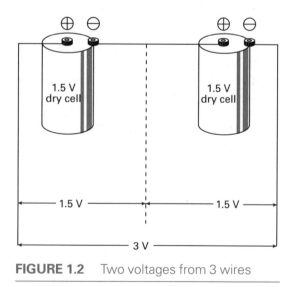

FIGURE 1.2 Two voltages from 3 wires

used to supply a house or group of houses is wired in the same manner.

Distribution Transformer

The local hydro utility uses a series of transformers to lower its voltages in stages from the power station to the residential street. Each locality may differ slightly in the actual voltage that arrives at the distribution transformer on the street. One common voltage is 2400 V. (See Fig. 1.3)

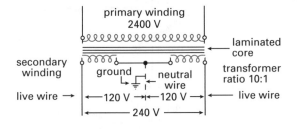

FIGURE 1.3   Distribution transformer circuit diagram

High-tension (high-voltage) *lines* supply the primary coil of the transformer with *2400 V*. This transformer reduces the voltage on a 10:1 ratio, giving a *secondary* output of *240 V*. A wire is connected to the midway point of the secondary winding, dividing its 240 V in half. This middle, or *neutral*, wire provides two voltages on a 3 wire system.

The two outer wires of the secondary winding are known as the *live* wires. It should be noted that in a residential wiring system, the *neutral* wire is *white* or *grey* in colour, while the *live* wires are usually *black*. For safety purposes, the *live* wires are switch controlled and have fuses connected in series with them. The *neutral* wire (also called the *grounded, identified conductor*) is grounded (connected to the earth) at

the transformer and in the residential *main switch* box.

Residential Overhead Supply System

Figure 1.4 shows the method used to supply power to many of the houses in a community. In some areas, a fourth wire is brought from the hydro pole to the house for the purpose of supplying a flat-rate, hot-water heater system. (This is covered in more detail in Chapter 15.)

Residential Underground Supply System

Modern community planning has led to the development of an underground supply system in which wires and cables are placed below ground. The main advantage of this system is to give the community an uncluttered look. Figure 1.5 shows one variation of this type of system.

Switching the Live Wire

On reaching the house, the hydro supply lines pass through the *kilowatt hour meter*. (The amount of power used is recorded on this meter in kilowatt hours.) The supply lines are then carried into the house by a *conduit*, which takes them directly to the *main disconnect switch*. As Figure 1.6 shows, the supply lines are connected at the top of the switch, which is standard practice. The upper portion of the switch is called the *line side*. Most manufacturers have the word *line* printed somewhere near the line terminals.

Should an electrical emergency arise, turning this switch off will disconnect the power supply to all parts of the house. Power may be restored by turning the main switch on again. Once the

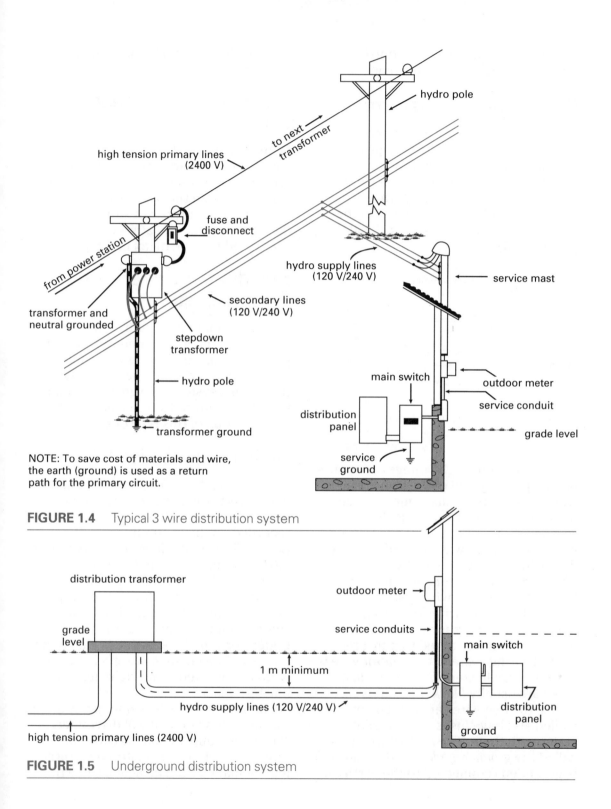

FIGURE 1.4   Typical 3 wire distribution system

high tension primary lines
(2400 V)

to next transformer

hydro pole

fuse and disconnect

from power station

hydro supply lines
(120 V/240 V)

service mast

secondary lines
(120 V/240 V)

transformer and
neutral grounded

stepdown
transformer

hydro pole

transformer ground

main switch

distribution
panel

outdoor meter

service conduit

grade level

service
ground

NOTE: To save cost of materials and wire,
the earth (ground) is used as a return
path for the primary circuit.

distribution transformer

outdoor meter

grade
level

service conduits

main switch

1 m minimum

distribution
panel

hydro supply lines (120 V/240 V)

high tension primary lines (2400 V)

ground

FIGURE 1.5   Underground distribution system

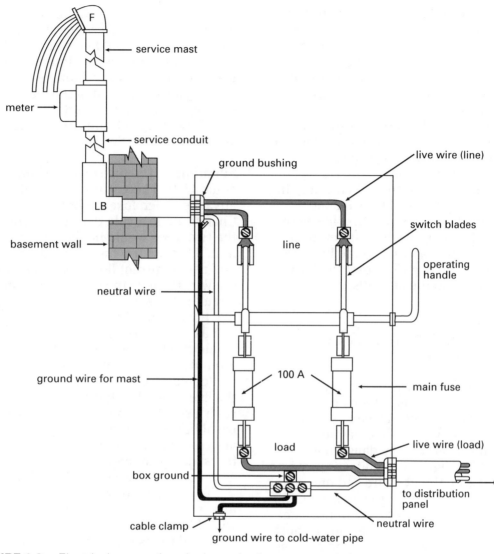

F

service mast

meter

service conduit

ground bushing

live wire (line)

LB

switch blades

basement wall

line

operating handle

neutral wire

ground wire for mast

100 A

main fuse

load

live wire (load)

box ground

to distribution panel

cable clamp

neutral wire

ground wire to cold-water pipe

FIGURE 1.6   Electrical connections in the main disconnect switch

main switch has been *pulled,* or turned off, all contents of the main switch box are safe to handle, with the exception of the two terminal connections that receive the incoming hydro supply lines. Fuses may be replaced or checked, repairs to the switch made, and any of the wires at the bottom, or *load*, side of the switch may be handled in safety.

Fusing the Live Wire

The purpose of any fuse is to limit the amount of electrical *current (amperes)* that can pass through a given wire. If more current is passed through the wire than it is designed to carry, the wire will heat up and eventually start to burn the insulation covering it. A fuse is designed to heat up and melt *before* the wire it protects is damaged.

The fuses in the main disconnect switch are designed to protect those conductors going from the *load* side of the switch to the *distribution panel*. The *voltage rating* of the fuse is matched to the voltage of the supply system, while the *current rating* of the fuse is matched to the current carrying capacity *(ampacity)* of the wire.

The amount of current entering the house on one of the live wires must leave the house on the other live wire and/or the neutral wire. Therefore, a fuse is needed on each of the live wires. Figure 1.6, on page 5, shows the location of the fuses in the main disconnect switch.

Why the Neutral Wire Is Not Fused

When the 3 wire distribution system first came into use, a fuse was placed in series with the neutral wire as well as with the live wires. This practice was soon discovered to be dangerous.

A 120 V lighting circuit was protected by a fuse on the live wire and one on the neutral wire. Since both wires were the same size, both fuses were of the same current rating. If a fault causing a short circuit condition occurred and excessive current started to flow, the fuse would open the circuit and prevent damage to the wire. Since both fuses in the circuit were the same size, several things could happen:

Situation A. *Both* fuses could blow at the same time and prevent any further current flow. Both the live and neutral wires would be safe to handle, and repairs could be made to the circuit without fear or danger of electrical shock. (See Fig. 1.7)

Situation B. The fuse on the *live* wire could blow. Both fuses are rated the same, but there could be some small difference in their construction that would make one fuse weaker than the other, causing it to open first. If this happened, the wires would be protected by the current flow being cut: the circuit would be safe to work on. (See Fig. 1.8)

Situation C. The fuse on the *neutral* wire could blow. There is no way to tell in advance which fuse might be weaker. Chance alone would determine which of the two fuses would be the one to blow first. If the neutral fuse burned out first, no further current flow would damage the wires. (Since the current entering a circuit is the same as the current leaving, the neutral fuse could open the circuit as well.) The live wire, however, would still be intact. A person (if grounded) attempting to repair the circuit could receive a 120 V shock from the live wire. The chances of receiving such a shock are high, because all electrical boxes in modern wiring systems are grounded and because plumbing fixtures, damp concrete floors, etc., are common in the home. (See Fig. 1.9)

To eliminate this danger of shock, a fuse is no longer installed in the neutral wire. Modern service installations are designed to fuse only the live wires.

Grounding the Neutral Wire

The majority of electrical service parts, wiring boxes, conduits, and similar fittings are made of metal. Sometimes, wire comes in contact with metal boxes— perhaps insulation on conductors was damaged during the initial installation of the circuit or became worn over a period of years. A grounded system is, therefore, needed.

Applications of Electrical Construction

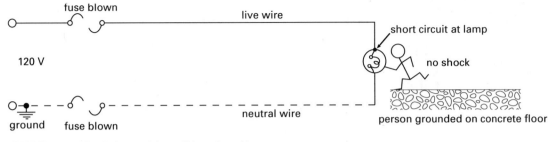

FIGURE 1.7   Both fuses blow (Situation A).

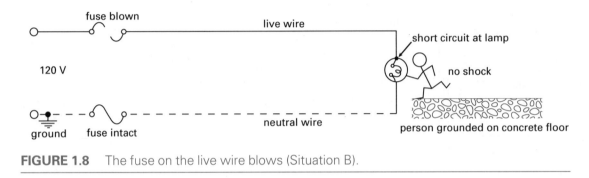

FIGURE 1.8   The fuse on the live wire blows (Situation B).

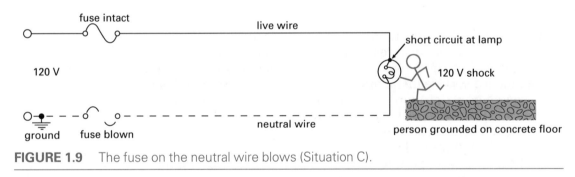

FIGURE 1.9   The fuse on the neutral wire blows (Situation C).

Service masts, which rise above the roofs of many single-storey houses, can be targets for lightning during thunderstorms. To prevent these metal conduits from attracting lightning, they and their boxes are grounded. Steel rods are driven into the earth or wire fastened onto cold-water pipes where they enter houses.

If one of the two live wires in the 3 wire system comes in contact with a grounded service box, many dangerous possibilities arise. Anyone can become grounded in a house. For example, if the second live wire of the 3 wire system comes in contact with the frame of a faulty power tool and if the first live wire touches the metallic service box, a grounded person could receive a 240 V shock—strong enough to kill. A person trying to fix a light over the kitchen sink could receive a 240 V shock by touching the sink and the second live wire in the fixture. Cleaning a laundry room fixture

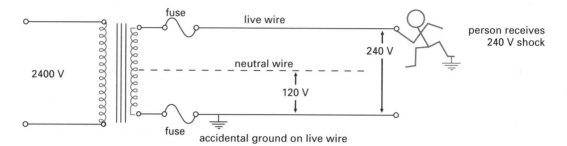

FIGURE 1.10   Accidental grounding of the live wire creates a 240 V shock hazard.

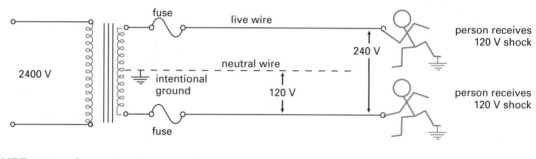

FIGURE 1.11   Grounding the neutral wire limits the shock hazard to 120 V.

with a damp cloth while standing on the basement floor might also result in a 240 V shock. (See Fig. 1.10)

In modern service installations, the main switch box is grounded and a *terminal block* is placed in the lower portion of the switch, where the neutral wire could also be intentionally grounded. With the neutral wire grounded, the maximum shock a person can receive from any one live wire is 120 V. This means that a *grounded person* working with power tools, repairing equipment, or cleaning fixtures can receive no more than the voltage between the neutral and a live wire, that is, *120 V*. Under certain conditions, this voltage can kill, but the danger is greatly reduced. (See Fig. 1.11)

If one of the live wires accidentally touches a metal box or fitting with the neutral wire grounded, a *short circuit*

results. The fuse on the live wire blows and protects the circuit. The neutral wire is safe to handle even when a person is grounded, since there is no voltage difference between it and ground.

Unbalanced System

When a distribution panel is installed in a house, some attempt is usually made to distribute the load *evenly* on each of the live wires supplying the panel. It is rare, however, for a dwelling to have the same number of lights or devices turned on for the load to be balanced on each side of the panel. The *neutral* wire in the system is important here, because it returns the unbalanced amount of current to the transformer.

If the neutral wire is broken or disconnected in any way, there is danger to the electrical equipment in the circuit.

Applications of Electrical Construction

Assuming that each individual load device draws the same current, the side of the system with the greatest number of devices turned on will have the greatest number of parallel circuit paths. In a *parallel circuit*, the more paths there are, the lower the total resistance of the circuit. The difference in electrical resistance between the two sides of the panel is determined by the number of devices operating on each side.

If the neutral wire is disconnected, there will no longer be a 120 V circuit. The two sides will now be *in series* with each other and have an applied voltage of *240 V*. Since voltage is divided in a series circuit, the side of the panel with fewer devices turned on (and, therefore, higher resistance) will receive a voltage much higher than normal. The other side of the panel, with its lower resistance, will receive the balance of the 240 V. Light bulbs will glow much more brightly with the increased voltage, and sensitive equipment, such as stereos and television sets, may be seriously damaged. When connecting service conductors, therefore, the *neutral* wire must be *disconnected last* and *reconnected first*, when power is to be restored.

The danger of an unbalanced condition is another reason for *never fusing*

the *neutral* wire. A blown fuse on the neutral wire will result in an unbalanced voltage situation.

As current passes through a circuit, each lamp forms part of the circuit and offers some resistance to the current flow. (See Fig. 1.12) Three factors are involved in current flow: line *voltage, amperage,* and *resistance*. The value of each factor is determined according to *Ohm's Law*:

$$\text{Current} = \frac{\text{Voltage}}{\text{Resistance}}$$

In terms of units, the Law can be expressed as follows:

$$\text{Amperes} = \frac{\text{Volts}}{\text{Ohms}}$$

As a mathematical formula, it is stated in this way:

$$I = \frac{E}{R}$$

To make calculation easier, each lamp shown in Figure 1.12 is the same size, receives 120 V, and draws 1 A of current. The resistance of *one* lamp is found by using Ohm's Law:

$$I = \frac{E}{R}$$

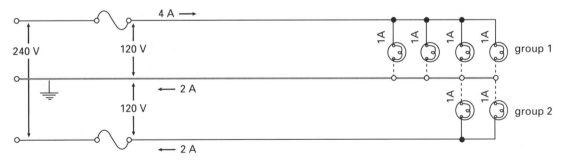

FIGURE 1.12 Each lamp is the same size (amperage). The current flows as shown by the arrows.

Transposed, this equals

$$R = \frac{E}{I}$$

since R represents Resistance, which is measured in Ohms.

$$R \text{ (Ohms)} = \frac{E \text{ (Volts)}}{I \text{ (Amps)}}$$

Therefore, the resistance of one lamp equals $\frac{120 \text{ V}}{1 \text{ A}}$

That is, R equals 120 Ω.

The resistance of one lamp is 120 Ω.

The *total resistance* of all lamps of the same size in a *parallel circuit* is equal to the resistance of one lamp *divided by* the number of lamps in the circuit.

Resistance of the lamps in Group 1:

$$R = \frac{120 \text{ Ω}}{4 \text{ (lamps)}} = 30 \text{ Ω}$$

Resistance of the lamps in Group 2:

$$R = \frac{120 \text{ Ω}}{2 \text{ (lamps)}} = 60 \text{ Ω}$$

When the neutral wire is broken, the lamps in Groups 1 and 2 are in *series* with each other. (See Fig. 1.13) The *total resistance* for the whole circuit is now the *sum* of the resistances in Groups 1 and 2 which is 30 Ω + 60 Ω = 90 Ω.

Since the neutral wire is broken, the line voltage is now 240 V, and the total current in the circuit is

$$I = \frac{E}{R} = \frac{240 \text{ V}}{90 \text{ Ω}} = 2.67 \text{ A}$$

The voltage in a series circuit is divided between the groups of lamps, according to Ohm's Law:

$$I = \frac{E}{R}$$

Transposed, this equals

$$E \text{ (V)} = I \text{ (A)} \times R \text{ (Ω)}$$

The voltage of Group 1 is

$$E = 2.67 \text{ A} \times 30 \text{ Ω}$$
$$= 80.1 \text{ V}$$

This low voltage will cause the lamps in Group 1 to glow dimly.

The voltage of Group 2 is

$$E = 2.67 \text{ A} \times 60 \text{ Ω}$$
$$= 160.2 \text{ V}$$

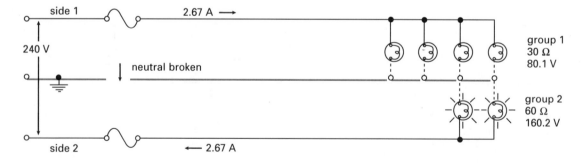

FIGURE 1.13 Each lamp is the same rating (wattage). The current flow is altered when the neutral wire is broken.

This higher voltage will cause the lamps in Group 2 to glow more brightly than usual and may cause them to burn out.

F o r   R e v i e w

1. What voltage is used in houses for portable equipment? for heavy-duty equipment?
2. Which electrical frequency is used most on the North American continent?
3. Explain how two separate voltages are obtained from the 3 wire distribution system. Include a diagram.
4. What does the term *neutral* mean when applied to a conductor?
5. What is the purpose of identifying the neutral wire?
6. What are the three differences between the neutral and live wires of a circuit, other than colour?
7. Explain why (a) the neutral wire is grounded, and (b) the neutral wire is never fused.
8. What method is used to ground the neutral wire in a house?
9. Why are the incoming hydro supply lines connected to the upper terminals of the main switch?
10. What voltage is available between (a) two live wires? (b) a live and neutral wire? (c) a live wire and ground? (d) the neutral wire and ground?

2

Lighting– Control Switches

The flow of electrical current in the various circuits of a building must be controlled. This is done by using a variety of switches capable of opening and closing the circuits.

Switches are divided into two main categories according to the method of installation:

Category 1. The *surface* type of switch is mounted on the face of the wall, with the entire body of the switch visible. (See Fig. 2.1)

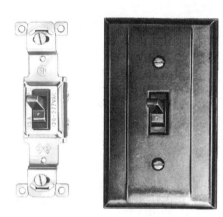

Courtesy Smith & Stone

FIGURE 2.2 Flush-mounted switch

Courtesy Smith & Stone

FIGURE 2.1 Surface-mounted switch

Category 2. The *flush* type of switch unit is mounted within an electrical box.

This box is recessed into the wall, so that only the operating handle of the switch is visible. This type of switch looks neater and is used more extensively. (See Fig. 2.2)

Operating Mechanisms

There are many different types of switches to control a wide variety of electrical devices. Just as there are many switches, there are also many different types of operating mechanisms to suit individual circuit needs. Figure 2.3 shows some of the more common operating mechanisms.

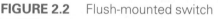

Applications of Electrical Construction

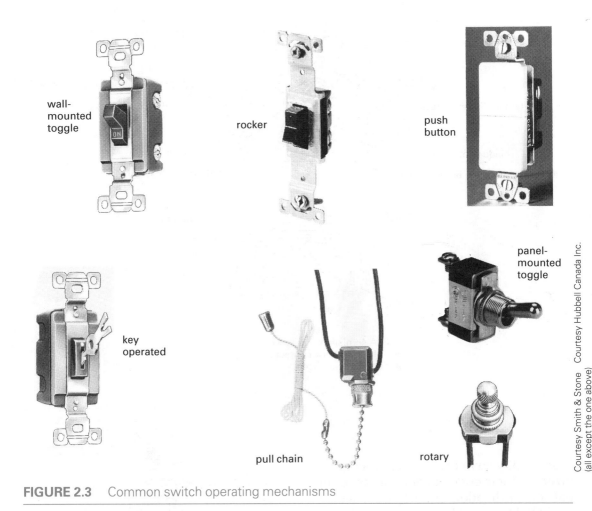

wall-
mounted
toggle

rocker

push
button

panel-
mounted
toggle

key
operated

pull chain

rotary

Courtesy Smith & Stone   Courtesy Hubbell Canada Inc.
(all except the one above)

FIGURE 2.3   Common switch operating mechanisms

Internal Construction

The basic function of the switch is to open and close a circuit. To do this the switch has a combination of *fixed* and *moveable* contacts, which are usually made of brass. Figure 2.4, on page 14, shows two types of designs and methods for moving the contacts.

Switches designed to carry a *high current* have larger and stronger contacts than those of switches designed to operate only one or two lights. Nearly all, however, have some form of *spring* inside to open and close the contacts quickly. If the contacts are slow to open and close, there is danger of the current forming an arc—a spark jumping across the contacts as they part—resulting in heat damage and possibly contact burning.

The main differences between a high-quality (expensive) and a low-quality switch are the size and strength of the contacts and spring. Take care to choose a switch capable of handling the intended load on the circuit.

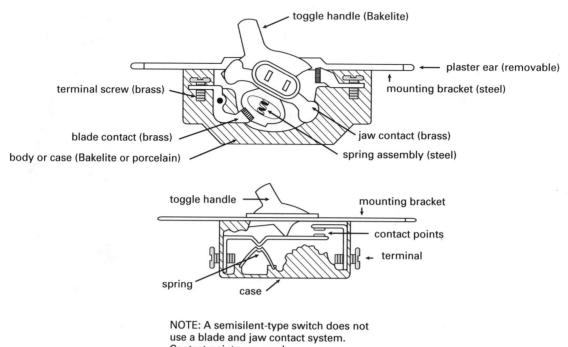

toggle handle (Bakelite)

plaster ear (removable)

terminal screw (brass)

mounting bracket (steel)

blade contact (brass)

jaw contact (brass)

body or case (Bakelite or porcelain)

spring assembly (steel)

toggle handle

mounting bracket

contact points

terminal

spring

case

NOTE: A semisilent-type switch does not use a blade and jaw contact system. Contact points are used.

FIGURE 2.4   Internal construction of a single-pole switch

Switch Terminals

High-quality switches are also recognized by their large-headed, brass terminal screws. In addition, they have well-designed terminal bases with good wire containment features. Some manufacturers provide a *push-in* type of terminal by which the terminal screw holds the wire in a vise-like grip. (See Fig. 2.5) Switches that make sole use of the push-in method of holding the wire may give trouble at their terminal connections in years to come, though. The installer should consider more than the ease with which the switches can be installed in the circuit.

Switch Ratings

Manufacturers usually place electrical

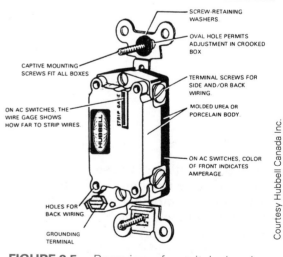

SCREW-RETAINING WASHERS.

OVAL HOLE PERMITS ADJUSTMENT IN CROOKED BOX

CAPTIVE MOUNTING SCREWS FIT ALL BOXES

TERMINAL SCREWS FOR SIDE AND/OR BACK WIRING.

ON AC SWITCHES, THE WIRE GAGE SHOWS HOW FAR TO STRIP WIRES.

MOLDED UREA OR PORCELAIN BODY.

ON AC SWITCHES, COLOR OF FRONT INDICATES AMPERAGE.

HOLES FOR BACK WIRING.

GROUNDING TERMINAL

FIGURE 2.5   Rear view of a switch showing 2 types of terminal connections

ratings on the mounting bracket of the switch. There are *six* possible ratings to assist a person in matching the switch to

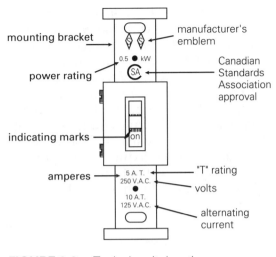

mounting bracket

manufacturer's emblem

power rating

Canadian Standards Association approval

indicating marks

amperes

"T" rating

volts

alternating current

FIGURE 2.6 Typical switch ratings

the job: voltage, amperes, test laboratory approval, "T" rating, AC or DC, and power. (See Fig. 2.6) The first three ratings *must* be placed on all switches. The second three—"T" rating, AC or DC, and power—are marked on switches designed for special applications.

Voltage. This rating refers to the electrical pressure the switch can control safely. Lighting switches should have a rating of 120 V to 125 V for control of *one* live wire and 240 V to 250 V for control of *two* live wires.

Amperes. This rating refers to the ability of the contacts to carry current. The *higher* the current rating, the *more* load can be controlled by the switch. The current ratings increase in units of five (for example, 10 A, 15 A, 20 A, etc.).

Test Laboratory Approval. All electrical products for sale or use in Canada must be submitted to the Canadian Standards Association (CSA) for testing and approval. Since Canadian

standards often differ from the standards of other countries, products made elsewhere must still be tested by the CSA. For example, products made in the United States and tested and approved by the American parent of Underwriters' Laboratories bear the mark *UL* or *Und. Lab.* If these products are to be sold or used in Canada, they must be retested by the CSA. If approved, both the CSA and Underwriters' approval marks may be placed on the products. The CSA mark is a guarantee to the user that the manufacturer's ratings are correct.

"T" Rating. Incandescent lamps (standard household lamps) have a tungsten filament. When the lamp is off and at normal room temperature, the filament's electrical resistance is very low. The low resistance of a cold filament allows a high inrush of current—eight to ten times' the normal operating current—to flow into the lamp for the length of time it takes to reach full brilliancy. Once the lamp is at operating temperature, filament resistance is high and normal current flows through the lamp. Switches designed to control a group of incandescent lights must have heavy-duty contacts and springs to withstand these operating conditions safely.

AC or DC. *Alternating current* (AC) is used for *residential* installations, and many switches produced for household use will be marked accordingly. Some *industrial* equipment, however, is designed to operate on *direct current* (DC). Switches for control of this equipment must have a direct current rating.

Once an arc is started between *direct*

current switch contacts, it will follow the contacts, continue to burn, and destroy the switch within seconds. (Arcing between alternating current switch contacts is usually less severe and damaging.) Switches designed for direct current use, therefore, must have strong contacts and springs, together with adequate insulation.

Power.   Switches are also used to control electric motors. A motor requires three to five times as much current when starting as it does when running. Therefore, when a large motor, such as that of a refrigerator or air conditioner, starts up, room lights will dim briefly.

To be capable of handling high-current situations, a switch must be equipped with sturdy contacts and a strong spring. Since motors are rated in watts, the power rating tells the installer the maximum size of motor the switch can control safely.

Dimmer Switch Rating

This type of switch consists of an electronic circuit sensitive to the amount of load placed on it. When installing a dimmer switch, take care not to exceed the wattage rating on the unit. The total wattage of the lamps being controlled by the dimmer must not be greater than the wattage rating marked on the switch.

Lighting–Load Calculations

To determine the ampere rating of a lighting circuit, add the wattage of all the lamps to be controlled by the switch; then, divide the total by the circuit voltage to calculate the amperage (current flow) for the circuit.

For example, if a room has six light fixtures equipped with 100 W lamps on a 120 V circuit, the total wattage is 600 W (6 lights × 100 W). The current flow is 600 W divided by 120 V which is 5 A. A 10 A, "T"-rated, 120 V switch would, therefore, be adequate for this installation. The current ratings for fluorescent lights, however, are usually marked on the ballast of each fixture. Adding these ampere ratings will determine the current flow without further calculation being necessary.

Switch Test Equipment

Switch mechanisms are usually completely enclosed, and the internal connections are hidden from view. Simple test equipment for locating the internal connections can be made, however. Both of the testers described in the following paragraphs may be checked by touching the test clips together.

Series-Lamp Tester.   Figure 2.7 shows a tester made from a lamp socket, light bulb, and line cord. If, after the clips are fastened to the switch terminals, the lamp is on, there is a complete, or closed, circuit.

Safety Note: This tester should *not* be used on a switch connected to a circuit with its own source of voltage supply. Also, the operator must be careful not to touch the metal portion of the test clips, since this unit operates on 120 V.

Buzzer Tester.   Figure 2.8 shows a tester made with a pair of dry cell batteries and a small buzzer or bell, which rings when the circuit is closed. This simple unit is portable and may be carried in a tool box.

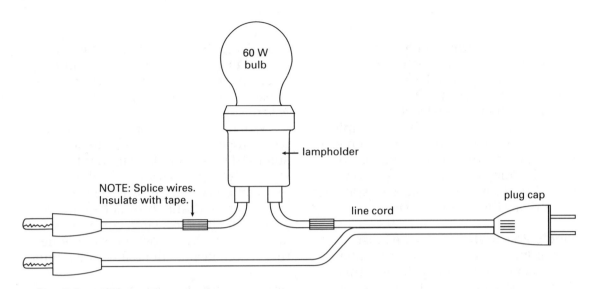

FIGURE 2.7 Series-lamp tester (120 V device)

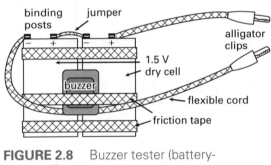

FIGURE 2.8 Buzzer tester (battery-powered)

Switch Wiring Symbols

A system of electrical symbols is often used to represent various types of switches on wiring diagrams. Figure 2.9 shows symbols commonly used on electrical drawings.

Switch Applications

Switches are available in a variety of types for a variety of uses. *Single-pole* and *double-pole* switches are used to control lights or equipment from one

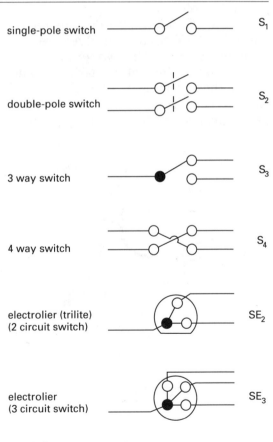

single-pole switch S_1

double-pole switch S_2

3 way switch S_3

4 way switch S_4

electrolier (trilite) (2 circuit switch) SE_2

electrolier (3 circuit switch) SE_3

FIGURE 2.9 Switch wiring symbols

location. *Three-way* switches are used to control lights in areas such as rooms with two entrances, hallways, or stairways where it is convenient to have control from either of two locations. *Four-way* switches are used where multiple-switch control is needed, for example, in large houses, apartment buildings, or office buildings that have large rooms with three or more entrances or stairways rising three or more storeys.

Single-Pole Switch. As the most commonly used switch for residential and commercial lighting circuits, this switch is also appropriate for fractional kilowatt motors, portable appliances, portable tools, etc.

The single-pole unit is designed to control a 120 V circuit (one live wire) from one location. The two terminal screws and the indicating marks (*on* and *off*) on the operating handle make it easy to recognize. (Some manufacturers use a small dot on the operating handle

to indicate the *on* position.) This switch is available in a variety of styles and operating mechanisms, with current ratings to suit any installation for which it was designed.

When installing a single-pole switch with a toggle or rocker mechanism, standard procedure is to mount the switch with the operating handle in the *up* position (when the switch is turned on). Figure 2.10 shows by schematic diagram a single-pole switch controlling two lights. Figure 2.11 shows how to connect this switch in a residential wiring circuit.

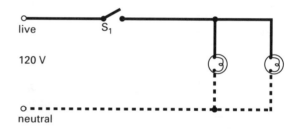

FIGURE 2.10   Schematic wiring diagram showing 2 lamps controlled from one location

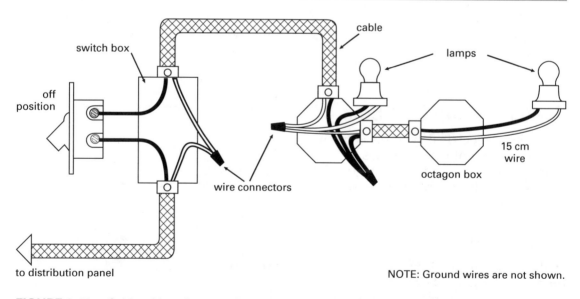

FIGURE 2.11   Cable wiring diagram showing 2 lamps controlled from one location

Applications of Electrical Construction

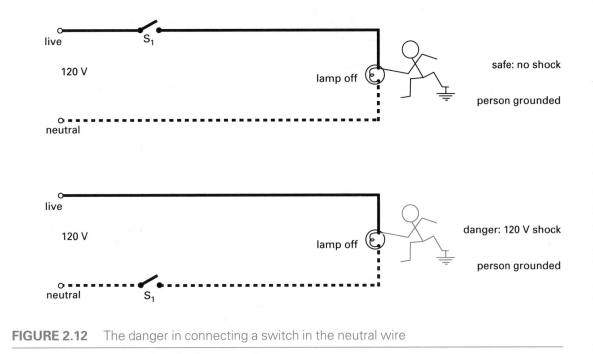

FIGURE 2.12   The danger in connecting a switch in the neutral wire

Remember that *only* the *live* wire is connected to the switch. (See Chapter 1) Figure 2.12 shows the danger of using the neutral wire, or grounded, identified conductor, to control the circuit. In either case, the light may be turned off, but the live wire is still dangerous at the light fixture.

Double-Pole Switch.   This switch is available in a variety of current ratings, but most often in the heavy-duty range. It is meant to control such energy users as electric heaters, air conditioners, and some motors operating on 240 V. Two live wires and a switch capable of opening both live conductors of the circuit simultaneously are required. The double-pole switch, much like a pair of single-pole switches in the same case, or body, is easily recognized by the four terminal screws and the indicating marks on the operating handle.

When installing a double-pole switch, take care to connect the circuit conductors to the proper terminals. This prevents short-circuit damage to the switch contacts. Figure 2.13 shows by schematic diagram a double-pole switch controlling an electric heater.

Figure 2.14 shows an improper connection, which will damage the switch badly the instant it is turned on.

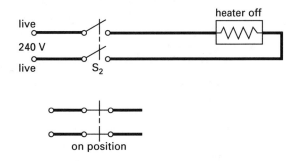

FIGURE 2.13   Schematic wiring diagram of a double-pole switch

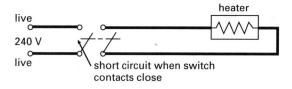

FIGURE 2.14 Contact damage results if a double-pole switch is connected improperly.

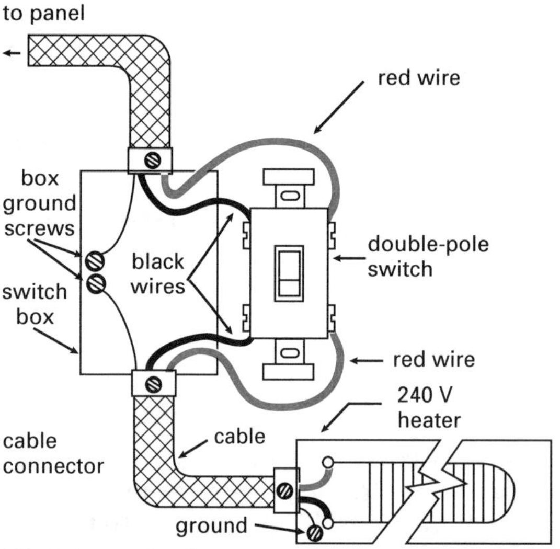

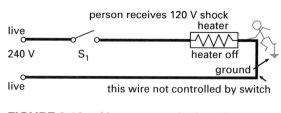

FIGURE 2.16 Never use a single-pole switch on a 240 V circuit.

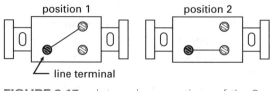

NOTE: Both wires in this circuit are alive and should be coloured accordingly. The modern cable containing one red and one black wire should be used.

FIGURE 2.15 Cable wiring diagram of a double-pole switch

Figure 2.15 shows how to connect this switch in a residential wiring circuit.

Safety Note: Remember that a single-pole switch must not be used on a 240 V circuit. Figure 2.16 shows the danger of such a connection.

Three-Way Switch. These switches are used in pairs, usually on 120 V circuits. They control one live wire in order to provide independent control of a light or group of lights from either of two locations. A 3 way switch is easily

recognized by its three terminals, one of which is marked as a *line* terminal, and by its lack of indicating marks. (The word *line* may be printed beside the terminal, or the terminal may be coloured for easy recognition.)

The internal design of the 3 way switch allows current to flow through the switch in either of its two positions. For this reason, the switch has no *on/off* marks on the operating handle. The circuit is controlled by using two switches as a team. Figure 2.17 shows the internal switch positions and also how one terminal—the common, or *line*, terminal—is used for both switch positions. Figure 2.18 shows by schematic diagram 3 way switch control of one light.

This switch is available in a variety of styles and operating mechanisms, with current ratings to match the load being controlled. Figure 2.19 shows how to connect this switch in a residential wiring circuit. The two conductors running between the switches are often called *travellers*, or *messenger wires*. If the need arises, this switch may also be

FIGURE 2.17 Internal connections of the 3 way switch

Applications of Electrical Construction

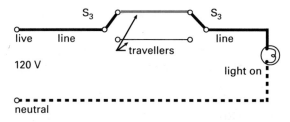

live line
S_3 S_3
travellers
120 V   line
light on
neutral

FIGURE 2.18   Schematic wiring diagram showing 1 light controlled from two locations

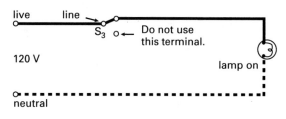

live   line
S_3 Do not use this terminal.
120 V   lamp on
neutral

FIGURE 2.20   Schematic wiring diagram showing how a 3 way switch can be used safely in place of a single-pole switch

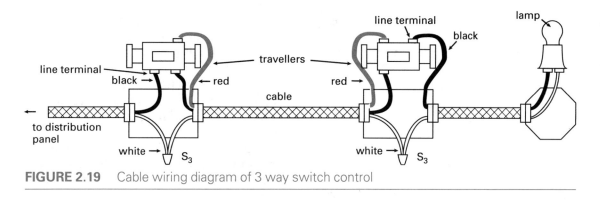

FIGURE 2.19   Cable wiring diagram of 3 way switch control

used safely as a single-pole switch. Figure 2.20 shows by schematic diagram such a circuit.

Four-Way Switch.   Multiple-switch control in large rooms or buildings is achieved by using 4 way switches in combination with a pair of 3 way switches in 120 V circuits. This combination is particularly useful in such areas as large rooms with three or more entrances and stairwells in buildings of three or more storeys. Control of the light from any of the switch positions may be obtained by operating any one of the switches in the group.

The 4 way switch is a *non-indicating* switch, because the lights can be controlled by any one switch in the circuit. It can be recognized by its four terminals and lack of indicating marks on the oper-

ating handle. Without proper test equipment, the only way to tell the difference between the 4 way and the double-pole switch is to remember that the 4 way has *no* indicating marks, while the double-pole switch does.

The 4 way switch is available in a variety of styles and operating mechanisms. It is thus able to suit any application. However, because the internal mechanism is more complicated and there is less consumer demand for it, the 4 way switch may be more expensive than other types.

Figure 2.21 shows the internal switch positions, and Figure 2.22 shows by schematic diagram control of a light from three locations. Figure 2.23 shows control from five locations, and Figure 2.24 shows how to connect this switch in a wiring circuit.

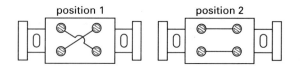

FIGURE 2.21   Internal connections of the 4 way switch

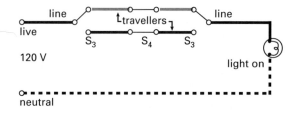

FIGURE 2.22   Schematic wiring diagram showing 1 light controlled from three locations

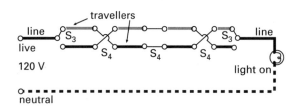

FIGURE 2.23   Schematic wiring diagram showing 1 light controlled from five locations

Electrolier Switches

Two-Circuit Electrolier.   The 2 circuit electrolier is more commonly known as the *trilite* switch. Used with the dual filament lamp, it provides three levels of light (low, medium, and high) for many table lamps, floor lamps, pole lamps, and hanging, or swag, lamps. (See Fig. 2.25) The switch unit and bulb in combination give this circuit its versatility.

The trilite switch has three terminals, or *leads*, and is almost always a *rotary* switch. It usually has no indicating marks, since the installer may use either of the two line terminals, depending on the choice of sequence. Trilite switches are used mainly for 120 V lighting units.

Figures 2.26 and 2.27 show the four positions of the trilite switch. One terminal, called the *common*, or *line*, terminal, is used in all three of the *on* positions. To turn the bulb off, the blades of the switch are simply moved *away* from the line terminal. A lamp with a switch as shown in Figure 2.26 operates in a high, medium, and low sequence. If the switch is connected with the line terminal as shown in Figure 2.27, however, the lamp will operate in a low, medium, and high sequence.

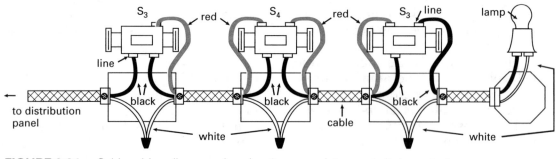

FIGURE 2.24   Cable wiring diagram showing 3 way and 4 way switch control

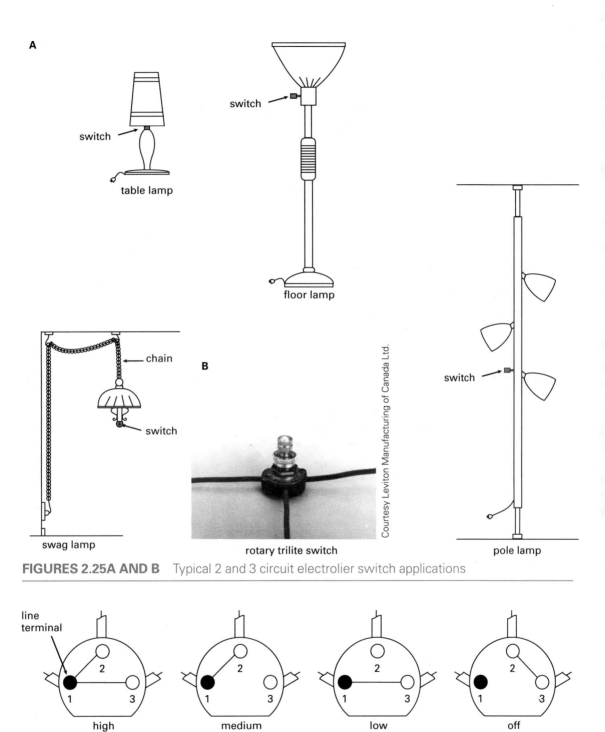

FIGURES 2.25A AND B Typical 2 and 3 circuit electrolier switch applications

FIGURE 2.26 Trilite switch positions for high to low sequence

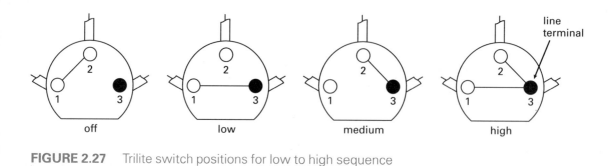

FIGURE 2.27   Trilite switch positions for low to high sequence

Trilite bulbs are available in three sizes, or wattage ranges. The largest has a *mogul* base equipped with 100 W and 200 W filaments. Both filaments operating together provide 300 W of light in the high position. The other two bulb sizes use a *medium* base and are available in 50 W-100 W-150 W and 40 W-60 W-100 W combinations. These units are popular for smaller light fixtures.

Figure 2.28 shows the switch and socket assembly wired as a working unit. Some table and floor lamps are manufactured with a switch built into the socket assembly. The only connections required are for the live and neutral wires in the lamp's cord. The switch may

be purchased as a separate unit for connection to two independent light sockets, as shown in Figure 2.29. This type of connection may be installed in fixtures using two standard light bulbs, rather than the trilite bulb.

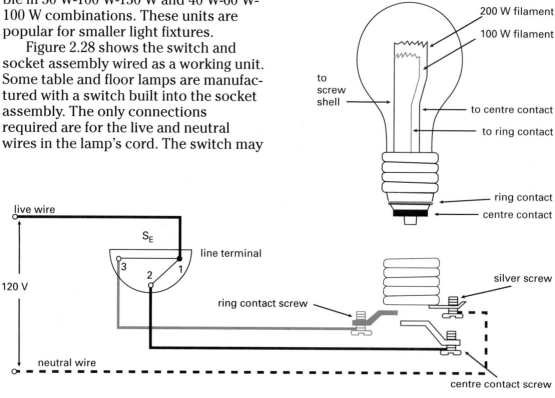

FIGURE 2.28   Schematic wiring diagram showing a 2 circuit electrolier switch

Three-Circuit Electrolier. There are two types of 3 circuit electroliers. The *alternate* type (See Fig. 2.30) puts one bulb on at a time and may be used for pole lamps or other decorative units. The *consecutive* type (See Fig. 2.31) is often fitted into pole or swag lamps. All three lamps may be on at the same time.

Dimmer Switches

Lighting not only illuminates, it can create moods, as well. For example, dimming lights in a dining area can create a more intimate atmosphere. Two types of dimmer switches, the *high-low* and the *infinite*, are commonly used in homes and function on *electronic* circuitry. Their operating methods are described below. Other dimmers functioning on *electrical resistors* alone produce a great deal of heat, which is difficult to dissipate, and so are not desirable for home use.

High-Low. This two-stage, flush-mounted, wall switch usually has a toggle action. When the operating handle is in the *up* position, the

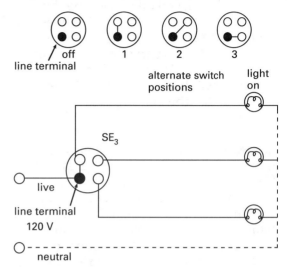

FIGURE 2.30 Schematic wiring diagram and alternate switch positions for a 3 circuit electrolier

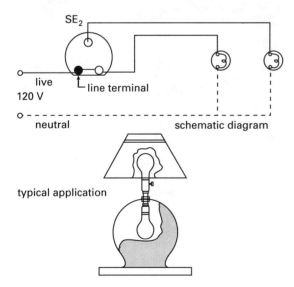

FIGURE 2.29 Schematic wiring diagram and typical application showing a 2 circuit electrolier switch using 2 standard light bulbs

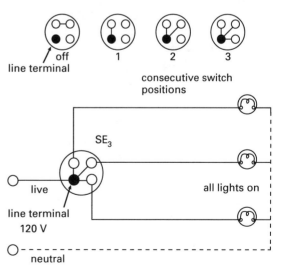

FIGURE 2.31 Schematic wiring diagram and consecutive switch positions for a 3 circuit electrolier

standard light bulb in the fixture glows at full brilliancy. When the operating handle is placed in the *down* position, the bulb glows at the lower level of light. The light is turned off by moving the handle of the switch to a centre position. (See Fig. 2.32)

This switch is wired in a way similar to any single-pole switch. Take care to install a switch of sufficient wattage rating to handle the lighting load of the circuit.

Infinite Dimmer Control. This unit mounts in a standard electrical switch box and can replace any standard flush-mounted switch. The majority are wall-mounted; however, units built into light sockets for table lamp and floor lamp applications are available. Both single-pole and 3 way units are made. The standard light bulbs used are turned on or off by pressing the control knob. (See Figs. 2.33 and 2.34) Any level of light from off to full brilliancy can be obtained by *rotating* the control knob.

The dimmer switch is usually equipped with leads for connection to the electrical circuit.

Safety Note: Take care to ensure that the total wattage of the circuit load does not exceed that marked on the switch.

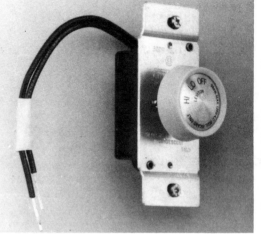

FIGURE 2.33   Infinite dimmer switch

FIGURE 2.34   A heavy-duty dimmer switch equipped with an aluminum heat-sink for proper dissipation of heat (right) compared to a standard 600 W dimmer switch. The heavy-duty dimmer switch can be used on lighting circuits up to 1000 W.

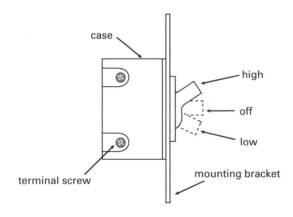

FIGURE 2.32   High-low dimmer switch

Applications of Electrical Construction

1. What is the purpose of a switch in a circuit?
2. List and describe the two main types of switches, according to method of installation.
3. List seven operating mechanisms used in lighting switches.
4. List the parts of a switch, and state the purpose of each.
5. Define *arcing*, and describe its effect on a switch.
6. List the six electrical ratings that may be marked on switches.
7. Why do manufacturers mark electrical ratings on switches?
8. Explain the function of the Canadian Standards Association and the American parent Underwriters' Laboratories.
9. An oil burner has a 560 W motor operating on 240 V. It it draws 6.9 A, what type of switch is needed to control the motor? Which electrical ratings should appear on the switch?
10. A room is equipped with twelve 120 V light fixtures, each having a 100 W bulb. What type of switch is required to control these lights? Which electrical ratings should be marked on the switch?
11. Explain how to use a series-lamp tester to locate a switch's internal connections.
12. Explain why the neutral wire should never be used to control a circuit.
13. Explain why a single-pole switch should never be used to control a 240 V circuit.
14. Can a double-pole switch be used safely to control a 120 V circuit? Explain, with the aid of a diagram.
15. Explain why care must be taken when connecting a double-pole switch to a 240 V circuit.
16. List three places in a home where 3 way switch control would be useful.
17. Show by diagram a 3 way switch being used to replace a single-pole switch.
18. What is the name of the two conductors that connect 3 way switches to one another?
19. List the types and quantities of switches required for stairway lighting with control from each storey of a six-storey building.
20. List three applications of 2 circuit electrolier (trilite) switch control in the home.
21. Which terminal of the electrolier switch must be located to determine lighting sequence?
22. Explain why you think it would be desirable to be able to change the level of light in a room.
23. Explain how a trilite lamp produces three levels of light.
24. What is the advantage an infinite-dimmer control switch has over a trilite switch?
25. Explain why care must be taken not to exceed the wattage rating on a dimmer switch.

Lampholders

There are many types of lampholders used for residential and commercial lighting installations. This chapter covers only the *screw-base* type.

Screw-Base Sizes

There are five standard screw-base sizes, each of which has a particular area of use. (See Fig. 3.1)

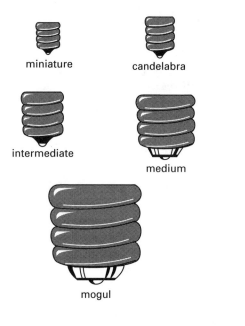

miniature

candelabra

intermediate

medium

mogul

FIGURE 3.1   Standard screw-base sizes

Mogul.   As the largest of the screw-base units, the mogul is used in residential trilite lamps and in commercial areas such as parking lots, service garages, and warehouses where larger light bulbs are employed.

Mogul-base light bulbs range in size from 300 W to 1500 W, which the Canadian Electrical Code has set as the maximum size for incandescent-bulb mogul sockets. Because 1500 W bulbs produce a great deal of heat as well as light, mogul-base sockets are usually heavy-gauge brass with a porcelain covering to withstand the heat.

Bulbs for the mercury vapour type of lamp exceed 1500 W. They are also equipped, however, with mogul bases.

Medium.   This is the most common residential socket size. Standard incandescent light bulbs from 7.5 W to 300 W are equipped with medium bases, as is the 125 V plug fuse for the distribution panel. Any commercial lighting installation using a bulb size of up to 300 W is also equipped with a medium-base socket.

The Canadian Electrical Code requires that no lamp in excess of 300 W be used in medium-base lampholders, *unless* the lampholder is made of heat-resisting material. The Code also requires that all medium-base lampholders have an electrical rating of 660 W and 250 V. This regulation is designed to protect the user by ensuring safe operation for the range of bulbs available.

Intermediate. This socket is used for such purposes as outdoor Christmas tree lights, showcase and aquarium lighting, sewing-machine lamps, and appliances such as electric stoves. The intermediate socket has an electrical rating of 75 W and 125 V.

Candelabra. As the smallest screw-base lampholder that may be connected directly to a 120 V circuit, the candelabra has an electrical rating of 75 W and 125 V. Decorative lighting takes full advantage of this small socket. The light bulbs used in it are produced in a wide variety of shapes, often simulating the flames of candles. They are ideal for indoor Christmas tree lights and crystal chandeliers.

Miniature. This is the smallest screw-base lampholder in the standard group. It is used for dial-illuminating panel lights in radio or television sets. One type of Christmas tree light string also uses these sockets in a 120 V series circuit. The voltage is divided among the number of lights in the string, so that, for example, if there are eight lights, each light receives approximately 15 V.

This lampholder does *not* carry a 120 V rating due to the limited space for contact separation inside the socket.

Adapters

An adapter is a device that can be used to reduce socket size so that the lampholder will accept a bulb with a smaller base. The adapter has an *external* thread to fit the socket being reduced and an *internal* thread to fit the smaller base of the bulb. Also, there are adapters that allow a lampholder to accept two bulbs rather than one and/or a plug from an extension cord.

Safety Notes: When using these devices, take care not to place a load on the lampholder in excess of its electrical rating. Also, when an extension cord adapter is being used, remember that no effective ground connection is available—the protection normally offered by a grounded tool or device is lacking. Remember, too, that it is dangerous to attempt to increase socket size by using adapters, because the heat and current from the larger bulb may damage the smaller socket.

Lampholder Construction

The basic components of all lampholders are similar (Fig. 3.2). The lampholder body, however, may be made of different materials, such as Bakelite, rubber, porcelain, brass, or aluminum. (See Fig. 3.3)

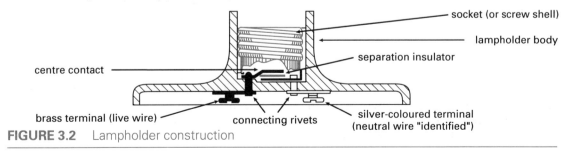

FIGURE 3.2 Lampholder construction

turn knob
socket

brass

push-
through
switch
socket

brass

pull-chain
switch
socket

pull-chain,
Bakelite
lampholder

keyless (no
switch)
socket

weather-
resistant
pigtail
socket

Courtesy Leviton Manufacturing of Canada Ltd.

FIGURE 3.3   Common medium-base lampholders

Lampholder Switch Mechanisms

Lampholders with built-in switch mechanisms are called *key* types. (See Fig. 3.4) This term came into use when lampholders had ornate, rotary switch handles similar to door keys. Lampholders without switches built into the socket assembly are called *keyless* types. (See Fig. 3.5)

FIGURE 3.4   A push-through switch socket

FIGURE 3.5   A keyless lampholder

Circuit Connection

The terminals on lampholders are colour-coded to make the connection of the circuit conductors easier. (Most of the lampholders mentioned in this chapter are designed for 120 V circuits and make use of this colour code.)

A 120 V circuit has a live and a neutral wire, each of which must be connected to the proper terminals at the socket. One terminal is a natural brass colour and is to receive the *black live* wire. (Inside, this brass terminal joins with the centre contact.) The *white neutral* wire is fastened to the "identified," *silver-coloured* terminal. This terminal screw is joined to the screw shell of the lampholder. (See Fig. 3.2)

Note: The Canadian Electrical Code requires that a terminal intended solely for the connection of a neutral wire *must* be identified by a tinned finish, a nickel-plated finish, or by means of some distinguishing mark.

It is important to use this method of connection in order to avoid the danger of shock. A person changing light bulbs or cleaning a light fixture with a damp cloth could easily touch the screw shell and become grounded on the bathroom or kitchen sink. If the screw shell is alive, the person could receive a 120 V shock. If, however, there are proper wiring connections, the *only* live part of the socket is the *centre contact*, which is not easily touched by accident.

Pull-Chain Insulators

The Canadian Electrical Code requires that lampholders with pull-chain switch mechanisms have insulating links in their chains. This regulation is designed to prevent a person in contact with ground from receiving a 120 V shock if a defective switch unit makes the chain alive. (See Fig. 3.6)

Location of Lampholders

Section 30 of the Canadian Electrical Code outlines the location of lampholders. These guidelines are subject to frequent change, because new products

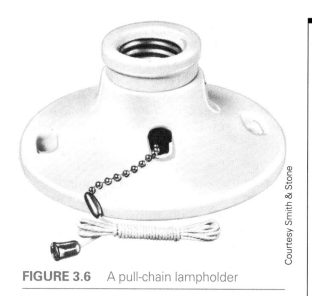

FIGURE 3.6 A pull-chain lampholder

Courtesy Smith & Stone

and materials are constantly being introduced. When planning an installation, make sure to obtain the latest edition of the Code.

1. List the five standard screw-base lampholder sizes, and give one practical application for each.
2. What is the wattage rating for incandescent lamps for the largest screw-base lampholder?
3. Which electrical ratings for screw-base lampholders are required by the Canadian Electrical Code?
4. Why is bulb size for the medium-base socket limited to 300 W?
5. List two types of adapters used with light sockets.
6. What precaution is necessary when using an adapter?
7. List the types of lampholder switch mechanisms. What is meant by the term *keyless*?
8. Explain how and why the conductors from a 120 V circuit must be connected to a lampholder.
9. Why is an insulating link required in the chain of one type of switch mechanism?
10. Where can information be found regarding the installation of lampholders?

4 Receptacles

The *receptacle* is the most widely used electrical device, because it is the point in a circuit at which power may be taken to supply lamps, appliances, or portable plug-in devices. The residential receptacle delivers 120 V to any electrical device plugged in to it. Every modern home is equipped with a number of duplex receptacles.

Receptacles are available in a variety of quality and duty ranges. *Light-duty* receptacles, frequently sold as *standard* or *residential grade*, may be used in living-rooms or bedrooms, where a table lamp or radio will be the largest current-demanding device plugged in. The slightly better quality *medium grade* receptacles stand up to somewhat more frequent use in kitchens and utility rooms. *Heavy-duty* receptacles, known in the electrical industry as *premium specification grade*, or often simply *specification grade*, are much better designed and constructed. They are most suitable for use in kitchen areas, workshops, and industrial/commercial settings. Electric frying pans, teakettles, and other fast-heating appliances requiring a current of 12 A to 15 A benefit from being plugged into heavy-duty receptacles: if a light-duty receptacle were installed for such appliances, it would heat up, the contacts would become soft and lose their ability to grip the plug firmly, and dangerous overheating would follow. The result would be damage to the receptacle, wiring, and plug.

Careful advance planning is necessary in order to prevent overloading of receptacles. For trouble-free service, the duty range of the receptacle should be matched to the type of appliances expected to rely on the outlet.

Number of Outlets on a Circuit

Section 26 of the Canadian Electrical Code outlines regulations for the installation of receptacles. The number of outlets per circuit, the location, and the number of outlets required for a given room or location are discussed in detail. With products and building standards constantly changing, *only* the Code can be relied upon to provide accurate, up-to-date information.

Receptacle Construction

There are many shapes and styles of receptacles, but each has basic similarities. Figure 4.1 shows the standard, duplex, "U" ground receptacle found in the modern home. This receptacle is

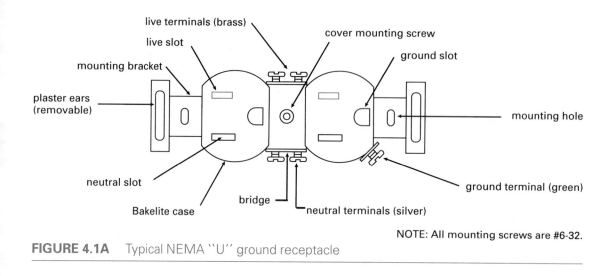

FIGURE 4.1A   Typical NEMA "U" ground receptacle

Labels: live terminals (brass), live slot, mounting bracket, plaster ears (removable), cover mounting screw, ground slot, mounting hole, neutral slot, bridge, Bakelite case, neutral terminals (silver), ground terminal (green)

NOTE: All mounting screws are #6-32.

Courtesy Hubbell Canada Inc.

FIGURE 4.1B   A 15 A, 125 V, specification grade, heavy-duty receptacle in NEMA "U" ground configuration

also widely used in office buildings, stores, and industry. It provides a firm electrical connection for 120 V equipment, together with ground protection for portable equipment.

Receptacle Combinations

Receptacles are available in *single*, *duplex*, and *three-plug* units. The most common of all is the duplex receptacle, which is designed to receive two electrical plugs. Some duplex units have one half as a "U" ground and the other half in

a different shape. Also, there are receptacles teamed with switches, which allow a light to be turned off either in the room or at the receptacle itself.

Receptacle and Plug Shapes

Figure 4.2 shows some of the configurations (shapes) of receptacles and plugs that are available. There are many others made for specialized use. This chapter discusses only the more common ones.

Old-Style Two Prong.   Many buildings constructed during the 1940s and 1950s were equipped with this unit. The main disadvantage is the lack of ground protection.

"U" Ground.   This receptacle accepts both the 2 prong plug and the 3 prong "U" ground plug. If the ground prong is removed from the 3 prong plug, it will still fit the receptacle in one direction only. Careful inspection of the receptacle

Applications of Electrical Construction

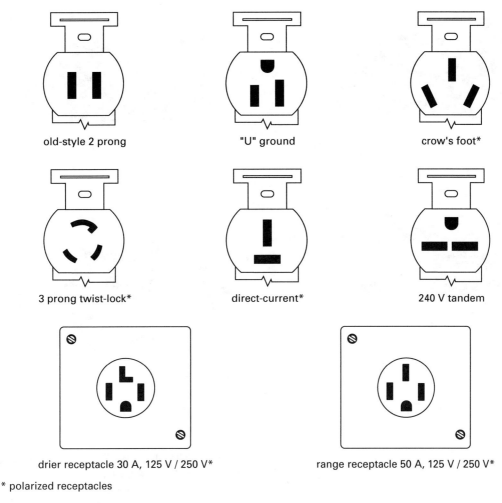

FIGURE 4.2   Receptacle and blade patterns (configurations)

will show that the live and neutral slots are different sizes to prevent interchanging the connections. (See Fig. 4.3)

Crow's Foot.   This plug was an early attempt to provide ground protection in a receptacle. Unfortunately, the shape of the 2 prong plug could be modified with pliers, allowing it to be fitted into the receptacle. Also, the crow's-foot plug often had its ground prong removed and

blades reshaped for use in standard 2 prong receptacles. Ground protection was lost in both cases. It became obvious a change in design was needed.

Twist-Lock.   This receptacle is available in 2, 3, and 4 prong units. Its major advantage is that the plug cannot be pulled out of the receptacle accidentally. The 3 and 4 prong units provide ground protection.

FIGURE 4.3   A specification grade "U" ground receptacle

FIGURE 4.6   A 250 V tandem "U" ground receptacle

FIGURE 4.4   A 250 V direct current "U" ground receptacle

FIGURE 4.7   A 250 V, 15 A, single tandem receptacle

FIGURE 4.5   A 125 V direct current "U" ground receptacle

unit. It is equipped with a ground prong and is used primarily on 240 V circuits, where its blade shape prevents 120 V units from being plugged in.
(See Figs. 4.6 and 4.7)

Range and Drier Receptacle.   Each time an electric range or clothes drier is connected to the cable in a house or apartment building, the cable is shortened slightly. Sometimes the cable is shortened to the point where it can no longer be used. The heavy-duty range and drier plug and receptacle were designed to prevent this from happening. They allow the range and drier to be pulled out from the wall for spring cleaning or simple removal.
(See Figs. 4.8 and 4.9)

Direct-Current.   The main use for this receptacle is to keep the positive and negative conductors from being interchanged in DC systems. Some models have ground protection.
(See Figs. 4.4 and 4.5)

Tandem.   Air conditioners, motors, and heaters of 240 V make use of this

Applications of Electrical Construction

FIGURE 4.8 A residential electric drier (30 A) receptacle

Courtesy Smith & Stone

FIGURE 4.9 A residential electric range receptacle, rated at 50 A

Courtesy Leviton Manufacturing of Canada Ltd.

Commercial and industrial applications often require voltages and currents other than those found in residential applications. To prevent accidental mismatching of cords and receptacles, a wide variety of receptacle configurations has been approved by the Canadian Standards Association (CSA). These can be seen in Figures 4.10A and 4.10B on pages 38 and 39.

Receptacle Grounding

Receptacles equipped with a *ground slot* are designed to provide safety for the person using the equipment connected to the receptacle. If a fault occurs in an electric drill, for example, the current will travel to the frame of the drill, then back along the ground conductor to the distribution panel. The circuit fuse in the panel will blow and prevent the drill's user from receiving a shock. If there is no ground protection, however, the user could receive a 120 V shock as the current flow passes from the tool's frame through the operator on its way to ground.

The ground-equipped receptacle has

a *green, hex-shaped terminal screw* for. connection to the grounding circuit of the electrical box that supports the receptacle. The ground wire in a non-metallic or armoured cable wiring system runs between the ground terminals of the distribution panel, those of the receptacle box, and the receptacle itself. A metal conduit system carries the ground connection from the distribution panel to the receptacle box within its metal casing. A flexible conduit system usually requires a separate, green, insulated conductor to be pulled into the conduit to realize the same purpose. These are Canadian Electrical Code requirements.

Isolated Grounding Receptacles

Some sensitive electronic equipment such as cash registers, computers, and medical instruments will perform poorly if any electromagnetic interference passes through their regular grounding circuits. However, special receptacles,

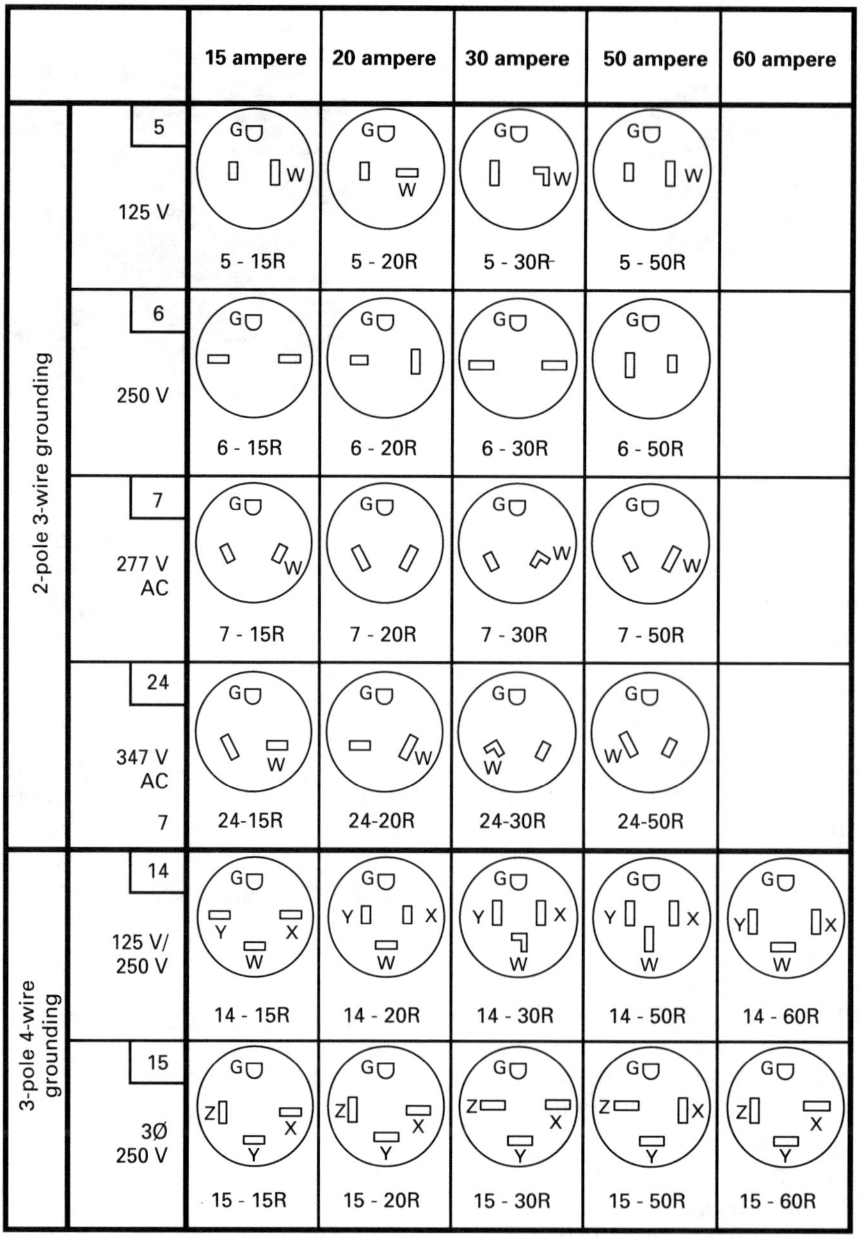

		15 ampere	20 ampere	30 ampere	50 ampere	60 ampere
2-pole 3-wire grounding	**5** 125 V	5 - 15R	5 - 20R	5 - 30R	5 - 50R	
	6 250 V	6 - 15R	6 - 20R	6 - 30R	6 - 50R	
	7 277 V AC	7 - 15R	7 - 20R	7 - 30R	7 - 50R	
	24 347 V AC 7	24-15R	24-20R	24-30R	24-50R	
3-pole 4-wire grounding	**14** 125 V/ 250 V	14 - 15R	14 - 20R	14 - 30R	14 - 50R	14 - 60R
	15 3Ø 250 V	15 - 15R	15 - 20R	15 - 30R	15 - 50R	15 - 60R

NOTE: 3Ø refers to a 3 phase system. The "Y" symbol signifies a 3 phase, 4 wire Wye-connected system.

FIGURE 4.10A CSA configurations for non-locking cord caps and receptacles

		15 ampere	20 ampere	30 ampere	50 ampere	60 ampere
2-pole 3-wire grounding	L5 125 V	L5-15R	L5-20R	L5-30R	L5-50R	L5-60R
	L6 250 V	L6-15R	L6-20R	L6-30R	L6-50R	L6-60R
	L7 277 V AC	L7-15R	L7-20R	L7-30R	L7-50R	L7-60R
	L8 480 V AC		L8-20R	L8-30R	L8-50R	L8-60R
	L9 600 V AC		L9-20R	L9-30R	L9-50R	L9-60R
3-pole 4-wire grounding	L14 125 V/ 250 V		L14-20R	L14-30R	L14-50R	L14-60R
	L15 3Ø 250 V		L15-20R	L15-30R	L15-50R	L15-60R
	L16 3Ø 480 V		L16-20R	L16-30R	L16-50R	L16-60R
	L17 3Ø 600 V			L17-30R	L17-50R	L17-60R
4-pole 5-wire grounding	L21 3Ø 208Y/120 V		L21-20R	L21-30R	L21-50R	L21-60R
	L22 3Ø 480Y/277 V		L22-20R	L22-30R	L22-50R	L22-60R
	L23 3Ø 600Y/347 V		L23-20R	L23-30R	L23-50R	L23-60R

NOTE: 3Ø refers to a 3 phase system. The "Y" symbol signifies a 3 phase, 4 wire Wye-connected system.

FIGURE 4.10B CSA configurations for locking cord caps and receptacles

with ground terminals electrically isolated from mounting straps/brackets, can overcome this problem. Such receptacles have separate, insulated ground wires that provide grounding paths separate from normal grounding circuits. Bothersome malfunction of equipment is thus prevented. These receptacles, marked by the manufacturer, are frequently produced in a bright orange colour for easy recognition.

Split Receptacles

Duplex units are designed so that the *bridge* linking the two terminal points on the one side can be removed. This separates electrically the upper and lower portions of the receptacle from each other. (See Fig. 4.11) Dividing the receptacle electrically has two major advantages.

One is that half of a receptacle placed, for example, in a living-room may be controlled by a wall switch. This allows a lamp to be turned on without entering the darkened room. The other half of the receptacle may be made alive all the time, for use with radios, televisions, or other appliances in the room. Figure 4.12 shows a diagram of such a circuit.

The second major advantage is for use of appliances in the kitchen area. Splitting the receptacle allows two separately fused circuits to be run to one receptacle at the counter. This permits two high current-consuming devices to be plugged in to one receptacle without overloading the circuit. Without a split receptacle, using two appliances, such as an electric kettle and frying pan, in the same outlet would normally blow a 15 A fuse. Adequate protection for the circuit conductors is not lost when the split receptacle system is used. Figure 4.13 shows a wiring diagram of such a circuit.

Replacement of Receptacles

While most receptacles deliver years of trouble-free service, replacements are sometimes necessary. When receptacles no longer hold the plug firmly or heat up during use, a new unit should be installed. Units are available in a variety of styles and colours to suit any room's decor, but attention must be paid to the receptacles' electrical ratings to ensure safe operation.

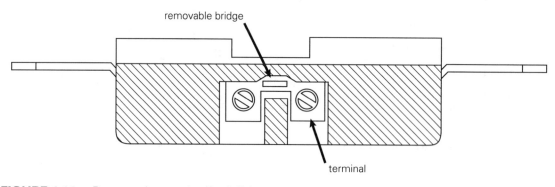

removable bridge

terminal

FIGURE 4.11   Receptacles can be "split" by removing the bridge.

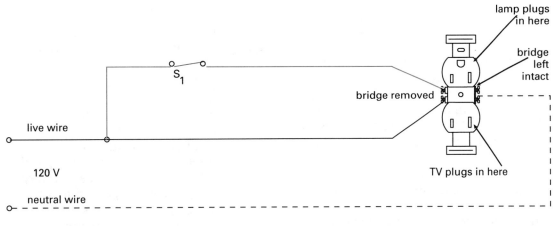

FIGURE 4.12 Wiring diagram showing a split-switched receptacle

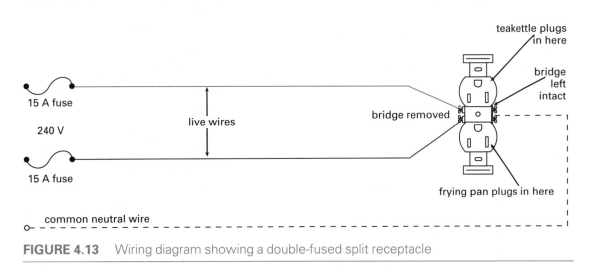

FIGURE 4.13 Wiring diagram showing a double-fused split receptacle

NEMA Receptacles

The National Electrical Manufacturers' Association (NEMA) is an American organization dedicated to standardizing electrical devices. Through its efforts, a radio of any make can now be plugged in to a receptacle in any town in North America and be expected to operate satisfactorily. Receptacles have been standardized in mounting technique and slot placement. Receptacles, wall plates, and receptacle boxes from most manufacturers are completely interchangeable.

The Electrical Electronics Manufacturers' Association of Canada (EEMAC) is the Canadian equivalent of NEMA.

Polarized Receptacles

The crow's-foot, DC, and twist-lock receptacles are examples of polarized units. They accept *only* their own style of plug and the plug can be inserted in *only one* way—a necessary feature in circuits where it is dangerous to interchange any of the circuit conductors.

Plaster Ears

Many receptacles are equipped with small extensions to the mounting bracket called *plaster ears*. When a receptacle is installed in a building with plastered walls or wood panels, plaster ears are very useful. Plaster around the receptacle box often crumbles, leaving a space with little or no support for the receptacle. The plaster ear extensions provide extra length and width to the mounting bracket, making possible a secure, flush mounting. (See Fig. 4.14)

Receptacle installations in surface wiring boxes, such as the utility box or FS fitting, do *not* require plaster ears. (See Fig. 4.15)

Removal of the ears is a simple matter. Bend them back and forth several times with a pair of pliers.

Ground Fault Interrupter Receptacles

Modern wiring regulations require outdoor receptacles to be of the *Ground Fault Interrupter* (GFI) type. Consult the Electrical Code for your area to determine exactly where these units must be installed. The GFI receptacles are designed to protect users of portable tools and equipment from electrical shock, which can occur if the tool or equipment becomes faulty. Each year, many people are killed or endangered by the portable electric devices that have developed internal defects or insulation breakdowns. It is not always obvious to the operator that this breakdown has occurred until the device is plugged in and a shock received.

Normal fuses and circuit breakers will not blow or trip unless the current flow to ground (earth) exceeds the ampere rating of the protection device. They are designed to protect a circuit's wire, NOT the person using the circuit. GFI receptacles are designed to detect small leakage currents when they develop and trip open the circuit to protect the person using that circuit. Currents as low as 5 mA (0.005 A) can be sensed by the unit when it is properly installed.

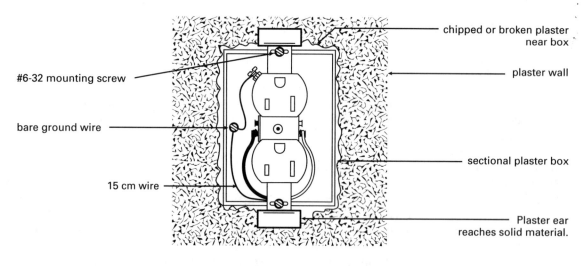

#6-32 mounting screw

bare ground wire

15 cm wire

chipped or broken plaster near box

plaster wall

sectional plaster box

Plaster ear reaches solid material.

FIGURE 4.14 Flush-mounting a receptacle using plaster ears

Applications of Electrical Construction

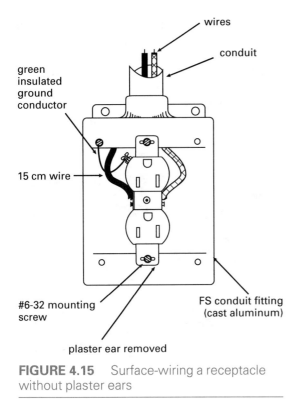

green
insulated
ground
conductor

wires

conduit

15 cm wire

#6-32 mounting
screw

FS conduit fitting
(cast aluminum)

plaster ear removed

FIGURE 4.15 Surface-wiring a receptacle
without plaster ears

Dampness in the tool, metal shavings, or rough handling can cause the initial breakdown in the insulation of the tool or equipment. Standing on a concrete floor, outdoors on grass or earth, or on metal plumbing or frame units of a building will place the operator in contact with the earth or ground.

An electrical shock can have serious effects on a person who is subjected to it. The human body does not allow a high current to flow due to its normally high electrical resistance. However, very small amounts of current can cause pain or upset bodily functions such as breathing and heart beat.

The unit of measurement for this discussion will be the milliampere. One milliampere of current is equal to 0.001 A. A person will feel a slight shock if 5 mA of current flow through the body—not enough to harm, but sufficient to know a shock hazard exists.

At between 10 mA and 15 mA of current flow, *muscular freeze* may occur, preventing the operator from letting go of the tool being used. At the 50 mA to 100 mA level, heart fibrillation and death occur. As can be seen by these figures, a normal 15 A circuit fuse would be of no value in protecting a person using the circuit. Chapter 16 explains the workings of ground fault devices in more detail.

One of the features of the Ground Fault Interrupter receptacle is that once connected to a circuit, any receptacle connected *after* the GFI unit will also offer ground fault protection to the circuit. Outdoor receptacles, garages, bathrooms, pool areas, workshops, laundry rooms, and numerous other locations can well take advantage of the protection offered by ground fault interrupters. (See Figs. 4.16 and 4.17 for typical GFI receptacles.) Great care must be taken to connect the leads of the receptacle according to the manufacturer's

Courtesy Smith & Stone

FIGURE 4.16 Front view of a GFI receptacle

instructions. If connected incorrectly, the GFI receptacle will not provide the protection desired. Figure 4.18 illustrates a typical wiring diagram.

Some manufacturers produce a GFI receptacle that uses terminal screw connections rather than connection leads.

The choice of connection method is then left to the installer. Figure 4.19 illustrates a typical commercial-duty, specification grade GFI, suitable for both vertical and horizontal mounting.

GFI receptacles are equipped with *test* and *reset* buttons. The unit should

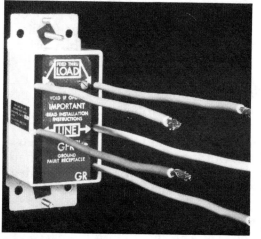

FIGURE 4.17   Rear view of a GFI receptacle, showing the connection leads

Courtesy Smith & Stone

FIGURE 4.19   A commercial-duty, specification grade ground fault protection receptacle and cover plate

Courtesy Hubbell Canada Inc.

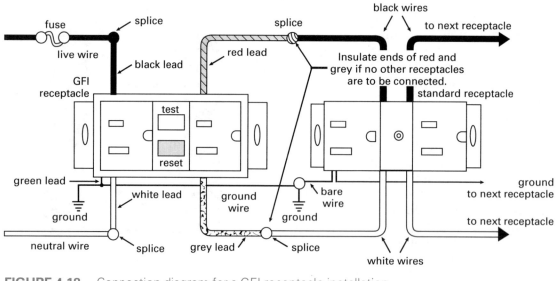

FIGURE 4.18   Connection diagram for a GFI receptacle installation

Applications of Electrical Construction

FIGURE 4.20   A typical electric shaver outlet

Courtesy Smith & Stone

green dot

Courtesy Leviton Manufacturing of Canada Ltd..

FIGURE 4.21   A 125 V "U" ground recepta-cle of hospital grade, identified by a green dot on the lower face

be tested at least once a month by press-ing the test button on the face of the receptacle. This simulates a ground fault in the circuit and the relay within the receptacle should react and open the circuit. The reset button will at this time protrude from the face of the receptacle, indicating successful operation of the unit. Pressing the reset button back into place will reactivate the relay in the receptacle for further use. This simple test assures the owner that the unit is in proper working order.

Prior to recent GFI developments, electric shaver receptacles that isolated the supply voltage from the shaver were available. This eliminated the shock haz-ard present when electric shaving equip-ment was used near plumbing fixtures. These receptacles were, however, too small (electrically) for use with modern hair driers. (See Fig. 4.20)

Hospital Grade Grounding Receptacles

Receptacles in hospitals and other medi-cal facilities are often subjected to severe use and mechanical abuse. When an emergency occurs or a life is at stake, time becomes important. Medical staff will frequently move plugged-in pieces of equipment about, causing unintentional abuse to both the cord connector and receptacle.

Hospital grade receptacles are made from a heavy-duty, abuse-resistant ther-moplastic, capable of withstanding impact damage, while resisting the tend-ency to crack or break. The possibility of short circuits is reduced by having a thick-walled, moulded base of similar material. A one-piece, integral grounding contact ensures proper grounding of the circuit and equipment being used. Hos-pital grade receptacles are identified by the *green dots* on their faces, visible even when their cover plates are installed. (See Fig. 4.21)

Receptacle Covers

Indoor receptacles use standard-sized metal or plastic covers when installed in residential circuits. (See Fig. 4.22) When covers are used outdoors or in other damp areas, they should be the type that provides easy access to the recepta-cle but keeps out weather-moisture and dirt. (See Fig. 4.23)

Receptacles

FIGURE 4.22   Indoor receptacle and cover

FIGURE 4.24   Combination switch and receptacle unit

device in another area of the room or building. See Figure 4.24 for a typical unit.

FIGURE 4.23   Weather-resistant receptacle cover

Combination Units

When installation space is limited, a unit that allows an installer to locate the switch and receptacle in the same single gang box is recommended. The switch can either control the receptacle portion of the unit, a light fixture, or another

F o r   R e v i e w

1. What duty range of receptacle should be selected for kitchen use? Why?
2. Where in the home may light-duty receptacles be safely installed?
3. Where can information be found regarding the number of receptacles required for a room?
4. Explain where the live, neutral, and ground wires are connected on a duplex "U" ground receptacle.
5. List two uses for a switch and receptacle combination unit.
6. What is the main disadvantage of the old-style 2 prong receptacle?
7. Would an extension cord fitted with twist-lock units be an advantage on a construction job site? Why?
8. Explain the purpose of the ground slot on a receptacle.
9. Where does the ground in a receptacle originate?

Applications of Electrical Construction

10. How is the ground circuit carried in a conduit system?
11. What is the danger in using a portable electric tool that has had the grounding prong removed from its plug? Explain.
12. List two areas for use of split receptacles in the home.
13. What are the advantages of using split receptacles?
14. What is a NEMA receptacle? Explain.
15. List three examples of polarized receptacles.
16. What are plaster ears? Where and why are they used?
17. What type of receptacle is recommended by the Electrical Code for use in outdoor areas?
18. What advantage has a GFI receptacle over other types of receptacles?
19. Why will a normal fuse or circuit breaker not protect a person using portable tools on a circuit?
20. Why is it important for a GFI unit to detect leakage currents as low as 5 mA?

5

Conductors

One of the most important parts of any electrical system is the conductor that makes up the wiring circuit. Most conductors are rarely checked after the installation has been approved. For this reason and others, the choice of conductor is an important feature of circuit design.

Conductor Materials

Conductors used for residential and industrial wiring circuits are usually made from copper, aluminum, or steel. If exposed to the air, these metals combine with the oxygen. This *oxidation* produces a layer of oxide on the surface of the conductor. If a conductor is allowed to oxidize at the terminal point or splice, the current-carrying ability of the terminal will be seriously reduced. That is because the oxide layer does not conduct electricity.

Copper. Conductors are often made of copper, because it is an excellent conductor, easy to work with and handle, and does not oxidize as much as aluminum or steel. Copper is also used in electronic circuits because it solders readily, ensuring a secure electrical connection. Copper has become expensive, however, and a substitute is being sought.

Copper wire *splices* must be made secure to prevent the oxide coating from reducing the current-carrying ability of the splice.

Aluminum. Aluminum is lighter than copper but not as good a conductor. To obtain the same current-carrying capacity, therefore, an aluminum conductor must be slightly larger than one made of copper. Also, aluminum oxidizes rapidly. The aluminum oxide acts as an insulator and reduces electrical current flow.

Using aluminum poses two other major problems. One is *electrolysis*, which is the chemical breakdown of two metals reacting with one another. Aluminum will react with some metals in this way. The process is accelerated by moisture and the flow of current in the conductors. Approved connection devices resist this process, and so approved terminal splice connectors should be used. Also, *antioxidant chemicals* to prevent oxidation of the wire are available. These chemicals are usually applied with a brush, forcing the chemical into the strands of the cable. Figure 5.1 illustrates a typical chemical available for this purpose.

The second major problem with aluminum conductors is that the connection

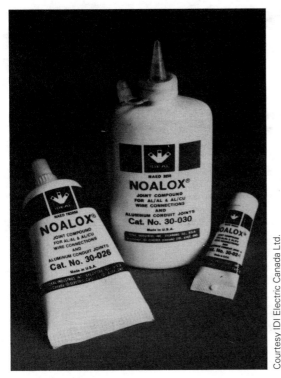

FIGURE 5.1 Antioxidant chemical for use on aluminum conductors

points lose their grip. A terminal screw can be tightened firmly, but in a few days, the aluminum will accept the new shape into which it has been forced. This reduces the pressure of the terminal screw on the wire, and the aluminum flows in to the new shape. This flow—called *cold flow*—results in a loosened connection. As a result, the terminal connection will overheat, causing damage to the connection and insulation.

If handled properly, aluminum is an effective conductor that provides many years of service. It is not of much use, however, for circuits where solder connections are necessary, because aluminum cannot be soldered simply.

Steel. Steel is the strongest of the three metals and is used mainly as a supporting material. For example, aluminum and copper cables are often wound around a steel centre core for outdoor wiring. Conductors made this way can withstand a great deal of stress. Also, electric railway or subway systems often use steel rails as the conductors in their supply system. The rails are sometimes made of special steels containing a small amount of copper to improve the current-carrying ability of the rails.

While the metal conduit and boxes are *not* used to carry current for the operation of electrical equipment, steel enclosures *are* used to complete the grounding circuit. In rural service installations, for example, two steel rods are often driven into the ground to provide a ground connection where there is no metal water supply system to the building.

Conductor Forms

Conductors used for residential and industrial wiring circuits are made in the form of wire, cable, and cord.

Wire. Wire, which is a single, solid strand, is the least flexible form of conductor.

Copper wire is usually made by *drawing* a soft copper rod, which is from 6 mm to 13 mm in diameter, through a series of doughnut-shaped metal blocks called *dies*. The hard centre of each die has a hole slightly *smaller* than the hole of the die through which the rod has just been drawn. As the rod passes through the dies, it is reduced in diameter and lengthened.

The wire tends to harden after passing through several dies. This *hard-drawn* copper is used for some outdoor wires because it is quite strong.

If the diameter needs to be made still

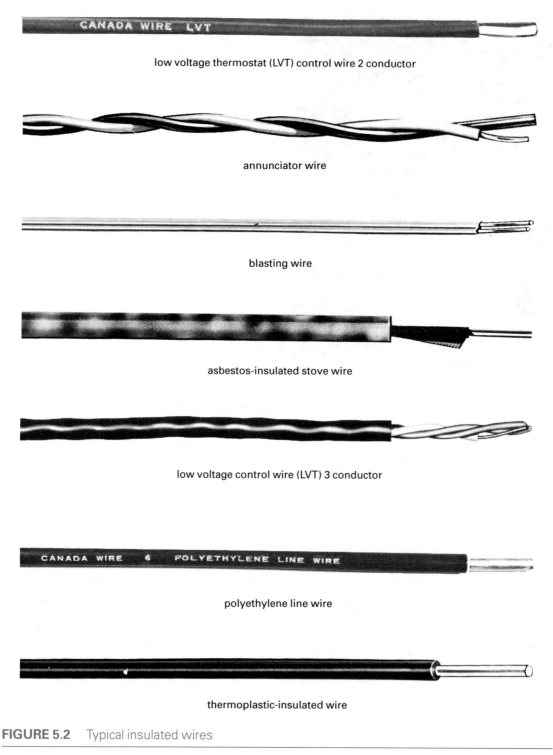

low voltage thermostat (LVT) control wire 2 conductor

annunciator wire

blasting wire

asbestos-insulated stove wire

low voltage control wire (LVT) 3 conductor

polyethylene line wire

thermoplastic-insulated wire

Courtesy Canada Wire & Cable Ltd.

FIGURE 5.2 Typical insulated wires

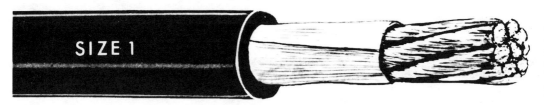

thermoplastic–insulated, weather- and heat-resistant cable

thermoplastic–insulated, weather- and heat-resistant cable

SIZE 1

flexible welding cable

EXELENE 600 VOLTS

service entrance cable

armoured cable

nonmetallic sheathed cable

FIGURE 5.3 Typical insulated cables

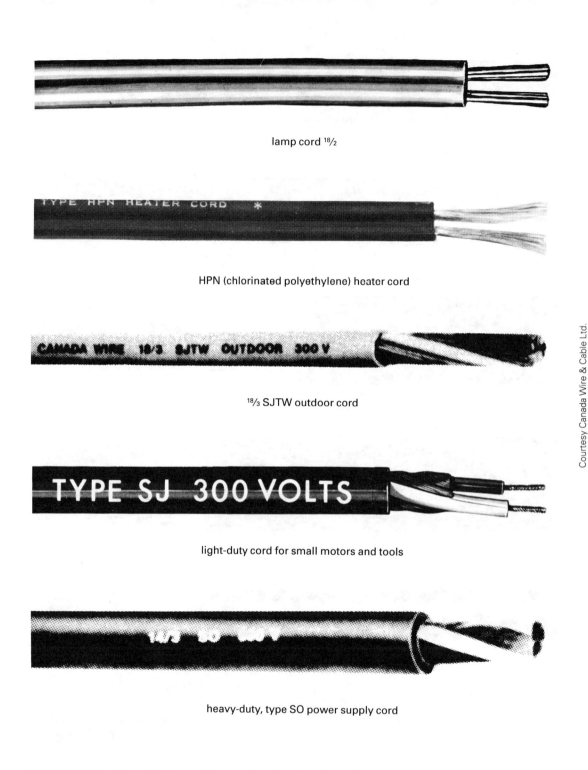

lamp cord $^{18}/_2$

HPN (chlorinated polyethylene) heater cord

$^{18}/_3$ SJTW outdoor cord

light-duty cord for small motors and tools

heavy-duty, type SO power supply cord

Courtesy Canada Wire & Cable Ltd.

FIGURE 5.4 Typical light- and heavy-duty insulated cords

smaller, the wire is first passed through an *annealing oven*, which heats and softens it. The wire can then be further reduced to any size required. The annealing process may have to be repeated several times. The wire may also be insulated with a variety of materials. (See Fig. 5.2 on page 50.)

Aluminum wire is produced in the same way.

Cable. A cable is a *compound* conductor made of a number of strands of wire. Some are twisted together to form a large conductor before being insulated. This type is usually used in circuits where there is a large flow of current. Others are assembled and placed under a common cover after each wire has been insulated. This type is used extensively for the wiring of buildings. (See Fig. 5.3 on page 51.)

A cable is more flexible than a wire of the same size.

Cord. The cord's conductors are made of many strands of fine wire twisted together. The cord is made of two or more separately insulated conductors assembled within an insulating jacket. It is the most flexible form of conductor and is used to supply current to hand-held appliances and portable tools, where freedom of movement is important. Appliance and tool cords sometimes have a third conductor (with green insulation), which is used for grounding. (See Fig. 5.4)

Conductor Sizes

Conductor sizes are measured and listed in two ways. One method is based on the American Wire Gauge. The other is based on area.

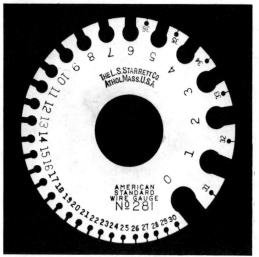

Courtesy Ron Ridsdill, Toronto

FIGURE 5.5 American Wire Gauge used to measure size of solid conductors (wires)

American Wire Gauge (AWG). This gauge is used to measure only *solid* conductors (wire). The outer edge of the gauge has slots, which are numbered. The *smallest slot* into which the wire will fit is the *gauge number* of the wire. (See Figs. 5.5 and 5.6) The AWG can measure wire from No. 36 (the smallest size) to No. 0 (the largest size).

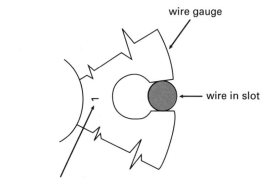

FIGURE 5.6 Using the American Wire Gauge

Wire produced for special applications can be as small as No. 44, however, and as large as No. 0000 (4/0). The small, hair-like wires are used in the windings of electric motors and similar equipment.

Area of Cross-Section. The size of a larger conductor is determined by calculating the conductor's cross-sectional area. When a compound conductor, such as a cable, is to be measured, calculate the area of one strand, then multiply the area of that strand by the number of strands in the cable. Traditionally, the area of cross-section has been referred to as the circular mil area. However, under the metric system, the cross-sectional area is measured in square millimetres. (The symbol is mm².) The cross-sectional area is obtained by using the formula $A = \pi \times r^2$, where A is area, r is the radius of the conductor in millimetres, and π is a constant (3.14).

Nonetheless, the term *MCM* still applies to particular sizes of wires. Used for large conductors (over 4/0 in size), it refers to thousands of circular mils. A conductor with a diameter of a thousandth (0.001 in.; 0.002 cm) of an inch has a diameter of 1 circular mil. (See Fig. 5.7A) Conductors larger than 1 mil in diameter must be measured with a micrometer to determine their diameters in mils. If, for example, a No. 6 AWG conductor has a diameter of 0.162 in. (0.411 cm), its diameter in mils is 162. Circular mil area is calculated by squaring the diameter in mils: this AWG conductor would be 26 240 circular mils.

Determining the circular mil areas of large conductors is more complicated, because such conductors are usually stranded to improve their flexibility. *Stranded* conductors are made up of

several rows of strands as follows. Row 1 is a single strand. Row 2 consists of 6 strands twisted over the top of row 1. The third row would contain 12 strands wrapped in the opposite direction to those of the second row. A fourth row would have 18 strands, wound in the opposite direction to those of the third row. Each new layer or row of strands added to the conductor will contain six more strands than the previous row. (See Fig. 5.7B) The number of strands must be calculated to arrive at the conductor's total circular mil area.

The next step is to determine the diameter of one strand, using the same method you would for small conductors.

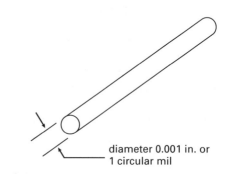

FIGURE 5.7A Diameter of conductor in circular mils

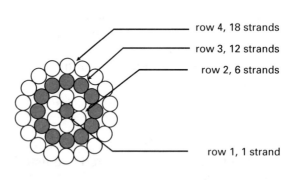

FIGURE 5.7B A multi-stranded conductor with 4 rows and 37 strands

Applications of Electrical Construction

Cable strands are frequently produced in diameters that suit cable size and design more than standard wire gauge sizes and dimensions. If a cable has a single-strand diameter of 0.116 in. (0.294 cm), the strand's diameter would be 116 mils. The circular mil area of the strand would be 13 456 circular mils (116 × 116). The size of the complete cable, as shown in Figure 5.7B, would be found by multiplying 13 456 × 37 strands. This 497 872 circular mil cable would be considered a 500 000 circular mil cable by Electrical Code tables—a 500 MCM cable.

Tables 5.1 and 5.2 show the metric sizes of various types of conductors. Bear in mind, however, that the current edition of the Canadian Electrical Code and much of the electrical industry still rely on the imperial system of measurement. Tables 5.3 and 5.4 show the imperial dimensions and sizes of bare copper wire in both solid and stranded configurations.

Wire Size Uses

Wire and cable for buildings are made in even gauge sizes, such as Nos. 14, 12, 10, 8, etc. Odd gauge sizes, such as Nos. 15, 13, 11, 9, etc., are not used for building wire and cable because there is not enough difference in current-carrying capacity (ampacity) to make production of these odd sizes worthwhile.

No. 10 gauge wire is the *largest single-strand* conductor allowed under a terminal screw by the Canadian Electrical Code. Larger conductors must be inserted into a compression-type fitting, called a *lug.* (See Fig. 5.8) Since wire and cable for buildings need to be flexible, conductors for this use are usually stranded when made in sizes larger than No. 10 gauge.

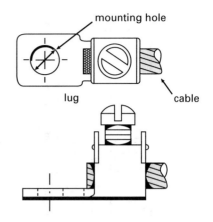

FIGURE 5.8   Method for installing cable in a lug

Wire for motors, transformers, and other magnetic equipment is made in both odd and even gauge numbers, because wire size for these uses is more critical. Conductors with large cross-sectional areas are used for industrial applications where a large amount of current must be carried.

Conductor Insulation

There are conductor insulators made for a wide variety of uses and situations. The more common types are thermoplastic, rubber and cotton, neoprene, asbestos, varnish, glass over thermoplastic, and cross link.

Thermoplastic.   This is one of the most common insulators used for residential and industrial wiring. Thermoplastic is an excellent insulator, but it is sensitive to extremes of temperature. At *high* temperatures, it melts. At *low* temperatures, it becomes brittle and cracks if handled roughly.

Weatherproof and heat-resistant thermoplastics are available, however. Thermoplastic weatherproof insulation is called *TW*. The *TW-75* type is

TABLE 5.1 Dimensions and related data for insulated copper and aluminum conductors

Copper and Aluminum Conductors   TW-75

Flameseal PVC Insulated Cables   600 V 75°C

Size AWG or MCM	Insulation Thickness mm	in.	Approximate Diameter mm	in.	Approximate Net Cable Mass — Copper Condr* kg/1000 m	lb./1000 ft.	Aluminum Condr* kg/1000 m	lb./1000 ft.	Ampacity (30°C Ambient) — Copper (amps) Free Air	Conduit	Aluminum Free Air	Conduit
SOLID												
14	0.8	0.030	3.2	0.13	30	20	—	—	20	15	—	—
12	0.8	0.030	3.6	0.14	35	25	25	15	25	20	20	15
10	0.8	0.030	4.2	0.16	60	40	30	20	40	30	30	25
STRANDED												
14	0.8	0.030	3.4	0.13	30	20	—	—	20	15	—	—
12	0.8	0.030	3.9	0.15	45	30	25	15	25	20	20	15
10	0.8	0.030	4.5	0.18	60	40	30	20	40	30	30	25
8	1.1	0.045	6.0	0.24	100	70	50	35	65	45	45	30
6	1.5	0.060	7.7	0.30	165	110	80	55	95	65	75	50
4	1.5	0.060	8.7	0.35	245	165	110	75	125	85	100	65
3	1.5	0.060	9.6	0.38	300	200	130	90	145	100	115	75
2	1.5	0.060	10.4	0.41	370	250	160	105	170	115	135	90
1	2.0	0.080	12.4	0.49	480	320	210	140	195	130	155	100
1/0	2.0	0.080	13.4	0.53	590	395	250	170	230	150	180	120
2/0	2.0	0.080	14.5	0.57	725	490	300	200	265	175	210	135
3/0	2.0	0.080	15.8	0.62	900	605	365	245	310	200	240	155
4/0	2.0	0.080	17.2	0.68	1 115	750	440	295	360	230	280	180
250	2.4	0.095	19.2	0.76	1 330	895	530	355	405	255	315	205
300	2.4	0.095	20.6	0.81	1 575	1 060	615	415	445	285	350	230
350	2.4	0.095	21.8	0.86	1 820	1 225	700	470	505	310	395	250
400	2.4	0.095	23.0	0.91	2 065	1 390	785	530	545	335	425	270
500	2.4	0.095	25.1	0.99	2 545	1 710	950	640	620	380	485	310

*"Condr" means "Conductor."

Colour—Standard colours are available.

TABLE 5.2 Dimensions and related data for bare copper and aluminum stranded conductors

Stranded Bare Copper and Aluminum Conductors

Conductor Size AWG	cmil	Area mm²	Area sq. in.	No. of Wires (a), (b)	Wire Diameter mm	Wire Diameter in.	Compressed Round mm	Compressed Round in.	Compact Round mm	Compact Round in.	kg/1000 m Copper	kg/1000 m Aluminum	lb./1000 ft. Copper	lb./1000 ft. Aluminum	Ω/1000 m Copper	Ω/1000 m Aluminum	Ω/1000 ft. Copper	Ω/1000 ft. Aluminum
14	4 110	2.08	0.003 23	7	0.61	0.024 2	1.80	0.071			18.9		12.7		8.61		2.63	
12	6 530	3.31	0.005 13	7	0.77	0.030 5	2.26	0.089			30.0	9.1	20.2	6.13	5.42	8.89	1.65	2.71
10	10 380	5.26	0.008 16	7	0.98	0.038 5	2.87	0.113			47.7	14.5	32.1	9.75	3.41	5.59	1.04	1.70
8	16 510	8.37	0.012 97	7	1.23	0.048 6	3.61	0.142	3.40	0.134	75.9	23.1	51.0	15.5	2.14	3.52	0.653	1.07
6	26 240	13.30	0.020 61	7	1.55	0.061 2	4.55	0.179	4.29	0.169	121	36.7	81.1	24.6	1.35	2.21	0.411	0.674
4	41 740	21.15	0.032 78	7	1.96	0.077 2	5.72	0.225	5.41	0.213	192	58.3	129	39.2	0.848	1.39	0.258	0.424
3	52 620	26.66	0.041 33	7	2.20	0.086 7	6.40	0.252	6.05	0.238	242	73.5	163	49.4	0.673	1.10	0.205	0.336
2	66 360	33.63	0.052 12	7	2.47	0.097 4	7.19	0.283	6.81	0.268	305	92.7	205	62.3	0.533	0.875	0.163	0.267
1	83 690	42.41	0.065 73	19(18|18)	1.69	0.066 4	8.18	0.322	7.59	0.299	385	117	258	78.6	0.423	0.694	0.129	0.211
1/0	105 600	53.51	0.082 91	19(18|18)	1.89	0.074 5	9.19	0.362	8.53	0.336	485	147	326	99.1	0.335	0.550	0.102	0.168
2/0	133 100	67.44	0.104 5	19(18|18)	2.13	0.083 7	10.31	0.406	9.55	0.376	611	186	411	125	0.266	0.436	0.0811	0.133
3/0	167 800	85.03	0.131 8	19(18|18)	2.39	0.094 0	11.58	0.456	10.7	0.423	771	234	518	157	0.211	0.346	0.0643	0.105
4/0	211 600	107.22	0.166 2	19(18|18)	2.68	0.105 5	13.00	0.512	12.1	0.475	972	296	653	199	0.167	0.274	0.0510	0.0836
250 kcmil		126.68	0.196 3	37(36|35)	2.09	0.082 2	14.17	0.558	13.2	0.520	1 149	350	772	235	0.142	0.232	0.043 2	0.070 8
300		152.01	0.235 6	37(36|35)	2.23	0.090 0	15.52	0.611	14.5	0.570	1 378	419	926	282	0.118	0.194	0.036 0	0.059 0
350		177.35	0.274 9	37(36|35)	2.47	0.097 3	16.79	0.661	15.6	0.616	1 609	489	1 081	329	0.101	0.166	0.030 8	0.050 6
400		202.68	0.314 2	37(36|35)	2.64	0.104 0	17.93	0.706	16.7	0.659	1 838	559	1 235	376	0.088 5	0.145	0.027 0	0.044 2
500		253.35	0.392 7	37(36|35)	2.95	0.116 2	20.03	0.789	18.7	0.736	2 298	699	1 544	469	0.070 8	0.116	0.021 6	0.035 4
600		304.02	0.471 2	61(58|58)	2.52	0.099 2	22.68	0.886	20.7	0.813	2 758	838	1 854	564	0.059 0	0.096 7	0.018 0	0.029 5
750		380.03	0.589 0	61(58|58)	2.82	0.110 9	24.59	0.968	23.1	0.908	3 447	1 048	2 316	704	0.047 2	0.077 4	0.014 4	0.023 6
1 000		506.71	0.785 4	61(58|58)	3.25	0.128 0	28.37	1.117	26.9	1.060	4 595	1 396	3 088	938	0.035 4	0.058 0	0.010 8	0.017 7
1 250		633.38	0.981 7	91	2.98	0.117 2	31.75	1.250			5 743	1 750	3 859	1 174	0.028 3	0.046 4	0.008 63	0.014 2
1 500		760.06	1.178	91	3.26	0.128 4	34.80	1.370			6 892	2 100	4 631	1 410	0.023 6	0.038 7	0.007 19	0.011 8
1 750		886.74	1.374	127	2.98	0.117 4	37.59	1.480			8 041	2 440	5 403	1 640	0.020 2	0.033 2	0.006 16	0.010 1
2 000		1 013.42	1.571	127	3.19	0.125 5	40.21	1.583			9 190	2 790	6 175	1 880	0.017 7	0.029 0	0.005 39	0.008 85

(a) Reduced number of wires for copper compact strandings shown in ( ) parentheses.

(b) Reduced number of wires for aluminum compact strandings shown in [ ] parentheses.

(c) Approximate weights and average DC resistances are considered to apply to all types of strands.

Conductor data and metric equivalents in this table are based where possible on EEMAC recommendations current at time of compilation, otherwise on published ICEA standards.

TABLE 5.3 Dimensions, Weights, and Resistance of Bare Copper Wire, Solid, AWG Sizes

Size AWG	Diameter mil	Area cmil	Weight lb./1000 ft.	Resistance Ω/1000 ft. 20°C Annealed Wire
0 000	460.0	211 600	640.5	0.049 0
000	409.6	167 800	507.8	0.061 8
00	364.8	133 100	402.8	0.077 9
0	324.9	105 600	319.5	0.098 3
1	289.3	83 690	253.3	0.124
2	257.6	66 360	200.9	0.156
3	229.4	52 620	159.3	0.197
4	204.3	41 740	126.3	0.249
5	181.9	33 090	100.2	0.313
6	162.0	26 240	79.44	0.395
7	144.3	20 820	63.03	0.498
8	128.5	16 510	49.98	0.628
9	114.4	13 090	39.62	0.793
10	101.9	10 380	31.43	0.999
11	90.7	8 230	24.92	1.26
12	80.8	6 530	19.77	1.59
13	72.0	5 180	15.68	2.00
14	64.1	4 110	12.43	2.52
15	57.1	3 260	9.87	3.18
16	50.8	2 580	7.81	4.02
17	45.3	2 050	6.21	5.05
18	40.3	1 620	4.92	6.39
19	35.9	1 290	3.90	8.05
20	32.0	1 020	3.10	10.1
22	25.3	640	1.94	16.2

thermo-plastic, weatherproof, and heat-resistant. Some of the heat produced by equipment such as electric heaters, stoves, and light fixtures travels back through the conductor, damaging the insulation. TW-75 is capable of withstanding such heat.

Rubber and Cotton. Homes wired with the knob and tube system have conductors covered first with a layer of rubber and then an outer braid of cotton for extra protection. Also, the copper wire is coated with tin to prevent premature oxidation of the copper, because of the sulphur content of the rubber. This form of insulation is rarely used for modern wiring systems.

Neoprene. This is a special type of rubber insulation used widely on heat-proof line cords for such

Applications of Electrical Construction

Size AWG or cmil	Overall diameter mil	Number of Strands, Class B Stranding	Weight lb./1000 ft.	Resistance Ω/1000 ft. 20°C Annealed Wire
TABLE 5.4 Dimensions, Weights, and Resistance of Bare Copper Wire, Stranded, AWG and cmil Sizes				
2 000 000	1 630	127	6 175	0.005 29
1 750 000	1 526	127	5 403	0.006 05
1 500 000	1 411	91	4 631	0.007 05
1 250 000	1 288	91	3 859	0.008 46
1 000 000	1 152	61	3 088	0.010 6
900 000	1 094	61	2 779	0.011 8
800 000	1 031	61	2 470	0.013 2
750 000	998	61	2 316	0.014 1
700 000	964	61	2 161	0.015 1
600 000	891	61	1 853	0.017 6
500 000	813	37	1 544	0.021 2
450 000	772	37	1 389	0.023 5
400 000	726	37	1 235	0.026 5
350 000	679	37	1 081	0.030 2
300 000	629	37	925	0.035 3
250 000	574	37	772	0.042 3
0 000	552	19	653	0.050 0
000	492	19	518	0.063 1
00	414	19	411	0.079 5
0	368	19	326	0.100
1	328	19	259	0.126
2	292	7	205	0.159
3	260	7	162	0.201
4	232	7	129	0.253
5	206	7	102	0.320
6	184	7	80.9	0.403
7	164	7	64.2	0.508
8	146	7	51.0	0.641
9	130	7	40.4	0.808
10	116	7	32.1	1.020

heat-producing appliances as teakettles, frying pans, and soldering irons. Since neoprene is also oil-resistant, it can be used on extension cords for service stations. The abbreviation for heat-proof neoprene insulation is *HPN*.

Asbestos. At one time, major appliances such as electric stoves used asbestos-insulated wires for the heat-proof insulation required for their internal circuits. For example, asbestos insulation was used for connections between elements and control switches. It also appeared in the cords of certain appliances such as soldering irons and toasters.

It has been established that asbestos fibres can harm persons who breathe them in over a period of time. As a result, manufacturers no longer produce asbestos-insulated wire, which can still be found in older equipment in both homes and industries.

Varnish. Copper wire with a baked-on varnish insulation is used extensively for motor windings. The high quality of the insulating varnish means that the insulation can be very thin, allowing space for the many windings required. This type of insulation has a temperature rating in excess of 200°C. Its abbreviation is *V*.

Glass over Thermoplastic. Conductors supplying recessed fixtures require a heat-resistant insulation because there is little, if any, air circulation to cool the conductors. *GTF* (glass and thermoplastic for fixtures) is used for this purpose.

Cross Link. This material is thermo-setting and will not melt. It has a higher temperature rating than TW and is now widely used for building wire.

Ampacity

The purpose of a conductor is to carry current from one place in a circuit to another. *Ampacity* refers to the ability of a conductor to carry current. (See Tables 5.5 and 5.6)

The ampacity rate of a conductor is determined by its material, size, and insulation.

Material. The material of which the conductor is made determines how easily it will carry current. For example, copper is a better conductor than aluminum and will therefore carry more current.

Size. The larger the conductor, the more current it will carry without heating. Since conductors are often enclosed in conduits and boxes, take care to use conductors of large enough size and to provide sufficient space for air circulation. By doing so, you will avoid conductor overheating and thus prevent damage to the insulation.

Insulation. A conductor with insulation capable of withstanding heat will have a higher ampacity rating than a conductor of the same size with a lower insulator temperature-rating. Wires with *heat-resistant covering* can be contained in a more confined area than normal conductors, without heat damage to the insulation.

The *voltage* of the circuit will also affect the choice of insulator. The conductors must be insulated from each other as well as from ground, and the insulation must be capable of withstanding the electrical pressure of the circuit.

Under high-current conditions, the current flows mainly along the surface of the conductor. This is known as the *skin effect*, and steel-core conductors can be used in such situations to combine strength and ampacity, because the outer section of the conductor carries the bulk of the current.

Conductors must be handled carefully to prevent surface nicks and scratches. The conductor will heat at the point of a surface nick, because the nick will have reduced the cross-sectional area. If the nick is deep enough,

TABLE 5.5 Allowable Ampacities for Not More Than 3 Copper Conductors in a Raceway or Cable

Based on Ambient Temperature of 30°C

	Allowable Ampacity					
	60°C	**75°C**	**85-90°C**	**110°C**	**125°C**	**200°C**
Size AWG MCM	**Type TW**	**Types RW-75, TW-75**	**Types R-90, RW-90, T-90 Nylon, Mineral-insulated cable, Paper**			
Col. 1	**Col. 2**	**Col. 3**	**Col. 4**	**Col. 5**	**Col. 6**	**Col. 7**
14	15	15	15	30	30	30
12	20	20	20	35	40	40
10	30	30	30	45	50	55
8	40	45	45	60	65	70
6	55	65	65	80	85	95
4	70	85	85	105	115	120
3	80	100	105	120	130	145
2	100	115	120	135	145	165
1	110	130	140	160	170	190
0	125	150	155	190	200	225
00	145	175	185	215	240	250
000	165	200	210	245	265	285
0 000	195	230	235	275	310	340
250	215	255	265	315	335	—
300	240	285	295	345	380	—
350	260	310	325	390	420	—
400	280	335	345	420	450	—
500	320	380	395	470	500	—
600	355	420	455	525	545	—
700	385	460	490	560	600	—
750	400	475	500	580	620	—
800	410	490	515	600	640	—
900	435	520	555	—	—	—
1 000	455	545	585	680	730	—
1 250	495	590	645	—	—	—
1 500	520	625	700	785	—	—
1 750	545	650	735	—	—	—
2 000	560	665	775	840	—	—

For full information of conditions that may change the values in this table, see the corresponding table in the Canadian Electrical Code.

Based on Canadian Electrical Code

TABLE 5.6 Allowable Ampacities for Not More Than 3 Aluminum Conductors in a Raceway or Cable

Based on Ambient Temperature of 30°C

	Allowable Ampacity					
	60°C	**75°C**	**85-90°C**	**110°C**	**125°C**	**200°C**
Size AWG MCM	**Type TW**	**Types RW-75, TW-75**	**Types R-90, RW-90, T-90 Nylon, Paper**			
Col. 1	**Col. 2**	**Col. 3**	**Col. 4**	**Col. 5**	**Col. 6**	**Col. 7**
12	15	15	15	25	30	30
10	25	25	25	35	40	45
8	30	30	30	45	50	55
6	40	50	55	60	65	75
4	55	65	65	80	90	95
3	65	75	75	95	100	115
2	75	90	95	105	115	130
1	85	100	105	125	135	150
0	100	120	120	150	160	180
00	115	135	145	170	180	200
000	130	155	165	195	210	225
0 000	155	180	185	215	245	270
250	170	205	215	250	270	—
300	190	230	240	275	305	—
350	210	250	260	310	335	—
400	225	270	290	335	360	—
500	260	310	330	380	405	—
600	285	340	370	425	440	—
700	310	375	395	455	485	—
750	320	385	405	470	500	—
800	330	395	415	485	520	—
900	355	425	455	—	—	—
1 000	375	445	480	560	600	—
1 250	405	485	530	—	—	—
1 500	435	520	580	650	—	—
1 750	455	545	615	—	—	—
2 000	470	560	650	705	—	—

For full information of conditions that may change the values in this table, see the corresponding table in the Canadian Electrical Code.

Based on Canadian Electrical Code

there may be serious heat damage to the surrounding insulation.

Electrical Resistance

Current does not flow through a conductor by itself. Electrical pressure (voltage) must be applied to *force* the current along the conductor. There is always some form of *resistance* to current flow in the conductor. This resistance is measured in *ohms*.

Resistance is determined by the conductor's diameter, length, temperature, and material. The smaller the diameter, the longer the length, and the higher the temperature, the greater the resistance will be.

Resistance has little, if any, effect on electrical performance in the average residential circuit. In tall buildings, sprawling industrial complexes, or street lighting, where conductors must travel long distances, however, resistance is an important factor. Considerable voltage may be lost in delivering the current to the end of the circuit. The result is dimming lights and less efficient electrical motors. *Voltage drop* is reduced by increasing the diameter of the wire used in the circuit. Tables 5.3 and 5.4 provide a comparison of the various resistances for solid and stranded wires.

F o r   R e v i e w

1. What two materials are most commonly used for electrical conductors? Why?
2. Define *oxidation*, and describe its effect on metal conductors.
3. Describe the process for producing wire.
4. Why are cables and cords made from many strands of wire?
5. What is the name of the gauge used to measure solid conductors? What sizes of wire can this gauge measure?
6. What method is used to measure the size of large cables?
7. A cable consists of 19 strands. Each strand has a metric diameter of 2.388 mm. What is the gauge number of the cable?
8. A cable consists of 37 strands, each having a diameter of 80.8 mil. What is the gauge number of the individual strand? Calculate the nominal size of the cable itself.
9. List five or more materials commonly used for insulating conductors.
10. Define *ampacity*, and list and explain the factors that affect it.
11. Why should circuit voltage be considered when selecting a conductor?
12. Explain why cuts or other damage to conductors should be avoided during installation.
13. List and explain the factors that determine a conductor's resistance.
14. What is the effect of conductor resistance on circuit voltage? Give examples.
15. Calculate the resistance of 152 m of No. 14 AWG solid bare copper wire.

6

Cord Fittings

Any electrical device that is designed to receive its electrical supply from a receptacle will be fitted with a line cord. Cord fittings attached to the line cord are used as a convenient method for connecting the device to the power source.

Male and Female Plug Caps

There are two main types of cord fittings, or *plug caps: male* and *female*. Both are produced with the same blade shapes, as described in Chapter 4. (See Fig. 4.2)

The *male* plug cap is designed to be *inserted* in to the slots of a receptacle. (See Fig. 6.1) It is usually attached to the end of line cords on electrical appliances, lamps, and power tools.

FIGURE 6.1   Male plug cap

Courtesy Smith & Stone

Courtesy Smith & Stone

FIGURE 6.2   Female plug cap (connector body)

The *female* plug cap is designed to *receive* the male plug cap. (See Fig. 6.2) It is attached to one end of an extension cord, which has a male plug cap on the other end. Usually, both have the same blade shape.

Uses for Heavy-Duty Cord Caps

Mobile homes (trailers), electric stoves, and clothes driers require heavy-duty cord caps for their power supply. In industry, cord fittings with current and voltage ratings other than the usual 10 A to 15 A, 120 V residential ratings are often required. Some examples are cord caps for welding machines, floor finishers, battery chargers, and marine shore lines.

Plug Cap Connection

Plug caps are often damaged by carelessness. For example, many people remove plugs from receptacles by tugging on cords. Figure 6.3 shows how line cords and plug caps are connected so that the strain from pulling on the cord will be absorbed by the *blades* rather than by the terminals. To prevent loose strands of the flexible cord from slipping out of the terminal connection, the strands are *twisted* together and then wound *clockwise* around the terminal screw. Soldering the groups of twisted strands before assembly also helps make the electrical connection secure.

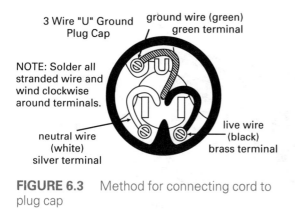

FIGURE 6.3 Method for connecting cord to plug cap

An *underwriter's knot* used to be tied in the cord to prevent the cord from being pulled out of the cord cap. But stress was placed on the cord's insulation. As a result, a knot is seldom used with modern plug caps in which conductor space is limited. Instead, industrial or heavy-duty plug caps are now equipped with metal or plastic *clamping devices* to hold the cords in the plug caps. (See Fig. 6.4) These devices are particularly useful on construction sites, where extension cords and caps are walked on, driven over, and treated roughly.

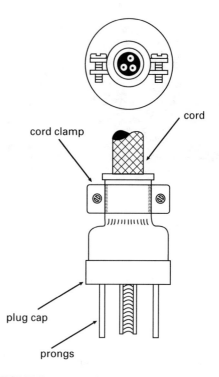

FIGURE 6.4 Industrial (heavy-duty) plug cap equipped with cord clamp

Dead-Front Plug Caps

To promote the personal safety of portable equipment users, the Canadian Electrical Code now requires that all new plug caps be of *dead-front* construction. A dead-front plug cap has prongs and terminals assembled as a removable unit. (See Figs. 6.5, 6.6, and 6.7) The cord connects to the rear of the unit. The front, which is the exposed portion of the plug cap, is free of terminals, conductors, and insulating disc.

The dead-front plug cap represents a significant improvement over the many older, plug cap models it is being used to replace. On older models, the *insulating disc* of the male plug cap fits separately over the prongs where the cord has been connected. The disc's purpose is to

Courtesy Hubbell Canada Inc.

prevent loose strands or terminals from coming in contact with the cover plate of the receptacle. But if the cover plate is made of metal and no insulating disc is present, a short-circuit flash can occur as the plug is inserted in to the receptacle. The hand of the person plugging in the cap may be burned.

The safer dead-front plug caps are produced in all cord cap configurations and in both residential and industrial grades.

FIGURE 6.5 Light-duty, dead-front plug cap in the "U" ground configuration, black neoprene body

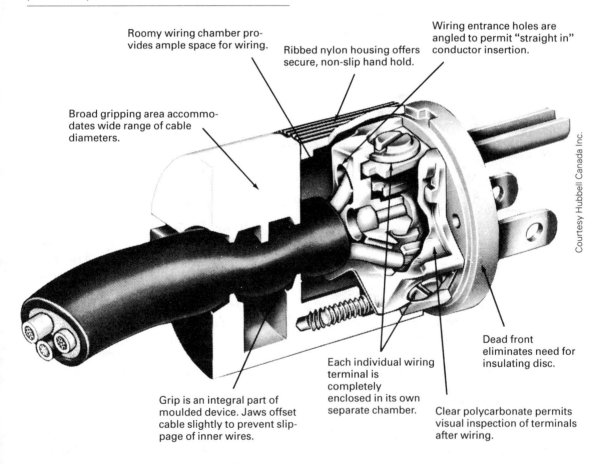

Roomy wiring chamber provides ample space for wiring.

Ribbed nylon housing offers secure, non-slip hand hold.

Wiring entrance holes are angled to permit "straight in" conductor insertion.

Broad gripping area accommodates wide range of cable diameters.

Grip is an integral part of moulded device. Jaws offset cable slightly to prevent slippage of inner wires.

Each individual wiring terminal is completely enclosed in its own separate chamber.

Dead front eliminates need for insulating disc.

Clear polycarbonate permits visual inspection of terminals after wiring.

Courtesy Hubbell Canada Inc.

FIGURE 6.6 Internal construction of a heavy-duty, dead-front cord cap showing cord and terminal connections

FIGURE 6.7 Female cord connector in dead-front design, constructed of black and white nylon

Appliance Plug Caps

Small portable appliances, such as tea-kettles, frying pans, and percolators, are often made with *removable* cord sets. These sets have plugs that are easy to grip, provide strain relief for the cord, and resist heat. Figures 6.8 and 6.9 show an appliance plug and its cord connections.

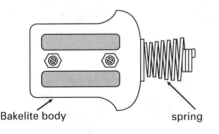

Bakelite body   spring

FIGURE 6.8 Appliance plug

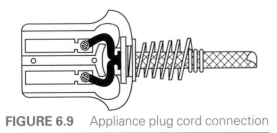

FIGURE 6.9 Appliance plug cord connection

Some large stationary appliances, such as electric stoves and clothes driers, are now being made with cord sets. This allows larger units to be pulled away from the wall for cleaning and quickly disconnected from the power source for servicing. Figures 6.10 and 6.11 show typical cord and cap sets for use with large appliances.

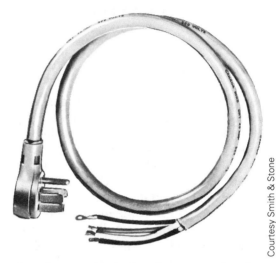

FIGURE 6.10 Typical 50 A range cord and plug cap

FIGURE 6.11 Typical drier cord and plug cap with a 30 A rating available in kit form

Electrical Ratings

Plug caps are rated in volts and amperes. The ratings of the unit must be matched to the requirements of the circuit. If a light-duty unit is installed where a heavy-duty unit is necessary, the plug will overheat. This can cause damage to the plug and/or receptacle and the conductor insulation. Look for the CSA stamp on the plug cap; this is a guarantee that the manufacturer's ratings are correct.

Grounded Cord Caps

As explained in Chapter 4, there is a need for a ground prong on a male plug cap and a corresponding slot in the matching female plug cap. These units are placed on any tool or device that could deliver a shock to the operator if the device is used in an area where the operator could become grounded. Faulty equipment often allows current to flow to the body or frame of the tool, then through the operator into the ground. This produces a serious electrical shock. The ground prong and ground wire can carry this stray or leakage current into the ground rather than have it go through the operator. This can prevent electrical shock to the operator. Figure 6.12 illustrates male plug caps with their rugged U-shaped ground prongs, recommended for all portable tools and equipment.

Twist-Lock Caps

Cords and connectors are often used in high traffic areas. Movement by persons walking past cords or movement of plugged-in equipment can cause cord caps to accidentally fall out. To prevent such inconveniences, a type of cord cap that requires twisting to lock the cap in place after insertion has been produced. Such *twist-lock* caps are used quite frequently in industry and business. They are available in many voltage and current ranges, each with a slightly different blade configuration. A typical 3 prong, 125 V, 15 A cord cap can be seen in Figure 6.13.

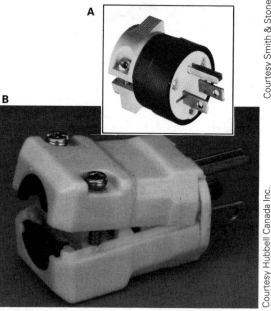

Courtesy Smith & Stone

Courtesy Hubbell Canada Inc.

FIGURES 6.12A AND B Male plug caps with ground prong in dead-front plug design

Courtesy Hubbell Canada Inc.

FIGURE 6.13 Twist-lock, 125 V, 15 A cord cap

Hospital Grade Cord Caps

Hospital grade cord caps are designed to meet the most demanding needs of hospital and health care facilities. Their nylon bodies resist impact, grease, oil, acid abrasion and ultraviolet radiation, while the cord grip action reduces strain on wiring terminals and helps prevent overtightening of assembly screws. Colour-coded faces are clearly marked with amperage and voltage ratings (15 A—blue, 20 A—red), and individual deep-funnelled wire wells accept up to No. 10 AWG conductors. Figure 6.14 illustrates hospital grade caps and connectors.

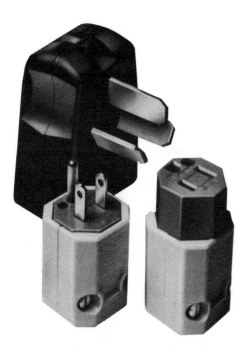

FIGURE 6.14 Hospital grade nylon plugs and connectors

Courtesy Leviton Manufacturing Co. Inc.

For Review

1. Describe the two main types of cord fittings.
2. List five common cord cap blade shapes.
3. What may happen if a plug is removed from a receptacle by pulling on the cord?
4. Why are the conductors taken around the blade of the plug cap before securing them to the terminal screw?
5. What is the purpose of soldering the cord strands before winding them around the terminal screw?
6. Why are heavy-duty plug caps equipped with clamping devices for the cords?
7. Describe the *insulating disc*, and explain its purpose.
8. What is the purpose of the dead-front plug cap?
9. List two reasons why large appliances are being fitted with cord sets.
10. List the electrical ratings that must be marked on plug caps.

7 Electrical Outlet Boxes

Modern wiring systems require an electrical outlet box at each point in the circuit where a switch, lampholder, receptacle, or splice is located. Section 12 of the Canadian Electrical Code provides an up-to-date summary of installation procedures.

Most electrical boxes are made of *galvanized* or *cadmium-plated* steel. Nonmetallic wiring systems may use boxes made of *Bakelite*. This heat-resistant material cannot be used for metallic wiring systems because it is quite fragile and not able to withstand the strain of metal box connectors.

Where there is much moisture, boxes may be made of *brass* or *everdur*. These nonrusting, high-corrosion-resistant materials prevent box damage from chemicals or moisture in the surrounding air.

There are several types of boxes for various uses: the octagon, pancake, square, sectional plaster, utility, and concrete-masonry-tile.

Octagon Box

This type of box usually *supports* light fixtures or serves as a *junction* point for wire splices. It can also be used with special covers as a supporting box for a switch or receptacle.

Dimensions. Octagon boxes are available in two diameters. The 10 cm diameter box is the most common. Boxes with diameters from 8 cm to 9 cm are available for applications where the box size is limited. (See Fig. 7.1)

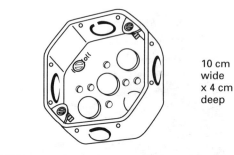

10 cm wide x 4 cm deep

FIGURE 7.1 Typical octagon box

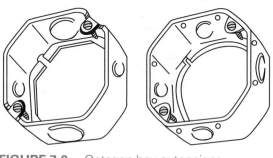

FIGURE 7.2 Octagon box extensions

Note: Check the Canadian Electrical Code requirements for your area.

Another important box dimension is depth. This measurement determines the amount of conductor space within the box. The most common depth is 4 cm. A box 5.5 cm in depth is available for installation where more conductor space is required.

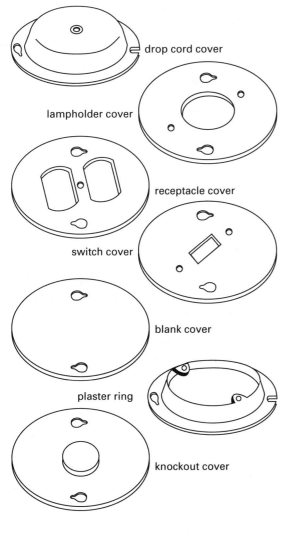

FIGURE 7.3   Octagon box covers

Box extensions are designed to mount directly on the top of an existing octagon box. They are simply octagon boxes without bottoms to provide an increase in conductor space when required. (See Fig. 7.2)

Covers.   There are many types of octagon box covers, allowing the box to serve many purposes. (See Fig. 7.3) Box covers and extension boxes are fastened with two No. 8-32 round-head machine screws, which are provided with each box.

Octagon Concrete Rings.   These rings are used in buildings where the boxes and wiring system are installed *before* the concrete is poured. (See Fig. 7.4) They are available in depths of 4 cm to 15 cm. There can be a cover on either end of the ring. Figure 7.5 shows a concrete ring installation.

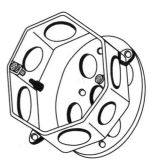

FIGURE 7.4   Typical 10 cm octagon concrete ring (Cover screw may be wax-protected.)

Knockouts.   Knockouts, or KOs, are removable and provide for the entrance of wire, cable, or conduit to the box. Described in Appendix G of the Canadian Electrical Code, they are made in common trade sizes, for example, 13 mm, 20 mm, or 25 mm. A 13 mm knockout is designed to accept a conduit with an internal diameter of 13 mm. The

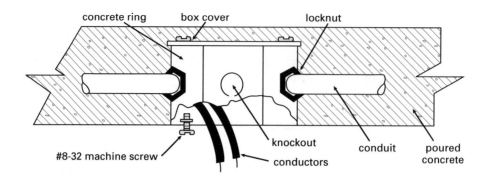

FIGURE 7.5 Concrete ring ceiling installation

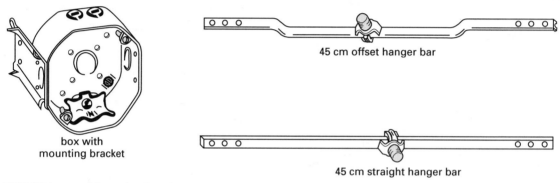

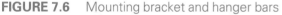

FIGURE 7.6 Mounting bracket and hanger bars

actual diameter of the knockout is approximately 22 mm.

Methods of Mounting. Octagon boxes can be used with either a surface or a concealed wiring system. There are screw holes in the bottom of the box for fastening the box on the surface of a wall or ceiling. Concealed wiring methods allow for several mounting techniques. Figure 7.6 shows the mounting-bracket and hanger-bar assemblies. Figure 7.7 shows three ways of supporting the boxes.

Pancake Box

This 10 cm round box is used primarily with an outdoor porch light. (See

Fig. 7.8) Its 1.3 cm depth allows a fixture to be mounted over it without the box being seen. Also, when a fixture must be installed on a finished wall, the use of a pancake box eliminates the need to make a hole in the wall. (See Fig. 7.9) Conductor space in the box is limited to two wires, and access is through knockouts in the back of the box.

Square Box

The square box is used primarily as a *junction* box for surface and concealed wiring systems. (See Fig. 7.10) There are special covers that permit this box to support switches, receptacles, and pilot lights. (See Figs. 7.11 and 7.12)

Applications of Electrical Construction

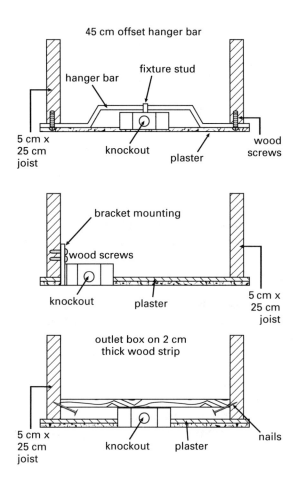

FIGURE 7.7   Methods for supporting octagon boxes

FIGURE 7.8   Typical 10 cm round pancake box

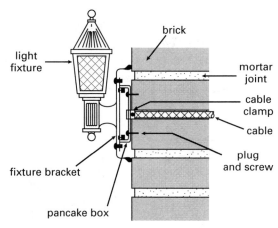

FIGURE 7.9   Installation of a fixture on a pancake box

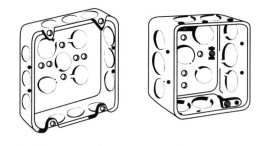

FIGURE 7.10   Square boxes

Dimensions.   Square boxes are made in two sizes. The 10 cm width is the most common. A box 12 cm wide is also available.

The standard square box depth is 4 cm. A box 5.5 cm in depth is also available.

When extra conductor space is needed, *extension rings* made for the square box are used. Box covers and extension rings are fastened with two No. 8-32 round-head machine screws. Knockouts in a combination of 13 mm, 20 mm, and 25 mm sizes are available. Figure 7.13 shows an extension box.

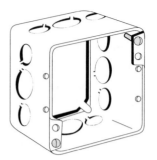

FIGURE 7.13   Typical square box extension

Methods of Mounting.   The square box is mounted in much the same way as the octagon box. (See Fig. 7.7)

Sectional Plaster Box

This box is used to *support* switches and receptacles in many *concealed* wiring systems. (See Fig. 7.14) With its irregular outer surface, it is not suitable for surface wiring.

The sectional plaster box was so named because it was designed for use in walls finished in plaster or masonry. It is used, however, with many types of wall construction.

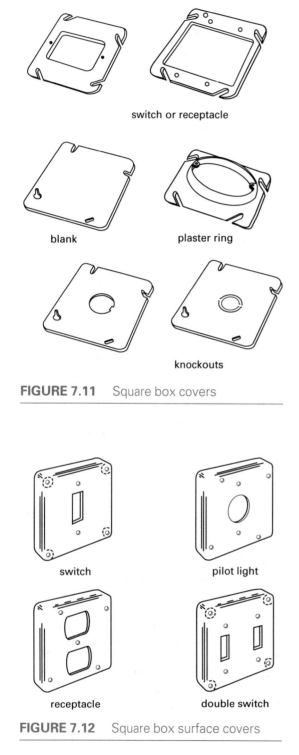

switch or receptacle

blank   plaster ring

knockouts

FIGURE 7.11   Square box covers

switch   pilot light

receptacle   double switch

FIGURE 7.12   Square box surface covers

5 cm wide x 7.5 cm high x
4 cm to 7.5 cm deep

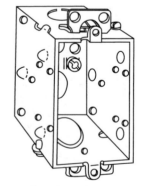

FIGURE 7.14   Sectional plaster box

Applications of Electrical Construction

The Canadian Electrical Code requires that the front of the box be flush with the surface of walls finished in combustible materials, such as wood panelling. (This is a precaution to prevent a flash fire in the box from spreading to the surrounding material.) The box may be recessed up to 6 mm in walls of plaster or masonry.

Construction. Most sectional plaster boxes are made of galvanized steel. Bakelite, or phenolic, boxes are available for use with *nonmetallic* wiring systems.

The sides of the metal boxes are easily removed for *grouping*, or *ganging* together, a series of boxes. (See Fig. 7.15) This feature allows an installer to quickly assemble a box capable of supporting any number of switches or receptacles. Sectional boxes made by one manufacturer may not link up with those made by another. When assembling a gang unit, therefore, take care to select boxes of the same type.

Dimensions. The standard sectional box is 5 cm wide and 7.5 cm in height. The depth of the box varies with the amount of conductor space required. The standard depth is 6.5 cm. Units

7.5 cm, 5 cm, and 4 cm in depth are also available.

Covers. Covers for this box are made primarily for switches or receptacles. A *blank* cover is used when the box is to serve as a junction point for splices in the wires. Plaster boxes have two No. 6-32 threaded mounting lugs spaced 8 cm apart, which accept any manufacturer's switch or receptacle. The covers are usually fastened to the switch or receptacle with No. 6-32 machine screws. (See Fig. 7.16)

FIGURE 7.16 Sectional plaster box covers

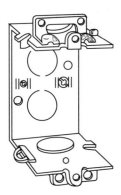
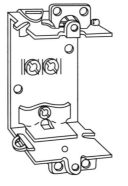
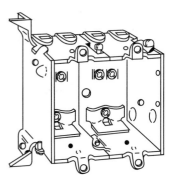

FIGURE 7.15 Sectional plaster boxes for ganging

Gang covers are used for multiple switch or receptacle units mounted in grouped boxes. Sectional boxes are sometimes used to enclose pilot lights, which indicate when a piece of electrical equipment is operating. Covers for this purpose are also available.

Bakelite, or phenolic, is usually used in the manufacture of covers. It is a good insulator, heat-resistant, easily moulded, and generally low in cost, making it ideal for the purpose. Brass, aluminum, stainless steel, and galvanized steel covers are produced where a stronger or more decorative cover is needed.

Methods of Mounting. Sectional plaster boxes can be supported in several ways. Figure 7.17 shows a group of boxes equipped with mounting brackets. Figure 7.18 shows how single or ganged units can be supported.

When a box must be installed in a finished wall, a special purpose unit can be used. This sectional box is equipped with an expanding bracket that will pass through a pre-cut hole in the wall, expand, and grip the plaster when a tension bolt is tightened. The cable must be fastened securely to the box before the box is inserted into the hole. Once the bracket has expanded, it cannot be removed easily. (See Fig. 7.19, page 78.)

A second method of mounting sectional boxes in existing walls depends on the use of the recently developed *swing arm*. After cutting a hole in the wall to accept the new box, the installer connects the cable to a built-in cable clamp in the usual manner and inserts the box into place in the wall. *Plaster ears* prevent the box from falling through the hole, while adjustment of a special screw brings the swing arm into a holding position on the inside of the wall. Further use of the special screw tightens the

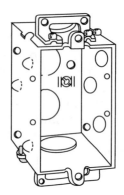

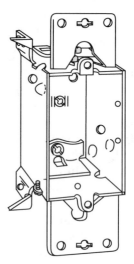

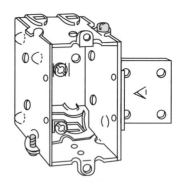

FIGURE 7.17 Sectional plaster boxes with mounting brackets

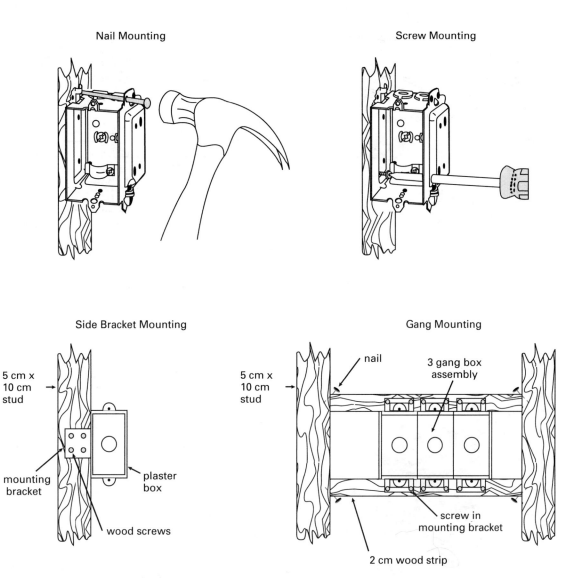

Nail Mounting

Screw Mounting

Side Bracket Mounting

5 cm x
10 cm
stud

mounting
bracket

plaster
box

wood screws

Gang Mounting

nail

3 gang box
assembly

5 cm x
10 cm
stud

screw in
mounting bracket

2 cm wood strip

FIGURE 7.18   Methods for mounting sectional plaster boxes

swing arm for secure holding of the box in the wall. (See Fig. 7.20)

Steel Stud Applications

Apartment buildings, office complexes, and other commercial buildings frequently contain partitions constructed with steel studding rather than the tradi-

tional wooden uprights. Special boxes have been designed for use in these areas. (See Fig. 7.21)

Sectional Box Accessories

Renovations to existing buildings often require the resurfacing of walls and partitions. When new dry wall or similar

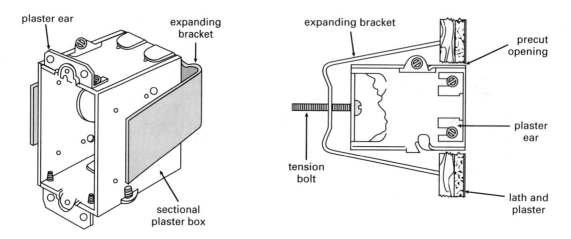

FIGURE 7.19   Method for mounting a sectional plaster box in a finished wall

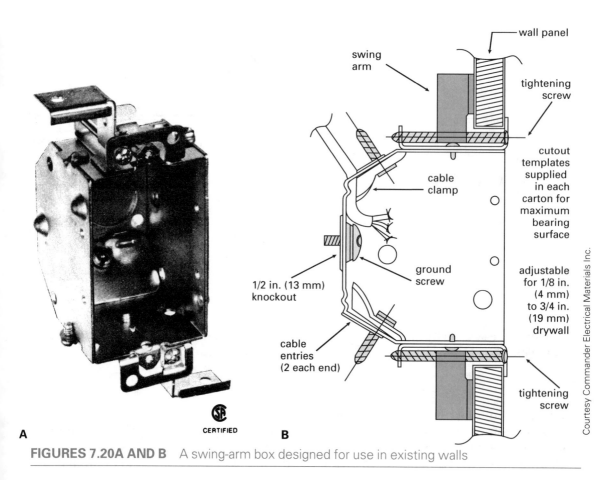

FIGURES 7.20A AND B   A swing-arm box designed for use in existing walls

Courtesy Commander Electrical Materials Inc.

Applications of Electrical Construction

Position the wrap-a-round box at required height and hold flat against the steel stud.

Fold both ends of the wrap-a-round bracket around the stud flanges.

Fold the wrap-a-round bracket into the steel stud using your fingers. No special tools required.

The wrap-a-round bracket and the steel stud flanges should be crimped together with pliers to secure the box in position.

Improve the installation by adding screws to the front and rear flanges.

Courtesy Commander Electrical Materials Inc.

FIGURE 7.21 Step-by-step mounting procedure for mounting steel stud boxes

material is added to the surface of a wall, the existing boxes are automatically recessed, making connections with new devices, such as switches and receptacles, awkward. A box extension that fits on the front of the previously installed box is now available. This extension provides the required metal barrier between the wall material and the conductors. It also provides a secure mount for new switches or other devices. (See Fig. 7.22)

Current regulations in the home insulation field require the installation of some form of vapour barrier around outlet boxes when they are mounted on outside walls. A fast, convenient way of providing this vapour barrier is to choose

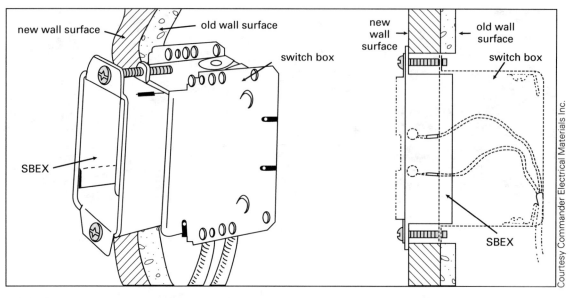

FIGURE 7.22 Installation of SBEX switch box extension when resurfacing old walls in existing buildings

something from the new line of transparent, tough, resistant plastic products that will stand up to extreme cold. The plastic is moulded into single and 2 gang box shapes. The larger, 2 gang units can also be used to enclose square or octagon boxes in the 4 in. (10 cm) size. (See Fig. 7.23)

Utility Boxes

This versatile box is used to *support* receptacles, switches, and pilot lights in *surface* wiring systems. Its rounded corners and smooth exterior design make it an ideal unit for surface wiring systems with cable or conduit.

Construction. The utility box is usually made of galvanized or cadmium-plated steel. (See Fig. 7.24)

Dimensions. Standard utility box widths vary between 5.5 cm to 6 cm

depending on the manufacturer. The length of the box is a standard 10 cm. The depth is usually 4 cm. A unit 5 cm in depth is available.

Covers. Utility box covers are usually made of plated steel. Bakelite covers should not be used because the sharp corners break easily when used with surface wiring materials.

Most utility box covers are made for receptacles and switches. However, pilot light and blank covers are also available. (See Fig. 7.25)

Methods of Mounting. There are several units with mounting brackets attached. (See Fig. 7.26) Holes are provided in the back of the box for bolts or screws when the box is mounted directly on a surface. The devices are fastened to the box with No. 6-32 machine screws.

Applications of Electrical Construction

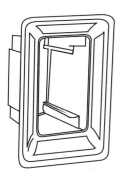

For use with any
5 cm x 7.5 cm
device box up to
7.5 cm

For use with all two
gang device boxes
up to 7.5 cm and 10 cm
square or octagonal
boxes in either shallow
or deep configurations

Installation Procedure

1. Remove the necessary pry-outs from the metal box. Place the metal box inside the vapour barrier. Adjust the front of the metal box at the required distance for flush mounting with drywall.

2. Attach the metal box with its vapour barrier to a stud with nails or screws.

3. Puncture the box's vapour barrier with a square head screwdriver where the pry-outs have been removed. Push the cable (nonmetallic sheathed cable) through the hole in the vapour barrier. Then, follow usual wiring procedures.

   a) Unskinned cable will penetrate a box's vapour barrier and metallic device boxes easier than skinned cable. Cable can be skinned after entered in the box and pulled back to its proper length.

   b) Insulation paper must be placed behind the flange of the vapour barrier so that the front surface of the flange will seal effectively.

Courtesy Commander Electrical Materials Inc.

FIGURE 7.23   Preshaped, plastic vapour barriers provide adequate protection from moisture trying to pass through an outlet box.

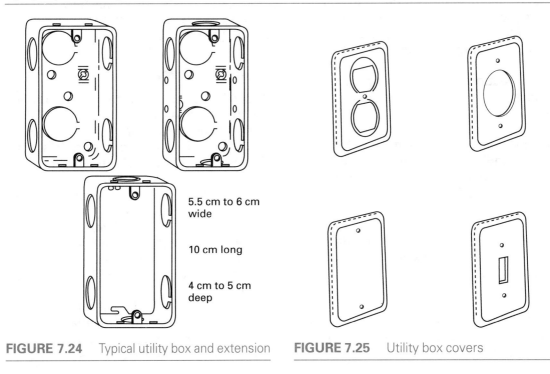

5.5 cm to 6 cm wide

10 cm long

4 cm to 5 cm deep

FIGURE 7.24   Typical utility box and extension

FIGURE 7.25   Utility box covers

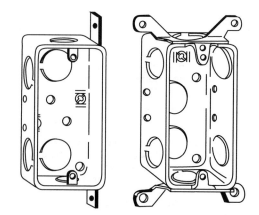

FIGURE 7.26 Utility boxes with mounting brackets

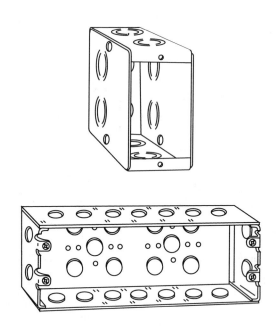

FIGURE 7.27 Concrete-masonry-tile boxes

Concrete-Masonry-Tile Boxes

Industrial and commercial buildings made with poured concrete or other masonry materials require a box that can be set directly into the material *during* construction. These boxes are intended to be used with *concealed metallic* wiring systems. (See Fig. 7.27)

Construction. Concrete-masonry-tile boxes are usually made of galvanized steel.

Dimensions. A *single-gang* unit is 5 cm wide and 9.5 cm in height. It is also available in a multi-switch unit capable of supporting five switches or receptacles.

Covers. Covers are made for single or group installation of switches, receptacles, and pilot lights, or combinations of all three.

Methods of Mounting. These boxes are held in place by the concrete or mortar of the masonry wall. When the building is made with poured concrete walls and floors, the boxes may be wired into position to hold them securely while the concrete is being poured.

Cable Clamps in Boxes

Electrical boxes provide access for conductors in two ways. The removable disc, called the *knockout*, allows conduit and/or cable connectors in trade sizes of 13 mm, 20 mm, and 25 mm internal diameter to be fitted to the box.

Boxes designed for use with nonmetallic or armoured-cable systems are available with *built-in clamps*. These cable clamps eliminate the need for separate connecting devices and also shorten the time required for this operation.

The removable disc designed for boxes equipped with cable clamps is called the *pry-out*. Appendix G of the Canadian Electrical Code describes the

Applications of Electrical Construction

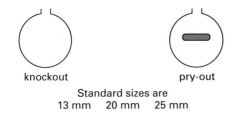

knockout   pry-out

Standard sizes are
13 mm   20 mm   25 mm

FIGURE 7.28   Typical knockout and pry-out

pry-out as "a knockout provided with a slot in order that a screwdriver may be inserted to pry out the knockout." (See Fig. 7.28)

There are several types of built-in box connectors. (See Fig. 7.29) Built-in cable clamps are usually found in octagon and sectional plaster boxes. (See Fig. 7.30)

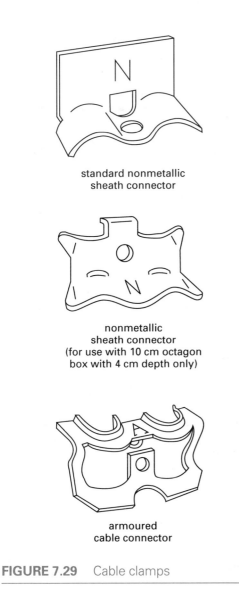

standard nonmetallic
sheath connector

nonmetallic
sheath connector
(for use with 10 cm octagon
box with 4 cm depth only)

armoured
cable connector

FIGURE 7.29   Cable clamps

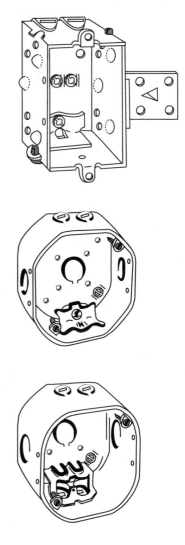

FIGURE 7.30   Boxes with built-in cable clamps

Box Grounding

A modern electrical wiring box provides one or two machine screws in the back of the box for grounding purposes. A nonmetallic cable system has a bare ground wire within the cable. The ground wire must be connected to one of the screws provided in the box. A receptacle should have a conductor joining its ground terminal with the ground screw in the box.

A metallic wiring system normally relies on a secure metal-to-metal connection with the box to complete its ground circuit.

Conductor Capacity of a Box

The number of current-carrying conductors contained in a box must be limited. When too many conductors are contained in a box, the conductors might be forced into a sharp edge or mounting screw within the box. The sharp edge might penetrate the insulation on the conductor, allowing the current to take other than its intended path. If the damaged conductor is a live wire, a short-circuit condition will occur and the circuit fuse will blow. If for some reason the box is not grounded, the box and any metal object in contact with it will become alive and dangerous.

A second, and equally important, reason for limiting the number of conductors in a box is overheating. Any conductor carrying current will produce some heat as a side effect. The more current passing through the conductor, the more heat will be produced. The more conductors there are in a box, the more heat will be accumulated within the box. Once the cover is placed on the box, there is little or no air circulation to cool the conductors. Modern conductor insulation is designed to withstand some heat, but it will become hard and brittle if overheated for an extended period of time.

Each size of conductor requires a certain amount of free air space for cooling. The Canadian Electrical Code lists the volume of air space required by some of the common conductor sizes. (See Table 7.1)

TABLE 7.1 Space for Conductors in Boxes		
Size of Conductor AWG Copper or Aluminum	**Usable Space Within Box for Each Insulated Conductor**	
	Cubic Centimetres	**Cubic Inches**
14	25	1.5
12	29	1.75
10	37	2.25
8	45	2.75
6	74	4.5

Based on the Canadian Electrical Code

The Canadian Electrical Code also lists the air space available in standard electrical boxes. (See Table 7.2) Mathematical calculation of the volume of a box (length × width × depth) will not give the same results as listed in Table 7.2. Because of the differences in tolerance during manufacture of boxes, the Electrical Code Committee decided on standard capacities. The actual internal dimensions of a sectional plaster box, for example, are slightly smaller than those listed in catalogues and tables.

Calculation of Box Capacity

Table 7.2 lists a device box 75 mm × 50 mm × 65 mm as having 205 cm³ of air

Box Dimensions Trade Size	Millimetres	Cubic Centimetre Capacity Copper or Aluminum	Maximum Number of Insulated Conductors Size in AWG				
TABLE 7.2 Number of Conductors in Boxes							
			14	12	10	8	6
Octagon	100 × 40	245	10	8	6	5	3
	100 × 55	345	14	12	9	7	4
Square	100 × 40	345	14	12	9	7	4
	100 × 55	490	20	17	13	10	6
	120 × 40	490	20	17	13	10	6
	120 × 55	690	28	24	18	15	9
Round	100 × 13	82	3	2	2	1	1
Device	75 × 50 × 40	130	5	4	3	2	1
	75 × 50 × 50	165	6	5	4	3	2
	75 × 50 × 55	165	6	5	4	3	2
	75 × 50 × 65	205	8	7	5	4	2
	75 × 50 × 75	245	10	8	6	5	3
	100 × 50 × 40	145	6	5	4	3	2
	100 × 55 × 40	165	6	5	4	3	2
	100 × 55 × 45	245	10	8	6	5	3
	100 × 55 × 48	230	9	8	6	5	3
	100 × 60 × 48	260	10	9	7	5	3
Masonry Box	95 × 50 × 65	230/gang	9	8	6	5	3
	95 × 50 × 90	345/gang	14	12	9	7	4
	100 × 55 × 60	330/gang	13	11	9	7	4
	100 × 55 × 85	365/gang	14	12	9	8	4

Extension rings to have the same value as the equivalent trade size box
Based on the Canadian Electrical Code

space. Table 7.1 lists a No. 14 gauge conductor as requiring 25 cm³ of air space. Therefore, the number of No. 14 gauge conductors allowed in this box will be:

$$\frac{\text{Air space in the box}}{\text{Air space of 1 conductor}} = \frac{205}{25} = 8.2$$

Of course, only 8 conductors can be put in the box.

Clamps, receptacles, switches, or other devices inside the box take up some of the free air space. Section 12 of the Canadian Electrical Code states that for *each* device in the box, *one* conductor must be *subtracted* from the total

listed (in Table 7.2). For example, the device box listed as having a capacity of 6 conductors would be limited to 4 conductors, if it contained a switch and built-in cable clamps.

Section 12 of the Canadian Electrical Code explains in detail how conductors entering and/or leaving a box must be counted to determine the total number of conductors allowed. Always use the latest edition of the Code, which is revised regularly.

Since the Canadian Electrical Code and much of the electrical industry use imperial measurements, imperial

TABLE 7.3 Imperial Dimensions and Conductor Capacities for Electrical Wiring Boxes

Box Dimensions Trade Size	Inches	Cubic Inch Capacity Copper or Aluminum	Maximum Number of Insulated Conductors Size in AWG				
			14	12	10	8	6
Octagon	4 × 1½	15	10	8	6	5	3
	4 × 2⅛	21	14	12	9	7	4
Square	4 × 1½	21	14	12	9	7	4
	4 × 2⅛	30	20	17	13	10	6
	4¹¹⁄₁₆ × 1½	30	20	17	13	10	6
	4¹¹⁄₁₆ × 2⅛	42	28	24	18	15	9
Round	4 × ½	5	3	2	2	1	1
Device	3 × 2 × 1½	8	5	4	3	2	1
	3 × 2 × 2	10	6	5	4	3	2
	3 × 2 × 2¼	10	6	5	4	3	2
	3 × 2 × 2½	12.5	8	7	5	4	2
	3 × 2 × 3	15	10	8	6	5	3
	4 × 2 × 1½	9	6	5	4	3	2
	4 × 2⅛ × 1½	10	6	5	4	3	2
	4 × 2⅛ × 1¾	15	10	8	6	5	3
	4 × 2⅛ × 1⅞	14	9	8	6	5	3
	4 × 2⅜ × 1⅞	16	10	9	7	5	3
Masonry	3¾ × 2 × 2½	14/gang	9	8	6	5	3
	3¾ × 2 × 3½	21/gang	14	12	9	7	4
	4 × 2¼ × 3⅜	20.25/gang	13	11	9	7	4
	4 × 2¼ × 3⅜	22.25/gang	14	12	9	8	4

TABLE 7.4 Wiring Boxes and Their Conductor Capacities for Use on 347 V Systems

Dimensions in Inches 347 V Boxes	Cubic Inch Capacity	Maximum Number of Conductors in Boxes Size in AWG			
		14	12	10	8
3 × 2¼ × 2½	15.7	8	7	5	3
4 × 2⅜ × 1⅞	16.5	9	7	5	4
4 × 2¼ × 2⅜	20.25/gang	11	9	7	5
4 × 2¼ × 3⅜	22.25/gang	12	10	7	6

measurements are provided in Table 7.3. This table lists electrical wiring boxes and their conductor capacities.

The trend toward using 347 V sup-plies for lighting circuits has affected electrical wiring boxes. Switches and other related devices must be somewhat larger to safely handle the higher volt-

Applications of Electrical Construction

TABLE 7.5 Wiring Boxes and Their Conductor Capacities for Use on 347 V Systems					
Dimensions in Millimetres 347 V Boxes	**Cubic Centimetre Capacity**	**Maximum Number of Conductors in Boxes Size in AWG**			
		14	**12**	**10**	**8**
75 × 55 × 65	268	8	7	5	3
100 × 60 × 47	282	9	7	5	4
100 × 55 × 60	330	11	9	7	5
100 × 55 × 85	467	12	10	7	6

age, and therefore the boxes supporting and enclosing these switches must be enlarged. By doing so, the required cooling air space is provided. Tables 7.4 and 7.5 list these boxes and their conductor capacities.

For Review

1. What materials are used for the construction of electrical boxes?
2. Where and why are brass boxes used?
3. What type of box is used for indoor light fixtures?
4. List the sizes in which octagon boxes are made.
5. Explain how switches or receptacles are fastened to the octagon box.
6. When are octagon concrete rings used?
7. Where and why are pancake boxes usually used?
8. Explain where and for what purposes square boxes are used.
9. How did the sectional plaster box get its name?
10. What is a gang box? What is it used for?
11. Explain where and how the sectional plaster box special purpose unit is used.
12. Why is Bakelite *not* recommended for utility box covers?
13. Which boxes are used for buildings made of poured concrete and block?
14. State two reasons for using boxes with built-in cable clamps.
15. What is a pry-out? How does it differ from a knockout?
16. How are boxes grounded in a conduit system? How are they grounded in a nonmetallic cable system?
17. State two reasons for limiting the number of conductors in a box.
18. Explain how the volume and the conductor capacity of a box are calculated.
19. Calculate the number of conductors allowed in a 10 cm octagon box, 4 cm in depth, to be used in a circuit with No. 12 gauge wires.
20. If the octagon box described in review problem 19 contains two built-in clamps and a fixture stud, how many wires are allowed, according to the Canadian Electrical Code?

Nonmetallic-Sheathed Cable Wiring

Nonmetallic-sheathed cable (NMSC) is used more often in residential wiring installations than any other wiring method. The Canadian Electrical Code permits this cable to be installed in a building made of combustible material or of wooden frame construction. It may *not* be used in other types of building construction without permission from the electrical inspection authorities.

Cable Construction

There are several basic types of nonmetallic-sheathed cable. (See Figs. 8.1, 8.2, 8.3, 8.4, and 8.5) Nonmetallic cable for *dry* locations *(NMD)* is used in normal residential circuits. Nonmetallic cable for wet locations *(NMW)* is used in farm buildings or similar structures, where there is usually more moisture. NMW cable can be buried directly in the earth, providing adequate protection is given to the cable.

Trade Names. NMSC was first produced by the Rome Wire and Cable Company, which named its new product *Romex*. This name is still used often in the trades. Each company producing NMSC, however, has its own product name ending in "ex", e.g., *Canadex* from Canada Wire and Cable Limited and *Philex* from Phillips Cables Limited.

Types of Insulation. NMD-3 cable has been used for several years for residential wiring. The number 3 indicates the maximum allowable temperature of the cable: 60°C.

This cable is not suitable, however, for use with modern light fixtures. Heat from the fixture bulbs often dries out the cable, making the insulation brittle and of little value. NMD-90 and NMD-90 XLPE cables with temperature ratings of 90°C have now replaced NMD-3 cable. These cables can also be used to supply electric heaters, stoves, and clothes driers.

Conductor Materials. NMSC is produced with copper or aluminum conductors. Since aluminum is not as good a conductor as copper, one wire size *larger* must be selected when using aluminum conductor nonmetallic-sheathed cable.

Conductor Sizes. NMSC with *copper* conductors is available in gauge Nos. 14, 12, 10, 8, 6, and 4. The smallest gauge of *aluminum* conductor cable produced, however, is No. 12. Cable for both dry and wet use is available with two or three insulated conductors and a bare ground wire.

FIGURE 8.1   NMD-90 cable

FIGURE 8.2   Nylon insulated NMSC

FIGURE 8.3   Moisture-proof NMSC with copper conductors (*Note*: Kraft paper omitted; extra TW applied to wires)

FIGURE 8.4   Typical 3 conductor, No. 8 gauge range cable (*Note*: Ground wire located between insulated conductors)

FIGURE 8.5   A 2 wire (black, red, and ground) nylon Heatex NMD-7 cable for use with electric heating circuits operating on 240 V

Cable Installation

Section 12 of the Canadian Electrical Code requires NMSC to be installed in a *loop system*. This means that all cables are run in continuous lengths between the electrical boxes. Joints or splices in the cable must be made in a box. (See Fig. 8.6) All electrical boxes in the system must be accessible for inspection or circuit repair after the building is completed.

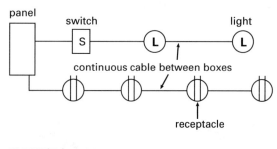

FIGURE 8.6   Loop system

The *knob* and *tube* wiring systems used before the loop system was developed allowed splices at almost any point in the circuit. This and the fact that the live and neutral wire of the same circuit did not always travel side by side made troubleshooting difficult even for the experts.

Cable Supports

NMSC must be fastened to the wooden members of a building by straps, staples, or other approved devices permitted by the CSA. (See Fig. 8.7) Staples take the least time, but the cable may be damaged if the staples are driven too deeply into the wood. (See Fig. 8.8) *Straps* may be secured with screws or nails.

The Canadian Electrical Code

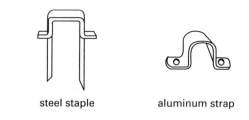

steel staple   aluminum strap

FIGURE 8.7   Cable supports

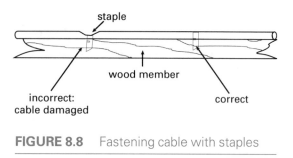

FIGURE 8.8   Fastening cable with staples

requires that a strap or staple be placed within 30 cm of every box. Doing so prevents any undue strain on the cable from pulling the conductors out of the box. Cable supports may be placed as far as 1.5 m apart on the runs between the boxes, but it is often a good idea to place them closer together. Cable installations are usually in service for many years, and so neat and secure installations are important.

Fastening a Cable to a Box

Safety Note: NMSC must be held securely by the clamp or connector at the box. Do not overtighten the clamp as this might crush the cable and create a short circuit.

Approximately 6 mm of outer sheath should extend *beyond* the clamp to protect the TW insulated wire from the clamp. (See Fig. 8.9) Also, a minimum of 15 cm of *free*, insulated conductor must be available for connection to devices in

Applications of Electrical Construction

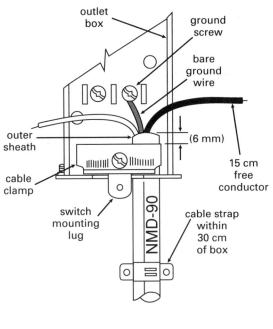

FIGURE 8.9 Method for connecting cable to outlet box

NMWU Buried in the Earth

The heavy layer of TW insulation on the NMWU conductors makes this cable suitable for use in underground runs supplying garden or post lights. Take care to protect the cable from garden tools. (See Fig. 8.10)

Cable Protection for Concealed Installations

The Canadian Electrical Code requires that NMSC be kept at least 3 cm from the outer edge of any wooden member. Otherwise, driven nails or screws supporting baseboard, plaster, wood panels, or other wall products may pierce the cable. When a cable has been pierced with a nail, the problem is usually not discovered until the building is completed and the circuit made alive. With a finished wall concealing the damaged

the box. The bare ground wire may be trimmed after connecting it to the box ground screw or, in the case of a receptacle, left long enough to connect to the receptacle ground screw as well. Take care to trim neatly excess kraft paper from the cable in the box. The trimming reduces any fire hazard.

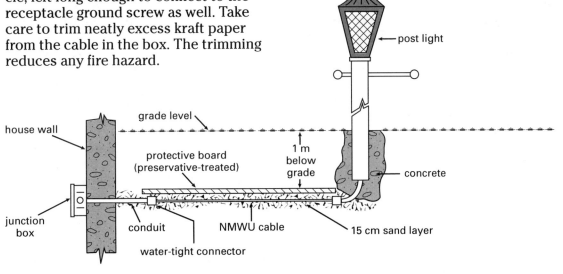

FIGURE 8.10 Post light installation using NMWU

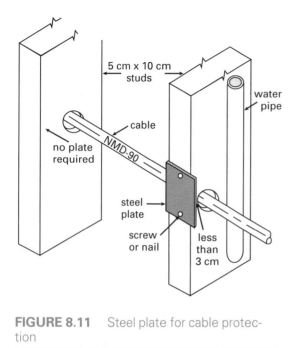

FIGURE 8.11 Steel plate for cable protection

cable, it is both difficult and costly to locate and repair the fault.

Normal procedure is to drill holes in the *centre* of the wooden members that the cable must pass through. Water and air circulation systems in the walls will often make it necessary to run the cable closer to the edge of a wooden member than the required 3 cm. In these cases, a *steel plate* is fastened to the wooden member in *front* of the cable to protect it. Often, the plate is a side of a sectional plaster box left over from a gang box assembly. (See Fig. 8.11) Take care not to damage the cable by locating it too near to a hot-water pipe or hot-air duct.

Residential Cable Applications

NMSC has a maximum rating of 300 V and will readily accept the 120 V/240 V supplied to a residence.

Most circuits in houses consist of No. .14 gauge NMSC and should be fused at a maximum of 15 A. Kitchen receptacles supplying current to frying pans, teakettles, or similar appliances can use No. 12 gauge NMSC fused at 20 A. This increase in cable size and ampacity provides a margin of safety to a circuit often operating very close to its capacity. Home electric heating systems often have circuits consisting of No. 12 gauge NMSC. Heat-sensitive fuses should be used to protect these circuits.

Clothes driers are supplied with 3 conductor, No. 10 gauge NMSC, commonly called *drier cable*. Fuse protection for this cable should not exceed 30 A. Heat-sensitive type fuses are the best. Electric stoves use a 3 conductor, No. 8 gauge *range cable* fused at 35 A to 40 A, depending on the size of stove. (See Fig. 8.4, on page 89.)

Cable Wiring Diagrams

An important part of any wiring system is preparation of a circuit diagram. This diagram helps determine the *sequence* in which the devices are to be connected and the *number* of wires required in the cable between the boxes.

Graphical symbols are used to simplify the drawing of electrical devices in a circuit. These universal symbols are listed in Appendix F of the Canadian Electrical Code. (See Table 8.1) Cables are represented by a single, solid line, with the number of insulated wires in the cable shown by short dashes across the cable line. For example, a 2 wire cable is shown as ——//——, and a 3 wire cable as ——///——.

Remember that nonmetallic-sheathed cable is available in 2 and 3 wire combinations. Any wiring circuit must be completed using *only* the white, black, or red NMSC wires available.

Applications of Electrical Construction

TABLE 8.1

ELECTRICAL SYMBOLS FOR ARCHITECTURAL PLANS

GENERAL OUTLETS

◯	─◯	Outlet
Ⓑ	─Ⓑ	Blanked Outlet
Ⓓ	─Ⓓ	Drop Cord
Ⓔ	─Ⓔ	Electrical Outlet; for use only when circle used alone might be confused with columns, plumbing symbols, etc.
Ⓕ	─Ⓕ	Fan Outlet
Ⓙ	─Ⓙ	Junction Box
Ⓛ	─Ⓛ	Lampholder
Ⓛ$_{PS}$	─Ⓛ$_{PS}$	Lampholder with Pull Switch
Ⓢ	─Ⓢ	Pull Switch
Ⓥ	─Ⓥ	Outlet for Vapour Discharge Lamp
Ⓧ	─Ⓧ	Exit Light Outlet
Ⓒ	─Ⓒ	Clock Outlet (Specify Voltage)

NOTE: Symbols on the left above refer to ceilings; those on the right above refer to walls.

RECEPTACLES

⊖	Duplex Receptacle
⊖$_{1,3}$	Other than Duplex Receptacle 1 = Single, 3 = Triplex, etc.
⊖	Split-Switched Duplex Receptacle
⊜	Three–Conductor Split-Duplex Receptacle
⊜	Three–Conductor Split-Switched-Duplex Receptacle
⊖$_{WP}$	Weatherproof Receptacle
⊜$_R$	Range Receptacle
⊜$_S$	Switch and Receptacle
⊖$_R$	Radio and Receptacle
⬤	Special Purpose Receptacle (Described in Specification)
⊙	Floor Receptacle

SWITCHES

S	Single Pole Switch
S$_2$	Double Pole Switch
S$_3$	Three Way Switch
S$_4$	Four Way Switch
S$_D$	Automatic Door Switch
S$_E$	Electrolier Switch
S$_K$	Key Operated Switch
S$_P$	Switch and Pilot Lamp
S$_{CB}$	Circuit Breaker
S$_{WCB}$	Weatherproof Circuit Breaker
S$_{MC}$	Momentary Contact Switch
S$_{RC}$	Remote Control Switch
S$_{WP}$	Weatherproof Switch
S$_F$	Fused Switch
S$_{WF}$	Weatherproof Fused Switch

SPECIAL OUTLETS

◯$_{a,b,c, etc.}$
⊖$_{a,b,c, etc.}$
S$_{a,b,c, etc.}$

Any standard symbol as given above with the addition of a lower case subscript letter may be used to designate some special variation of standard equipment of particular interest in a specific set of architectural plans.

When used they must be listed in the Key of Symbols on each drawing and, if necessary, further described in the specifications.

PANELS, CIRCUITS, AND MISCELLANEOUS

▬	Lighting Panel
▨	Power Panel
──	Branch Circuit; Concealed in Ceiling or Wall
─·─	Branch Circuit; Concealed in Floor
----	Branch Circuit; Exposed
⟶	Home Run to Panel Board Indicate number of circuits by number of arrows.

NOTE: Any circuit without further designation indicates a two-wire circuit. For a greater number of wires indicate as follows:
+++ (3 wires)
+++ (4 wires), etc.

architectural symbols

wiring diagrams

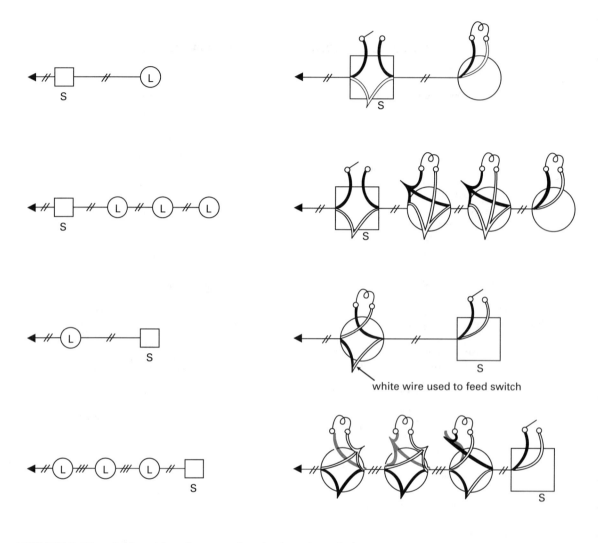

white wire used to feed switch

FIGURE 8.12   Cable wiring diagrams for single-pole switch

Figures 8.12, 8.13, and 8.14 compare *architectural symbols* with wiring diagrams of basic lighting circuits. Figures 8.15 and 8.16 provide more complex circuits designed to develop wiring skills. You will develop a greater understanding of these skills if you take time to draw and complete the circuits on note paper. Use coloured pencils to indicate the wires. Since a cable system of wiring requires that all splices and connections be in a box, it is unnecessary to show the individual conductors between the boxes. A single line is used to

Applications of Electrical Construction

architectural symbols   wiring diagrams

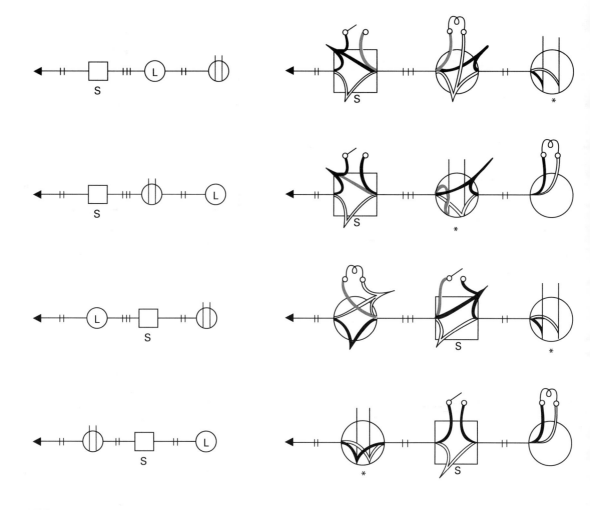

*NOTE: Receptacles alive at all times. Switch controls only the light.

FIGURE 8.13 Wiring diagrams with receptacles

represent the cable in these circuits. (See Fig. 8.12)

Figure 8.13 shows simple lighting circuits with a duplex receptacle added. Receptacles are considered to be alive at all times, unless otherwise marked on the diagrams. Three-conductor cable is required for some of these circuits.

Three and 4 way switch-control circuits are shown in Figure 8.14. Three-conductor cable is often used in this type of circuit.

architectural symbols   wiring diagrams

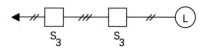

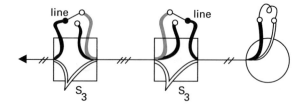

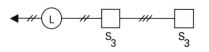

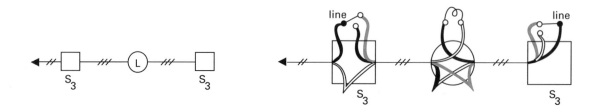

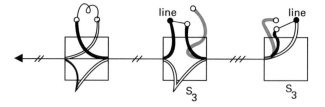

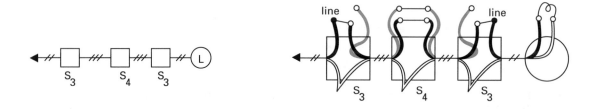

FIGURE 8.14   Wiring diagrams showing 3 and 4 way switch control

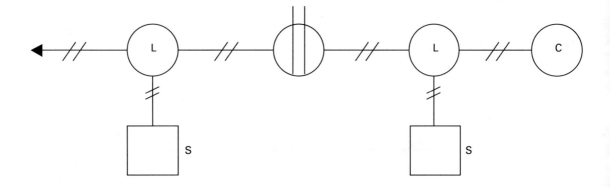

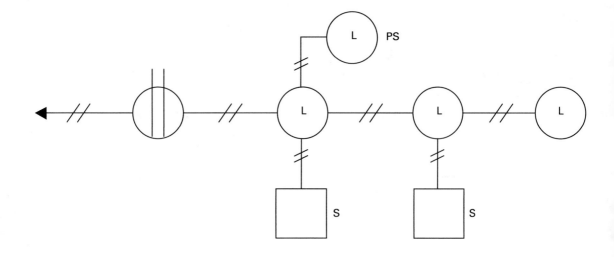

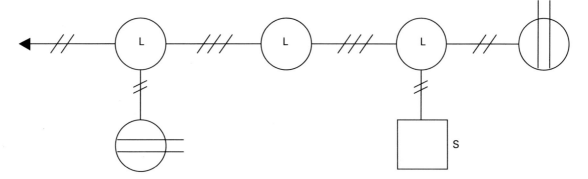

FIGURE 8.15 Wiring diagrams using architectural symbols to show 2 and 3 wire cables

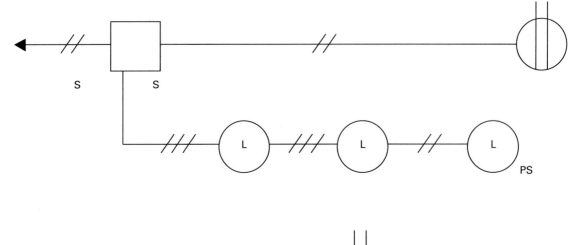

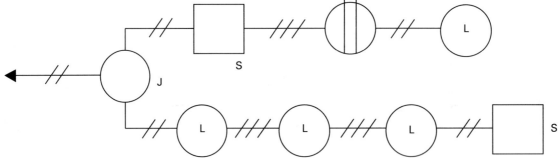

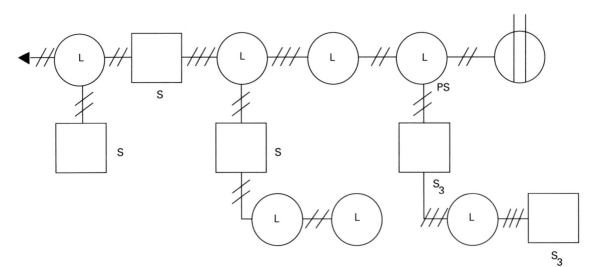

FIGURE 8.16 Complex wiring diagrams using architectural symbols to show combinations of lights, receptacles, and assorted switching methods.

NMSC Accessories

Not all electrical boxes are equipped with built-in clamps. Distribution panels and outlet boxes with 13 mm knockouts require *cable connectors*. (See Figs. 8.17 and 8.18)

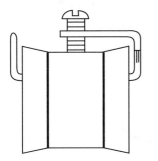

FIGURE 8.17   Typical 3030 style NMSC connector (expansion type)

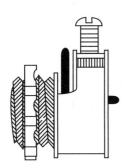

FIGURE 8.18   Aluminum die-cast connector (with locknut)

Some 15 cm of the cable's outer sheath is easily split with a *cable ripper*. (See Fig. 8.19) This simple metal tool saves time and prevents damage to the cable during the ripping process. Modern NMD-90 cables are smaller and more compact than the older cables with braided outer sheaths. Special cable rippers have been developed just for them, but great care must be taken not to damage the nylon sleeve over the PVC

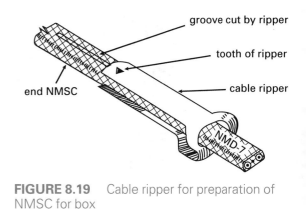

FIGURE 8.19   Cable ripper for preparation of NMSC for box

insulated wires. The need for care is most obvious when ripping 3 wire cable. *Cutting pliers (diagonal cutters)* are used to trim off the loose ends of kraft paper and outer sheath.

F o r   R e v i e w

1. With which types of building materials may NMSC be used?
2. What are the two basic types of NMSC, and where are they used?
3. What is the purpose of the bare wire in NMSC?
4. Describe the *loop* system used with NMSC wiring.
5. Why does the Canadian Electrical Code require that all joints or splices be made in a box?
6. List the devices that may be used to support NMSC. How should these devices be spaced, and why?
7. When fastening a cable to a box, how much conductor should be left free? Why?
8. Describe how NMW-10 is run under ground and connected to an outdoor post light.
9. When running NMSC through wooden members, what precautions should be taken? Why?

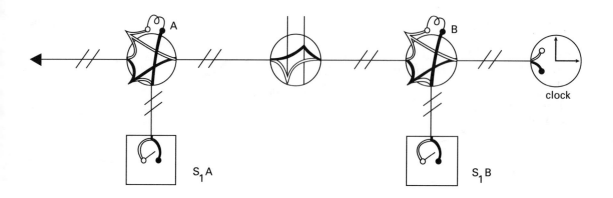

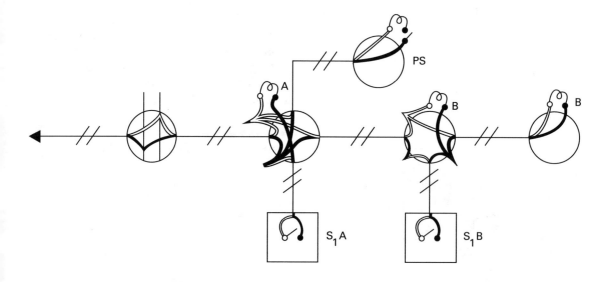

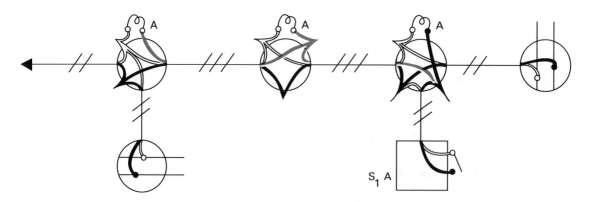

Answers to FIGURE 8.15 Wiring diagrams showing 2 and 3 wire cables

Applications of Electrical Construction

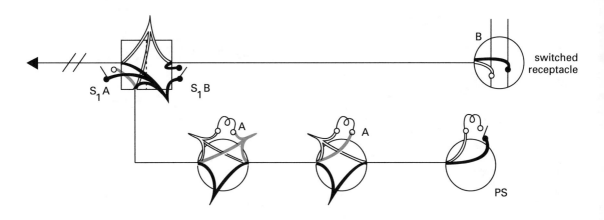

B

switched
receptacle

S_1A S_1B

A   A

PS

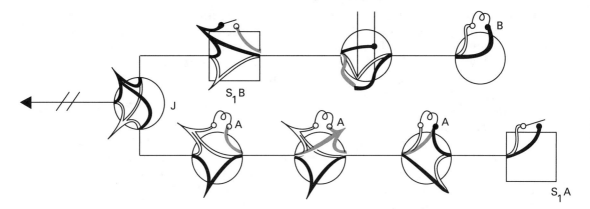

B

S_1B

J

A   A   A

S_1A

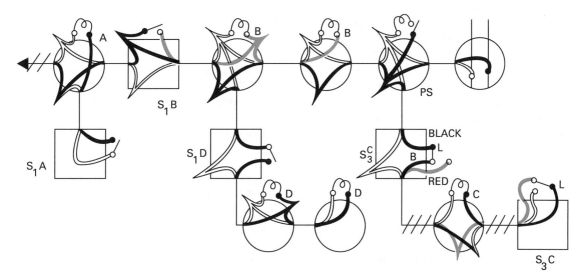

A   B   B   PS

S_1B

S_1A S_1D S_3^C BLACK   L   B   RED   C   L S_3C

D   D

Answers to FIGURE 8.16   Complex wiring diagrams showing combinations of lights, receptacles, and assorted switching methods.

9

Solderless Connectors

The Canadian Electrical Code once required that joints and splices in insulated conductors be soldered, and then covered with an insulating tape equivalent to the insulation on the conductor. Soldering the splice prevented the weakening of the electrical connection by the action of oxidation. But it was also time-consuming and could create other problems. When large cables were connected, they were heated to a temperature that often damaged the insulation near the splice. Also, to disconnect the conductors, more heat was needed to melt the solder. The number of tools required and the cost of one-time-use materials made this system unsatisfactory. The need for a convenient method of making electrical splices resulted in the development and subsequent CSA approval of solderless connectors.

Solderless Wire Connectors

One of the first solderless wire connectors consisted of a tapered porcelain cap with an internal screw thread. While porcelain is still used in the production of some heat-proof units, Bakelite and/or nylon are now most often used in the construction of wire connectors.

There are three main types of solderless wire connectors: twist-on, set-screw, and compression. The names for each refer to the system used to apply the unit.

Twist-on Connector. This connector has a cone-shaped, metal spring that threads itself around the conductors as the connector is rotated. Several manufacturers produce this type of connector, but the operating principle is the same. (See Fig. 9.1) The internal spring design takes advantage of leverage and vise action to multiply the strength of a person's hand. Thus, the conductors are forced into a solid, effective splice. (See Fig. 9.2)

Using the twist-on connector is one of the quickest ways of splicing and insulating electrical conductors. It is suitable for use with solid and/or stranded conductors operating at 600 V or less. Some models have been approved for use at 1000 V.

One variety of twist-on connector features *built-in wings* to increase the torsion achieved by the installer when joining conductors in the No. 12, 10, and 8 AWG sizes. (See Fig. 9.3)

Besides such special features, the twist-on connector is made in a range of sizes for splicing conductors from No. 18

Applications of Electrical Construction

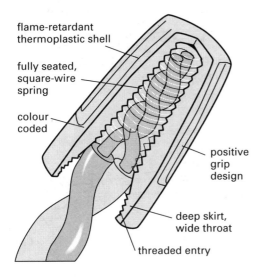

flame-retardant
thermoplastic shell

fully seated,
square-wire
spring

colour
coded

positive
grip
design

deep skirt,
wide throat

threaded entry

Courtesy IDI Electric Canada Ltd.

FIGURE 9.1 Cutaway view of a twist-on connector showing the connector's internal spring and its effect on the conductors when installed

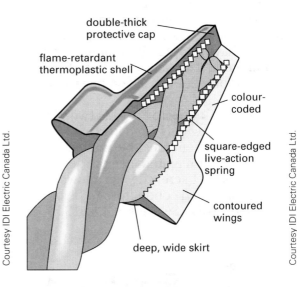

double-thick
protective cap

flame-retardant
thermoplastic shell

colour-
coded

square-edged
live-action
spring

contoured
wings

deep, wide skirt

Courtesy IDI Electric Canada Ltd.

FIGURE 9.3 Wing-nut style of wire connector for extra torque, or twisting power, while splicing larger conductors in the No. 12 to No. 8 AWG range

Step 1. Bare wires and cut to length.

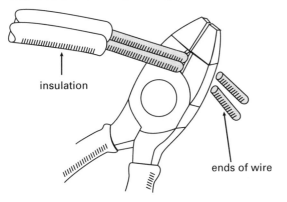

insulation

ends of wire

Step 2. Insert wires and rotate cap.

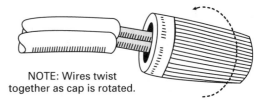

NOTE: Wires twist
together as cap is rotated.

Step 3. Tighten cap fully.

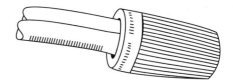

NOTE: No bare wire when cap in place

FIGURE 9.2 Method for installing a twist-on connector

gauge up to No. 8 gauge. Some manufacturers use a numbering system to identify the different sizes of their product. Others have adopted both number and colour codes for easy, visual product recognition. (See Fig. 9.8 on page 106)

Set-Screw Connector. This two-piece connector is widely used in circuits where equipment must be changed frequently for maintenance purposes. The simple set-screw design allows solid and/or stranded conductors to be interchanged easily. (See Figs. 9.4 and 9.5)

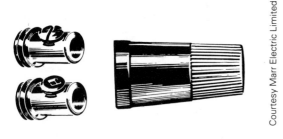

Courtesy Marr Electric Limited

FIGURE 9.4   Set-screw connector

Set-screw connectors are made for use on circuits up to 600 V, with approval for use on some lighting circuits of 1000 V. They are designed to splice conductors from No. 18 up to No. 10 gauge in size.

Figure 9.6 shows typical applications of both twist-on and set-screw connectors. Both are for use in electrical boxes or enclosures only.

Compression Connector. Unlike the mechanical (set-screw) connector, this two-piece compression connector requires a special *crimping* (compression) *tool* to install the conductor-retaining sleeve. The *retaining sleeves* are made of copper and/or zinc-plated steel. An *insulating cap* of plastic or nylon is fastened over the sleeve after

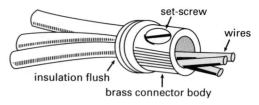

Step 1. Insert wires. Do not twist.

set-screw

wires

insulation flush

brass connector body

Step 2. Tighten set-screw.

threads for insulating cap

Be sure insulation does not slip into connector body.

Trim ends of wire.

Step 3. Install cap.

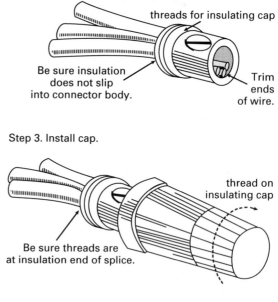

thread on insulating cap

Be sure threads are at insulation end of splice.

FIGURE 9.5   Method for installing a set-screw connector

the conductors have been crimped firmly. (See Fig. 9.7) This type is for *permanent* installation, because the conductor cannot be easily removed from the retaining sleeve.

These units are made for use on circuits of 600 V and may be used on lighting fixtures up to 1000 V. The retaining sleeves are made for splicing conductors No. 18 up to No. 6 gauge. The zinc-plated steel sleeves can be used *only* on copper conductors, due to the possibility of electrolysis acting on these sleeves.

Some compression connectors are

Applications of Electrical Construction

FIGURE 9.6 Typical twist-on and set-screw connector applications

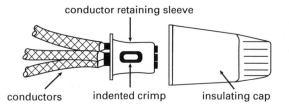

FIGURE 9.7 Method for installing a compression connector

conductor retaining sleeve

conductors   indented crimp   insulating cap

used to simplify the connection of stranded conductors to a terminal screw. These units prevent loose strands from slipping out from under the terminal screw and reducing the current-carrying capacity of the connection. (See Figs. 9.9 and 9.10) Special *crimping pliers* are needed to install these connectors on conductors ranging from No. 18 to No. 10 gauge.

Courtesy IDI Electric Canada Ltd.

FIGURE 9.8 Colour-coded wire connectors are available in a range of sizes from grey (small) to blue and orange (medium) and to yellow and red (large).

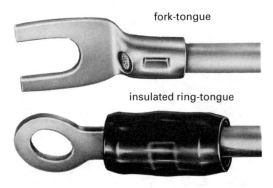

fork-tongue

insulated ring-tongue

Courtesy Burndy Canada

FIGURE 9.9 Fork-tongue and insulated ring-tongue compression lug installations

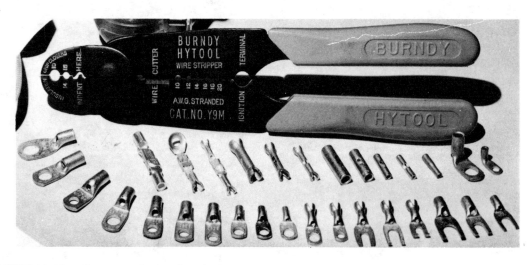

Courtesy Burndy Canada

FIGURE 9.10 Compression tool and crimp-on terminal lugs

Applications of Electrical Construction

Mechanical Cable Connectors

Connecting cables poses several difficulties: all strands must be held securely without damage to any, because damage will reduce the connection's conductivity; the connection must maintain a firm grip on the cable, compressing the strands into a solid group that will not loosen after a time; and the connector must be made of a metal that will not encourage electrolysis between itself and the cable. Mechanical connectors, called *lugs*, meet these requirements. (See Fig. 9.11)

There are several kinds of mechanical cable connectors (see Figs. 9.12, 9.13, and 9.14), and service entrance equipment for buildings makes extensive use of them. Figure 9.15, on page 108, shows neutral blocks from a distribution panel. This neutral block is capable of connecting the main neutral cable to the neutral wires of every circuit within a building.

Cable connectors are made in sizes for conductors of No. 14 gauge up to 1000 MCM. Connectors for cables larger than 1000 MCM may be obtained from some manufacturers by special order.

Courtesy Burndy Canada

FIGURE 9.12   Typical 3 conductor mechanical connector

Preparation of the Cable.   The following procedure is for installing copper and aluminum conductors in lugs.

Step 1.   Whittle off the insulation with a knife, taking care not to nick any of the

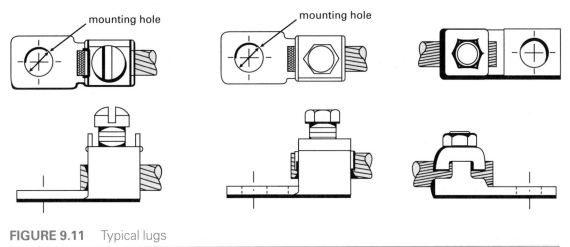

FIGURE 9.11   Typical lugs

FIGURE 9.13 Split-bolt connector

FIGURE 9.14 A single conductor mechanical connector

strands. Do not circle the cable with the knife, since this usually nicks the wire strands, which reduces their strength and conducting capacity.

Step 2. Remove any *oxide coating* (a dark dull coating) that is visible on the bared portion of the cable. This is an important procedure when using aluminum cable, because it oxidizes rapidly. Apply *antioxidant chemicals* to aluminum conductors at the same time. Use a wire brush to apply the chemical and remove oxidation.

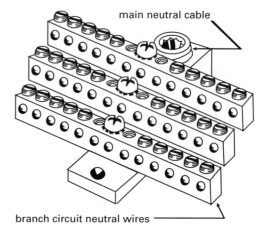

main neutral cable

branch circuit neutral wires

FIGURE 9.15 Neutral blocks for distribution panels

Step 3. Tighten the holding screw once the cable has been fully inserted into the lug. Allow several minutes to elapse, and then *retighten* the holding screw. The strands will have settled in place, making a second tightening necessary for a secure connection.

Compression Cable Connectors

These solderless connectors are made from a one-piece tubular form for installation with a hand- or hydraulic-powered compression tool. (See Fig. 9.16) The tubular forms are made from a high-conductivity, electrolytic *copper*. One end (the *tongue*) is flattened and drilled for fastening to a terminal block. The connectors are often electro-tinplated to minimize corrosion.

The strands of the cable are *compressed* within the copper tube by the compression tool until they form a solid mass of copper. This process ensures long life and maximum current-carrying capacity for the terminal con-

Applications of Electrical Construction

FIGURE 9.16 Compression type solderless terminal and splice

nection. There are three types of compression tools: the *hand-operated mechanical* type; the *hand-operated hydraulic* type; and the *motor-driven hydraulic* type. (See Figs. 9.17, 9.18, and 9.19)

Aluminum cables require special attention when terminated with a compression-type solderless lug. Figure 9.20, on page 111, shows the correct procedure for terminating aluminum cables. Figure 9.21, on page 112, shows the effect of a compression tool on a cable.

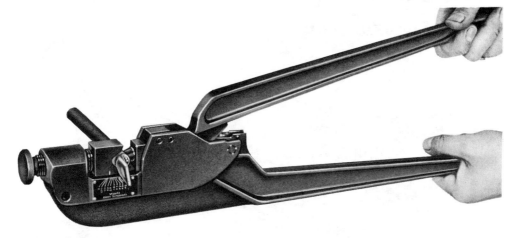

FIGURE 9.17 A hand-operated, mechanical type compression tool

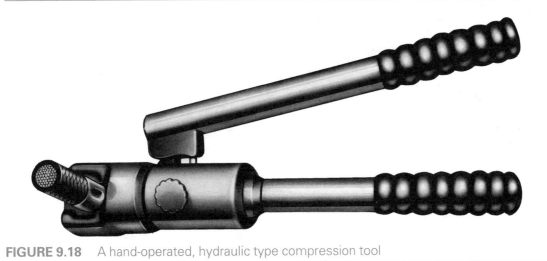

FIGURE 9.18 A hand-operated, hydraulic type compression tool

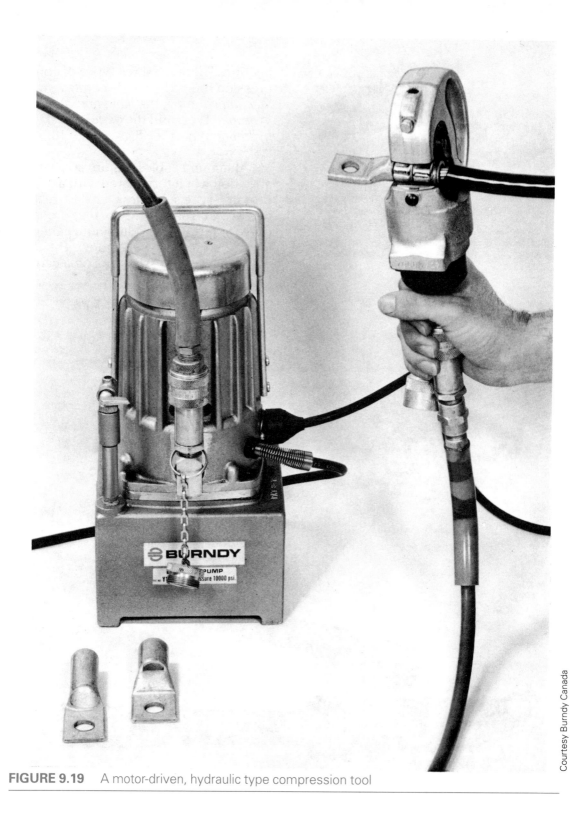

FIGURE 9.19   A motor-driven, hydraulic type compression tool

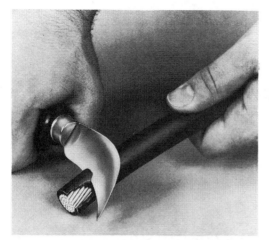

STEP 1. Carefully remove insulation without nicking conductor.

STEP 2. Wire-brush conductor to remove any oxide.

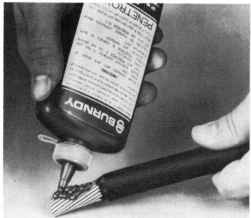

STEP 3. Apply antioxidant to prevent formation of surface oxide.

STEP 4. Tighten mechanical connectors securely.

STEP 5. Crimp compression type connectors with proper die and tool the recommended number of times.

FIGURE 9.20 Procedure for terminating aluminum cables

Courtesy Burndy Canada

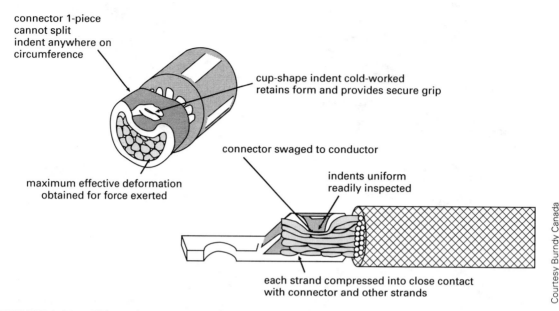

connector 1-piece
cannot split
indent anywhere on
circumference

cup-shape indent cold-worked
retains form and provides secure grip

connector swaged to conductor

indents uniform
readily inspected

maximum effective deformation
obtained for force exerted

each strand compressed into close contact
with connector and other strands

Courtesy Burndy Canada

FIGURE 9.21   Effect of compression tool on cable and lug

Figure 9.22 shows two outdoor applications for solderless compression connectors. In both cases, installing solder connections would be difficult and inconvenient.

Insulating Tapes

Once installed, solderless cable connectors require an *insulating tape* or other covering to replace the insulation removed during the making of the splice. (See Fig. 9.23) This tape or covering material must be capable of withstanding both the circuit voltage and the normal wear and tear on the cable. In other words, the covering material must have the same voltage rating and physical properties as the cable's original insulation.

Some splices require several types of insulating material to provide adequate electrical protection. Common coverings are friction tape, vinyl plastic, fibreglass, and insulating putty.

Friction Tape.   This basic cotton tape has an insulating compound impregnated into the weave. A low-quality insulator, it should not be used on circuits above 120 V.

Vinyl Plastic.   This excellent insulating product, available in assorted types and colours, comes in several thicknesses and widths. It adheres readily to most surfaces. A general purpose vinyl tape as seen in Figure 9.24 is approximately 7 mils thick and resists abrasion, sunlight, moisture, alkalis, and many acids. It has CSA approval for use on cable splices up to 600 V and fixture and wire splices up to a maximum of 1000 V at 105°C.

A more specialized type of vinyl tape can be used where cold temperatures are experienced. It is slightly thicker, at 8.5 mils, and has extra pliability for application at extremely low temperatures. It also remains easy to handle at normal temperatures. Figure 9.25 illustrates an application of this high-quality,

Applications of Electrical Construction

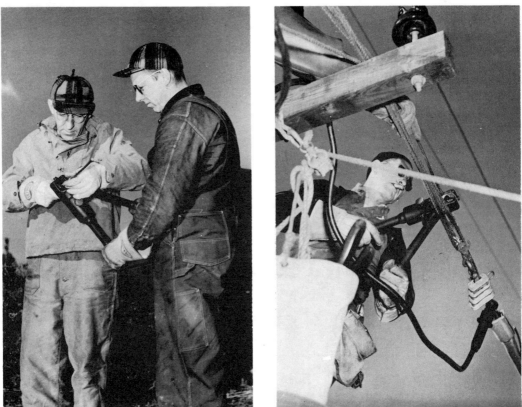

FIGURE 9.22   Typical outdoor uses for hand-operated, hydraulic compression tools

Courtesy Burndy Canada

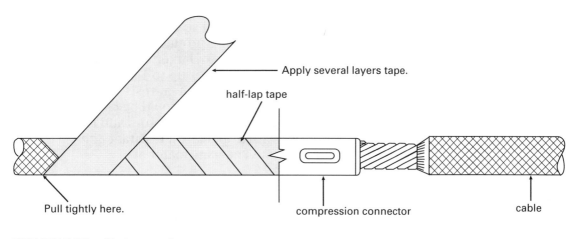

Apply several layers tape.

half-lap tape

Pull tightly here.

compression connector

cable

FIGURE 9.23   Taping a splice

FIGURE 9.24 General purpose vinyl plastic electrical tape

FIGURE 9.26 High-quality, all weather tape

FIGURE 9.25 Application of vinyl plastic electrical tape suitable for cold weather

all weather tape. Figure 9.26 shows a typical form of it.

Sturdy vinyl 10 mil tapes can be used where more abrasion resistance and mechanical strength are required. Wider widths are produced to speed up the insulating of larger splice areas. Where temperatures up to 105°C are encountered, a more heat-resistant tape is available. This tape could be used in and around electric motors and is equipped with a special oil resistant adhesive and vinyl backing material.

EPR Tape. Ethylene propylene rubber tape is a 30 mil, nonvulcanizing material that can be used for low voltage as well as high voltage applications up to 69 000 V. It can be stretched upon

Applications of Electrical Construction

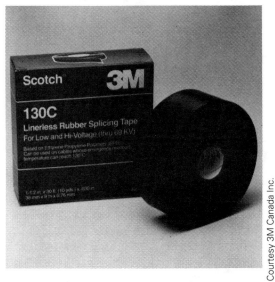

FIGURE 9.27 High and low voltage, EPR, linerless rubber splicing tape with a 130°C rating

application to ten times its normal length, thus forming a moisture-tight layer over the splice. (See Fig. 9.27)

Recent developments in tape technology have led to the production of an EPR tape that does not require a separation liner between layers on the roll. The liner prevented older-style tape from sticking together on the roll. It was often awkward to handle and messy to clean up after taping.

The new linerless tape has a unique ability to dissipate any heat from the splice. Its stable properties improve the chemical, electrical, and physical characteristics of the tape up to a maximum operating temperature of 130°C. This self-bonding, flame-retardant tape is self-extinguishing and suitable for use in areas previously restricted to specialty products.

Fibreglass. This tape is used on splices where the temperature may reach 130°C.

It has a pressure sensitive, thermo-setting adhesive that makes it suitable for use in such high temperature applications as furnace connections, water heaters, etc.

Insulating Putty. Large connectors require good quality insulation that will fill voids and pad any irregular shapes produced by the connector. An electrical grade, rubber-based, self-fusing elastic type putty is available in tape form for the insulating of large connectors. It should be used on low voltage circuits (600 V or less) where it will resist aging and not dry out. Figure 9.28 illustrates the proper method of insulating a large connector.

Another form of insulating putty is the vinyl mastic pad. (See Fig. 9.29 on page 117.) These pads consist of a self-fusing, rubber-based compound with a strong adhesive. They mould around difficult shapes and have excellent resistance to alkalis, acids, moisture, and varying weather conditions. Figure 9.30 illustrates the proper method of using them.

Resin Splicing Kits. A resin splicing kit can be obtained with sufficient materials included to complete one splice. A plastic mould, funnels, insulating and sealing compound, and pouring resin are included in the kit. Figure 9.31, on page 118, illustrates the pouring of the resin into the mould. These unique "field splicing kits" can be used for overhead, underground, or direct burial applications up to 5000 V. Figure 9.32 illustrates a cutaway of a resin splice after the mould has been removed. The kits are available in a variety of forms to handle numerous shapes and types of splices. Since they are produced in a variety of voltage

STEP 1. Apply first piece of insulating putty.

STEP 2. Overlap a second piece of insulating putty.

STEP 3. Press and form putty to shape of connection.

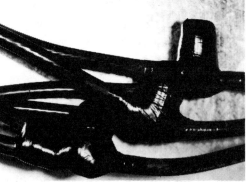

Courtesy 3M Canada Inc.

STEP 4. Complete insulating process with layer of vinyl plastic tape.

FIGURE 9.28 Insulating a splice using self-fusing elastic putty

ratings, take care to select the proper kit for the splice or connection to be insulated. If properly installed, the kit forms a moisture- and water-tight cover over a splice that could take consider-able time and material to insulate in any other manner.

A resin-pressure system of insulating splices is capable of insulating cables up to a capacity of 8000 V. To insulate in this manner, apply an open-weave, spacer tape around the splice. Next, tape

an injection fitting into place, and do a final taping of vinyl plastic. A liquid-tight mould forms. Now use a resin-pressure gun to pump the insulating resin into the wrapped splice. A tough, moisture-proof insulation on the splice will result. Figure 9.33, on page 118, illustrates the steps in making such a splice.

Electrical Coating.   Occasionally a splice that has been taped or been in service for a while may require a little

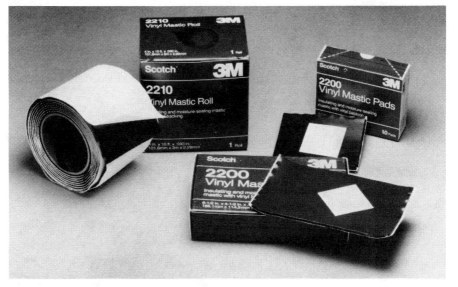

FIGURE 9.29   Vinyl mastic insulating material in roll and pad format

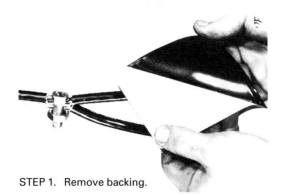

STEP 1.   Remove backing.

STEP 2.   Position pad.

STEP 3.   Wrap around.

STEP 4.   Insulation completed.

FIGURE 9.30   Proper technique for insulating with vinyl mastic pads

FIGURE 9.31   Pouring resin into a splicing mould

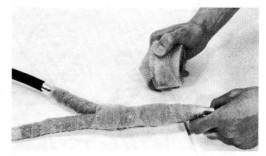

STEP 1.   Apply an open weave, spacer tape to splice.

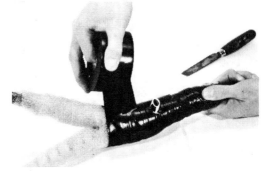

STEP 2.   Install injection fitting and cover. Splice area with plastic vinyl tape.

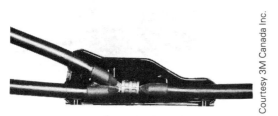

FIGURE 9.32   Cutaway view of a completed resin insulated splice

extra moisture proofing or insulating. A liquid, brush-on coating is available to provide this extra protection when required. Figure 9.34 shows this product being applied to a taped splice.

Coloured Tapes.   Vinyl tape is available in eight, fade-resistant colours: red, yellow, blue, green, white, orange, brown, and grey. Tape in these colours

STEP 3.   Force resin into splice with resin-pressure gun. Avoid excessive pressure to prevent bulging of tape.

FIGURE 9.33   Insulating a splice using the resin-pressure system

Applications of Electrical Construction

FIGURE 9.34 Application of liquid insulating coating

Courtesy 3M Canada Inc.

Heat-Shrink Tubing

Special plastic tubing that protects splices is now produced by the plastics industry. This tubing has a programmable memory which allows it to be shrunk or reduced in diameter when heated. Tubing of somewhat larger diameter than the splice to be insulated is easily slipped over the connection area. It is then heated with a hot-air blower and shrinks into a secure, one-piece plastic layer over the splice.

Chemical Make-up. A virgin plastic material, such as *polyethylene*, is used as the base for heat-shrink tubing. Such a material must possess mechanical strength and the capacity to resist certain fluids and ultraviolet light. It must also be a high-quality electrical insulator. The molecular structure of the plastic is modified by blending *additives* into it. These additives enhance the existing qualities of the plastic and add a few features, as well.

One new feature is that the tubing may soften under heat, but not turn into liquid. Under adverse high temperature conditions, such as a fire or short circuit, the plastic will not run off the splice. Too much heat can, of course, destroy the tubing, but it will remain in a rubber-like state until the point of destruction.

The second and most desirable feature is the *perfect elastic memory* produced by the radiation cross-linking of the plastic molecules. The tubing, supplied in an expanded (deformed) condition, will shrink tightly over irregularly shaped splices or objects when heated.

Different blends of the plastic produce a variety of tubing, making it useful in higher temperature areas, in cold weather applications, and in or near corrosive materials or liquids.

can be used to replace coloured insulation removed from a conductor before splicing or to identify and mark various circuit conductors. It is approved for use on 600 V at 80°C and has similar qualities to the other vinyl tapes available.

Taping a Splice

Figure 9.23 shows how to tape a splice. Manufacturers provide instruction sheets covering application methods for more specialized materials.

When removing insulation, take care not to nick the cable, something which would produce sharp edges or burrs in the strands. Burrs are potential weak areas that might puncture the tape or establish electrical stress in the splice. The higher the circuit current and voltage, the greater the danger of splice damage from electrical stress.

Types of Heat-Shrink Tubing

Two main types of this tubing are used extensively in industry. *Polyolefin* tubing, a simple, heat-shrink tubing, is one of the most popular types for covering splices or electrical connectors. (See Figs. 9.35 and 9.36) A second type of tubing, known as *dual-wall* tubing, has an outer tube of polyolefin and an inner tube of adhesive-type plastic which will form a perfect seal around the splice or connection. This seal is capable of keeping out all dirt, moisture, vapours, etc. The inner sealing wall is simply a different blend of polyolefin having adhesive/sealing properties. (See Figs. 9.37 and 9.38)

Courtesy Raychem Canada Limited

FIGURE 9.37   Dual-wall polyolefin, heat-shrink tubing, placed over a multi-conductor splice

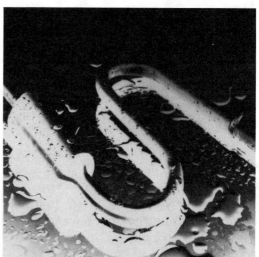

Courtesy Raychem Canada Limited

FIGURE 9.38   Dual-wall polyolefin tubing is capable of waterproofing splices and connections to components.

Courtesy Raychem Canada Limited

FIGURE 9.35   Polyolefin, heat-shrink tubing placed over a crimp-on connector

Courtesy Raychem Canada Limited

FIGURE 9.36   When heated the tubing will shrink to form an insulating barrier over any shape of connector or splice.

Heat Source

The approved method of heating the tubing is with a hot-air convection heat blower as shown in Figure 9.39. Shrink temperatures range from 80°C to 150°C, depending on the blend of polyolefin being used. In the lower heat ranges, a standard hair dryer can shrink the tubing.

When emergency situations arise and a blower is not available, a flame from a match or other low-level flame

Applications of Electrical Construction

FIGURE 9.39 Hot-air gun in use with heat-shrink tubing to activate the elastic memory

Courtesy Raychem Canada Limited

source can be used. Relying on a flame should not be done in general practice however: care must be taken not to exceed the plastic's temperature rating. (See Fig. 9.40)

FIGURE 9.40 Low temperature flame being used to heat-shrink tubing where power is not available

Courtesy Raychem Canada Limited

Applications

Due to the simple application process and neatness of the completed job, many uses have been found for polyolefin tubing. There are ten major colours, as well as clear plastic, available for the identification of conductors or connections. Colours with stripes are also available to provide an even greater number of identification combinations. This high-quality insulating product can hold conductors in groups for the separation of circuits and easily cover complex connectors or terminals. Heat-shrink tubing, with its versatility, has been used extensively in mass transit vehicles such as trains, military ships, aircraft and land vehicles. Special versions of the tubing are also being produced for use in space satellites. (See Figs. 9.41, 9.42, 9.43, 9.44, 9.45 and 9.46)

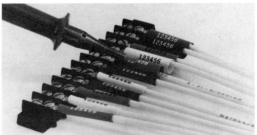

FIGURE 9.41 Flexible, general purpose tubing for identification of cables

Courtesy Raychem Canada Limited

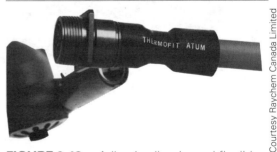

FIGURE 9.42 Adhesive-lined, semi-flexible, thinwall tubing with a high shrink ratio and flame-retardant jacket

Courtesy Raychem Canada Limited

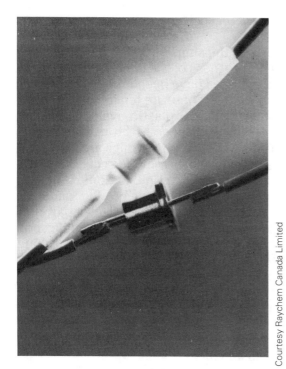

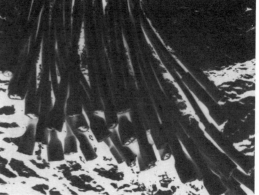

Courtesy Raychem Canada Limited

FIGURE 9.45   Heat-shrink tubing used to enclose crimp-on terminal connectors

FIGURE 9.43   Heat-shrink tubing used to insulate and protect an electronic component

Courtesy Raychem Canada Limited

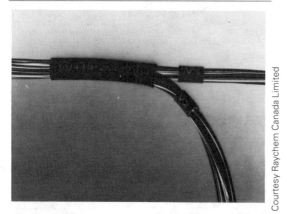

Courtesy Raychem Canada Limited

FIGURE 9.46   Bus-bars insulated and colour coded with heat-shrink tubing

FIGURE 9.44   Heat-shrink tubing used to identify circuits and individual conductors in a circuit

Courtesy Raychem Canada Limited

Ratings and Approvals

The tubing has both CSA and Underwriters' Laboratories approval for use in the electrical industry. Various thicknesses form protection for the many residential and industrial voltages in use, while temperature ratings for continuous-duty use range from -55°C to 135°C. Polyolefin tubing would appear to be a great boon to the electrical industry.

Applications of Electrical Construction

For Review

1. List two disadvantages of soldering and taping wire splices.
2. List three types of solderless wire connectors. What is the maximum circuit voltage for each?
3. Why are crimp-on connectors installed on small stranded conductors fastened under terminal screws?
4. What are the disadvantages of compression connectors?
5. Give three reasons why lugs are useful.
6. List the steps in fastening a copper cable to a terminal lug.
7. List the steps in fastening an aluminum cable to a terminal lug.
8. Why should nicks on the strands of a cable be avoided when removing insulation?
9. Why must a mechanical connector be retightened several minutes after the first tightening?
10. Describe the methods for crimping copper and aluminum compression connectors to a cable.
11. Which two types of protection must electrical tape provide?
12. Why is it necessary to observe the temperature ratings of electrical tape?
13. List three common types of electrical tape.
14. Describe how large, irregularly shaped cable connectors are insulated.
15. What are the advantages of insulating putty when covering large or irregularly shaped splices?
16. What advantages do resin splicing kits have over more conventional methods of insulating a splice?
17. Why must care be taken to select an insulating product that has the proper voltage rating?
18. What are the advantages of using coloured insulating tapes?
19. Why is it desirable to pull or stretch the tapes as they are being applied to the splice area?
20. Under what conditions would a brush-on, liquid insulating material be used?
21. Name the two main types of heat-shrink tubing in use throughout the electrical industry.
22. What advantage has heat-shrink tubing over other forms of electrical insulation?
23. List four different applications of heat-shrink tubing in the care and protection of electrical wiring circuits.
24. How can heat-shrink tubing be used in the identification of circuit terminals and conductors?
25. What methods are used to reduce the tubing to its final size after installation on splices or terminations?

10 Heat-Control Switches

Electrical cooking appliances, such as ovens, ranges, hot plates, and commercial coffee-makers used in restaurants, use a variety of heavy-duty switches to control their heating elements. One common method for providing different levels of heat is to use two elements and connect them in series/parallel combinations across 120 V and 240 V. The switches discussed in this chapter are capable of providing the series/parallel connections required for this method of heat control.

Three-Heat, Single-Pole

This simple heat-control switch is used primarily for rangettes and hot plates operating at 120 V. Heat ranges are provided by connecting two elements of the same size (wattage) as follows:

Low: 2 elements in series on 120 V;
Medium: 1 element on 120 V;
High: 2 elements in parallel on 120 V.

Figures 10.1 and 10.2 show the switch, its internal connections, and wiring by schematic diagram. The switch's connections can be tested with a series-lamp tester (described in Chapter 2).

Three-Heat, Double-Pole

This switch uses the same series/parallel combinations of the two elements as the single pole version. Switches controlling 240 V (two live wires) must be capable of opening both live wires at the switch. When the switch is turned to *off*, there must be no voltage present at the elements.

Figures 10.3 and 10.4 show the switch, its internal connections, and wiring by schematic diagram.

Five-Heat

More variations in heat are available by using the 5 heat, double-pole switch to control two elements of the same size. Two elements, each rated at 600 W on 240 V, will provide heat as follows:

Low: 2 elements in series on 120 V provide 75 W;
Low-Medium: 1 element on 120 V provides 150 W;
Medium: 2 elements in parallel on 120 V provide 300 W;
Medium-High: 1 element on 240 V provides 600 W;
High: 2 elements in parallel on 240 V provide 1200 W.

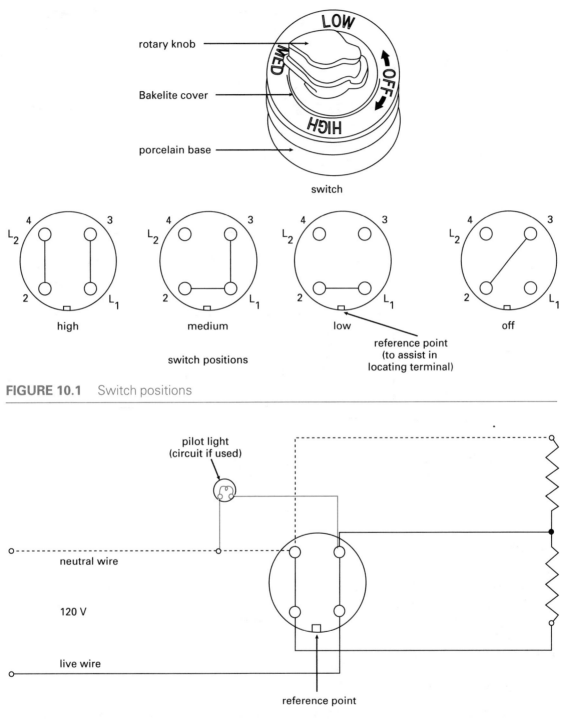

FIGURE 10.1　Switch positions

FIGURE 10.2　Schematic wiring diagram for a 3 heat, single-pole switch

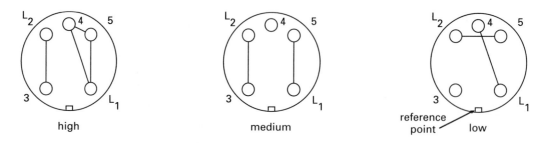

FIGURE 10.3   Switch positions for a 3 heat, double-pole switch

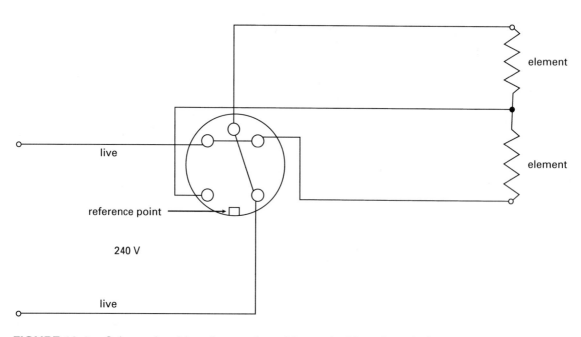

FIGURE 10.4   Schematic wiring diagram for a 3 heat, double-pole switch

Figures 10.5 and 10.6 show the switch's internal connections and its wiring by schematic diagram.

Heat variations other than those listed above can be obtained by using elements of two different sizes. Also, a safety pilot light indicates when the switch is on in one of its five heating positions. Take care to use a 120 V lamp for the pilot light. Lamps with lower voltage ratings will burn out if connected to this circuit. Lamps with higher voltage ratings may be too dim for effective use.

Seven-Heat

This switch looks much like the 5 heat switch, but has the advantage of two extra heat-levels. Because of this greater

Applications of Electrical Construction

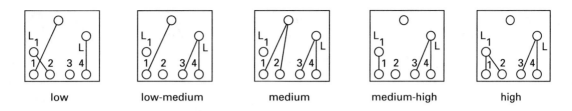

low   low-medium   medium   medium-high   high

FIGURE 10.5   Internal switch connections (rear view) for a 5 heat switch

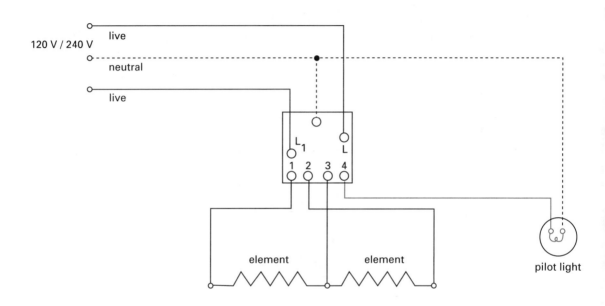

FIGURE 10.6   Schematic wiring diagram for a 5 heat switch

range of heat selection, the heat switch has gradually replaced the older 5 heat switch design.

The seven heat-levels are obtained by using two different elements, each rated at 240 V. A 600 W and an 800 W element in combination provide heat as follows:

Simmer: Both elements in series on 120 V provide approximately 87 W.
Position 6:   One 600 W element on 120 V provides 150 W;
Position 5:   One 800 W element on 120 V provides 200 W;

Position 4:   Both elements in parallel on 120 V provide 350 W;
Position 3:   One 600 W element on 240 V provides 600 W;
Position 2:   One 800 W element on 240 V provides 800 W;
High: Both elements in parallel on 240 V provide 1400 W.

Figures 10.7 and 10.8 show the switch's internal connections and its wiring by schematic diagram.

Most modern ranges have a pilot light terminal at the rear of the switch to

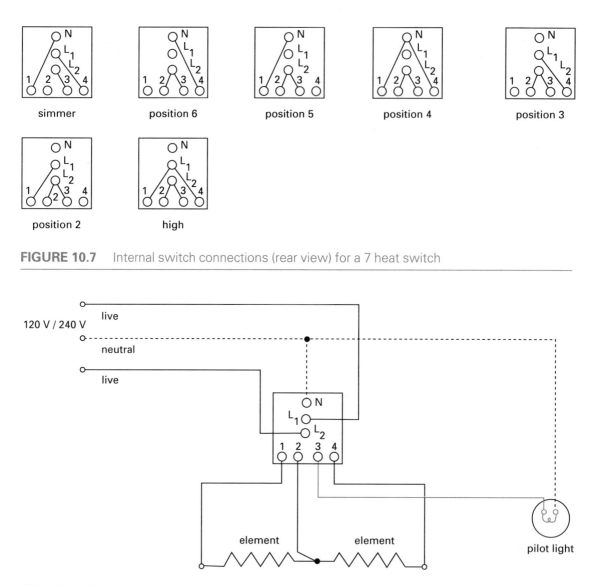

FIGURE 10.7   Internal switch connections (rear view) for a 7 heat switch

FIGURE 10.8   Schematic wiring diagram for a 7 heat switch

show whether the element is on. An element that has been left on may not glow red but still be hot enough to burn a person touching it.

Infinite-Heat

Improvements in the design of electric ranges created a need for a greater variety in heat levels. The infinite-heat switch, as the name implies, offers an unlimited variation in heat level. Although this switch is often more expensive than other heat control switches, a single element requiring less wiring offsets this initial cost.

Figure 10.9 shows by schematic wiring diagram the internal workings of a typical infinite-heat switch.

Applications of Electrical Construction

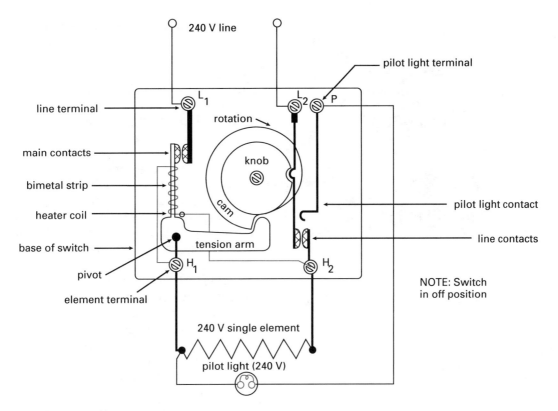

240 V line

pilot light terminal

line terminal

L_1

rotation

L_2 P

main contacts

knob

bimetal strip

cam

pilot light contact

heater coil

line contacts

base of switch

tension arm

pivot

H_1

H_2

element terminal

NOTE: Switch
in off position

240 V single element

pilot light (240 V)

FIGURE 10.9   Schematic wiring diagram showing infinite-heat switch parts and circuit

Operation of Infinite-Heat Switch.

A *cam* is fastened to the *knob* of the switch. One of the *main* contacts is on a *bimetal strip* fitted in to a *tension arm*. As the knob is turned slightly, the cam actuates the tension arm. The arm rotates on the pivot and closes the main contacts, thus heating the element.

A *heater coil* is wound around the bimetal strip and connected in *parallel* with the element. The heat from the heater coil warms the bimetal strip and causes the strip to bend outwards. This opens the circuit, and the heater cools. The strip contracts and closes the contacts. This activates the element and heater again.

The more the knob is turned, the more it causes the tension arm to exert

pressure on the contacts. Therefore, it takes longer for the heater to become hot enough to actuate the bimetal strip and open the contacts.

The single element operates at full power each time the contacts are closed, and the heat level is regulated by the speed at which the main contacts open and close. The number of operating cycles in a given period of time and the length of time the element is on in each cycle is controlled by the amount the knob on the switch is rotated.

A pilot light terminal simplifies the connection of a light to indicate the *on* position. Figure 10.10 illustrates the internal layout of a typical infinite-heat switch.

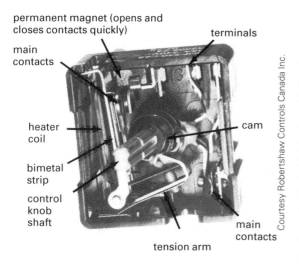

permanent magnet (opens and closes contacts quickly)

terminals

main contacts

heater coil

cam

bimetal strip

control knob shaft

main contacts

tension arm

Courtesy Robertshaw Controls Canada Inc.

FIGURE 10.10   Internal layout of an infinite-heat switch unit

Oven Control

The switch controlling the heat level in a modern oven also provides *infinite-heat*

control. A liquid expansion system is used to regulate the *on* and *off* cycles of the switch. Cycling of the switch is further regulated by an adjustable threaded shaft that operates like a cam. The timing of the heating cycle determines the amount of heat in the oven at a given time. The liquid expansion system replaces the heater coil found in the infinite-heat switch. (See Fig. 10.11)

The oil-like material used in the bulb of the liquid expansion system senses the oven temperature, expands accordingly, and operates a diaphragm device inside the switch. Both the expansion and contraction of the liquid and the movement back and forth of the diaphragm open and close a set of contacts inside the switch. These contacts switch the oven *on* and *off* at the proper rate of cycling to obtain the temperature selected by the person using the oven.

The oven switch controls two elements inside the oven. The upper broil element is turned on by rotating the

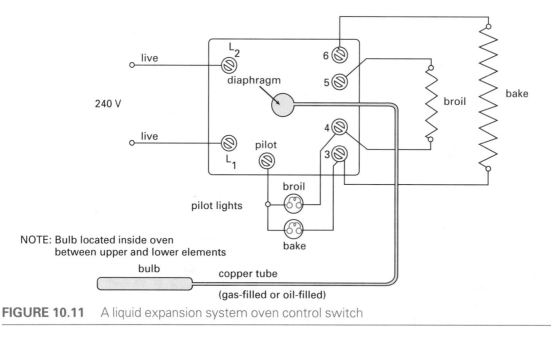

FIGURE 10.11   A liquid expansion system oven control switch

Applications of Electrical Construction

knob of the switch to the broil position. Oven manufacturers design the oven door so that it will remain in a slightly open position for this operation. They recommend that broiling be done with the oven door partly open to ensure that oven temperature does not get hot enough to cycle the contacts. The broil element will then remain *on*, necessitating careful food watching if burning is to be avoided. It is also common to have an oven with an automatic preheat process. When the oven is being preheated, the broil element will come on with the bake element. The broil element will be automatically turned *off* when oven temperature is approximately 40°C below the knob setting of the switch.

Oven temperature will then be brought to its selected heat level by the bake element and kept there as the switch cycles *on* and *off* during the cooking process. Having both elements *on* together provides a lot of heat and gets the oven up to temperature rather quickly. The purpose of the broil element then is to preheat the oven and/or brown food quickly. The lower bake ele-ment is used for normal heating of the oven and is under the control of the liquid expansion system at all times.

Pilot lamps are normally included in the oven circuit to indicate when the elements are *on* and producing heat in the oven. Figure 10.12 illustrates the internal layout of a typical oven control switch.

Self-Cleaning Ovens

Ovens with a self-cleaning feature do not require spilt or burnt food to be removed from the walls and floor of the oven through the use of strong cleaning materials. They thereby eliminate a tedious and often messy process.

Continuous Cleaning Type. This oven uses its normal cooking temperature to gradually reduce the burned food particles on its inner surfaces. An oven cleaned this way does not appear to be as clean as one of the second type and the process takes much longer.

Pyrolytic Self-Cleaning Type. This oven, with its more efficient cleaning

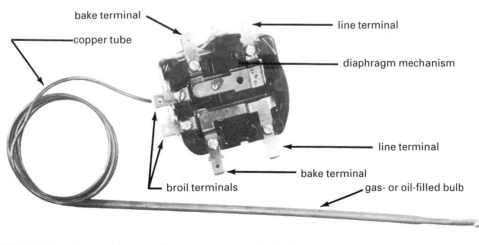

bake terminal
copper tube
line terminal
diaphragm mechanism
line terminal
bake terminal
broil terminals
gas- or oil-filled bulb

Courtesy Robertshaw Controls Canada Inc.

FIGURE 10.12   Internal layout of an oven control switch

system, uses a high oven temperature over a shorter period of time to reduce food particles to an easily removable ash.

A pyrolytic oven relies on a control switch-and-circuit that operates in much the same manner as the temperature-sensitive expansion system used in regular oven switches. However, the control switch-and-circuit has more functions to perform than the expansion system. The oven temperature reaches approximately 480°C during its cleaning cycle, and as mentioned above, reduces food spill-overs to a small amount of ash (much like a cigarette ash) in an hour or two. This ash can then be easily removed, leaving the oven clean and ready for future use.

Due to the much higher temperatures in this type of oven, the oil-like material in the expansion system is replaced with helium gas. A larger bulb accommodates the amount of helium required to operate the diaphragm in the switch.

This type of control switch may have up to four different sets of contacts, all of which are set to operate at different temperatures and open and close the various circuits as the gas-filled expansion system dictates. One set of contacts would be for baking and broiling, while a second set would be calibrated at the factory to operate the oven at high temperature during the cleaning cycle.

Although oven temperatures are high during the cleaning cycle, only a little electricity is used. That is because the elements are connected in such a manner that they operate at 120 V, producing less than their full wattage, but staying on throughout the cycle. The elements are not cycled on and off as frequently as they are during the cooking operation, but are allowed to produce heat for a longer period of time. Extra

insulation is usually put into a self-cleaning oven to assist the elements in obtaining the proper temperature and prevent excessive heat from passing through to the outer surfaces of the stove itself.

A third set of contacts can be used to engage a latching mechanism in the oven door. This prevents the door from being opened during the high temperature cleaning cycle. The oven door latching mechanism is set to keep the door closed whenever oven temperature exceeds 320°C for two safety reasons. First, burns to hands or face can be severe when oven surface and air temperatures reach this level. Second, a sudden inflow of oxygen from the air as the oven door is opened can cause food particles in the oven to burst into flames when the temperature is in the high range used for cleaning.

A fourth set of contacts on the switch can be used to hold the food at a warm temperature, after the baking process, until you are ready to eat the food. This temperature is approximately 80°C.

Most modern ovens have useful timer circuits built into their control panels. Ovens can be set to come on at a predetermined hour of day or night, to cook for a preset length of time, and to shut off or keep food warm until needed. Elements, receptacles, and minute minders can all be timed and/or controlled.

Figure 10.13 illustrates a circuit diagram for a self-cleaning oven. All switches contained within the dotted lines are controlled by the gas-filled expansion system. The door-lock switch operates the door-lock solenoid, or electrically powered, magnetic latch assembly, once the clean cycle has been selected and prevents the door from being opened during the cleaning operation. The door's electrical interlock (5 & 6) prevents the solenoid from releasing

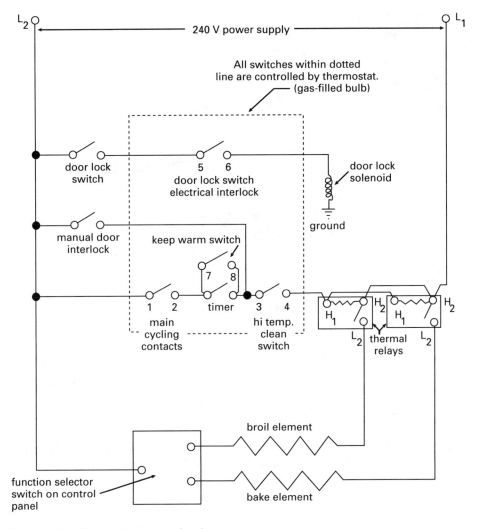

FIGURE 10.13 A self-cleaning oven circuit

the door once the oven temperature reaches cleaning level.

The manual door interlock switch by-passes the normal cycling contacts, thus removing the expansion system from the circuit for the cleaning operation and allowing the oven to reach 480°C. A hi-temp clean switch prevents the oven from rising above the desired cleaning temperature.

The function switch located on

the control panel of the stove/oven is used by the owner to select bake, broil, preheat, etc., functions for the oven. Both bake and broil elements are further controlled by a thermal relay (part of the gas expansion system) which controls their cycling to obtain proper oven temperature as selected by the operator. Figure 10.14 illustrates the internal layout of a self-cleaning oven switch, pyrolytic type.

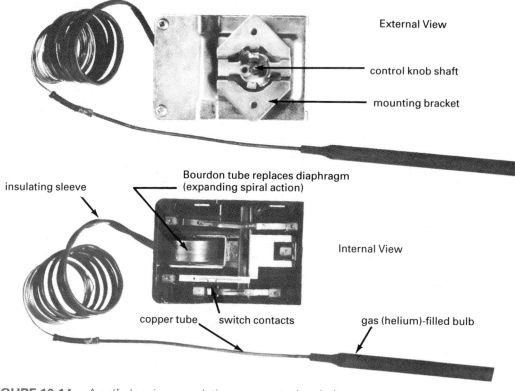

External View

control knob shaft

mounting bracket

insulating sleeve

Bourdon tube replaces diaphragm
(expanding spiral action)

Internal View

copper tube   switch contacts   gas (helium)-filled bulb

Courtesy Robertshaw Controls Canada Inc.

FIGURE 10.14   A self-cleaning, pyrolytic oven control switch

Recent developments in such switches have brought about the use of a sodium/potassium-filled tube and bulb expansion system. The sodium/potassium mixture turns into a liquid at oven operating temperatures, expanding to operate a bellows device in the switch. A higher degree of sensitivity is obtained with the new expansion system, but the electrical contacts operate in much the same manner as previous switch mechanisms.

Rotisserie.   Many modern ovens are equipped with a motor driven unit to rotate foods within the oven while cooking. Juices from the food being cooked fall from the rotating unit and can often land on the lower bake element. To prevent smoking of these greases, the broil element is used by many manufacturers for the rotisserie cooking cycle. Its location in the upper part of the oven prevents dripping particles from landing on it, thus reducing the smoke problem.

When selecting the cooking temperature for rotisserie cooking, the switch knob is rotated clockwise to the broil position, and then backed off to the desired cooking temperature. This engages the broil element only, through a mechanical selection system built into the front of the switch, and allows the broil element to heat the oven while being controlled by the gas expansion system.

For Review

1. List four appliances that use heat-control switches.
2. Explain how three levels of heat are obtained with the 3 heat switch.
3. Of what use is the reference point on the base of a 3 heat switch?
4. Why must heat-control switches operating on 240 V be capable of opening both live wires?
5. Why has the 7 heat switch gradually replaced the older 5 heat switch?
6. Why is a pilot light an important feature on heat-control switches?
7. List two advantages the infinite-heat switch has over other heat-control switches.
8. Explain in your own words the operation of the infinite-heat switch.
9. Where is the oil bulb for an oven-control switch located?
10. What precaution must be taken when using the broil element? Why?
11. State two advantages of the self-cleaning oven.
12. Which type of self-cleaning oven cleans the best? Explain why.
13. How is the high temperature obtained for the cleaning cycle of a self-cleaning oven?
14. Explain why extra insulation is required around a self-cleaning oven.
15. What are the advantages of an electrically timed oven circuit?

11 Armoured Cable

Armoured cable (*A/C*) can be used for both open and concealed wiring systems. Unlike NMSC, it may be run on the surface of walls and ceilings in buildings of masonry construction.

Armoured cable is widely used as a quick method for distributing power throughout new industrial plants, adding to existing plants, and relocating machinery. It can also be used in public and commercial buildings where the possibility of physical damage to the cable makes NMSC unsuitable. Armoured cable is more flexible than rigid conduit systems. It can be installed in long continuous runs without need for joints and splices. Conduit systems, however, require a box or fitting after each 30 m of conduit and/or after an accumulation of 360° of bend to ease the pulling in of the conductors.

Electrical Ratings

Unlike NMSC, A/C has a continuous protective metal covering, and so is approved for use on circuits up to 600 V maximum. It is available in 1,2,3, and 4 conductor combinations, ranging in size from No. 14 to No. 4/0. Larger conductors may be made to order.

Individual conductors in the cable used to be covered with a cotton-braid-over-rubber insulation which was rated for use at a maximum temperature of 60°C. Changes in the Canadian Electrical Code have resulted in improvement in the insulation on these conductors. Modern A/C conductors are insulated with a durable material called *Cross-Link* (*X-Link*). X-Link has a maximum temperature rating of 90°C.

Trade Names

Armoured cable was originally known as *armoured bushed cable (ABC)*, because a small, anti-short bushing was inserted in to the end of each cable termination. However, installers often refer to A/C simply as BX or BXL. *BXL* refers to the type that has an *internal lead sheath* over the conductors.

Cable Construction

Figure 11.1 shows the materials used in the construction of the older style of A/C. Figure 11.2 shows modern cable construction.

One type of armoured cable has a lead sheath placed between the conductors and the armour. This sheath prevents moisture or chemicals from entering the conductor portion of the cable. The sheath makes the cable suitable for outdoor use or direct burial in the earth. (See Fig. 11.3) The *aluminum* armour on most A/C, however, corrodes severely when placed under ground. Therefore, *steel* is used as a replacement on the cable for outdoor and underground use (*ACL*).

Cable Termination

The first step in fastening A/C to a box or fitting is to remove the armour with a

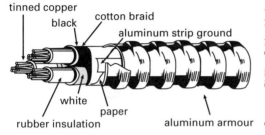

FIGURE 11.1   Older style 3 conductor flexible armoured cable

Courtesy Phillips Cables Limited

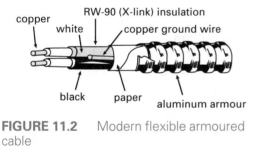

FIGURE 11.2   Modern flexible armoured cable

Courtesy Phillips Cables Limited

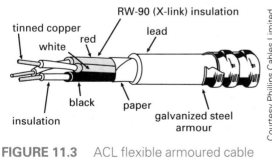

FIGURE 11.3   ACL flexible armoured cable

Courtesy Phillips Cables Limited

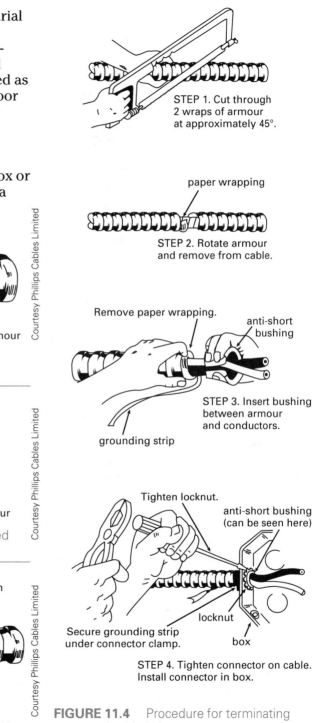

STEP 1. Cut through 2 wraps of armour at approximately 45°.

STEP 2. Rotate armour and remove from cable.

STEP 3. Insert bushing between armour and conductors.

STEP 4. Tighten connector on cable. Install connector in box.

FIGURE 11.4   Procedure for terminating armoured cable in box

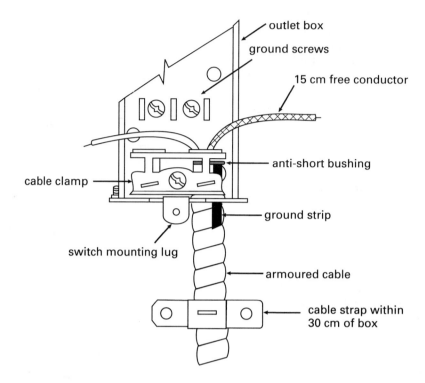

outlet box

ground screws

15 cm free conductor

anti-short bushing

cable clamp

ground strip

switch mounting lug

armoured cable

cable strap within 30 cm of box

FIGURE 11.5 Old-style armoured cable secured to box by built-in clamp

hacksaw. (See Fig. 11.4) Take care to remove the correct amount of armour the first time. It is often difficult to hold a small length of cable for a second cut. Take care, too, to protect the insulation on the conductors from being damaged by the saw during the cutting. Trim the paper wrapping around the conductors as close to the armour as possible.

After removing the armour and trimming the paper wrapping, fold back the aluminum grounding strip so that it is out of the way beside the cable. Insert a fibre or plastic anti-short bushing between the armour and the conductors. This anti-short bushing prevents the jagged edges of the armour from cutting into the insulation around the conductors. Once the anti-short bushing is in place, insert the cable in to an approved *box connector.* The box con-

nector holds the anti-short bushing in place, secures the cable to the box or fitting, and clamps on to the grounding strip. Use of it ensures that the ground circuit is complete. (See Figs. 11.5 and 11.6)

Cables now being produced have a *grounding conductor* instead of an aluminum grounding strip. This feature eliminates the need for folding back and clamping a grounding strip.

Cable Supports

As does NMSC, armoured cable *must* be supported by an approved strap or staple within 30 cm of every box. A *maximum* of 1.4 m between the supports on the cable run is allowed. If the cable is run through wooden joists, it *must* be kept back at least 3 cm from the face.

Applications of Electrical Construction

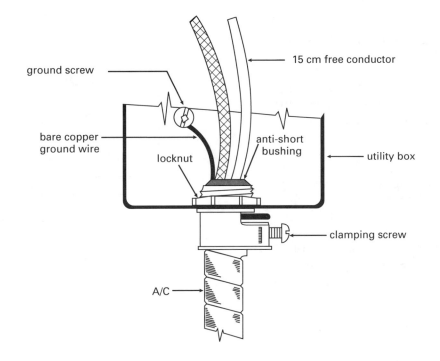

ground screw

15 cm free conductor

bare copper
ground wire

anti-short
bushing

locknut

utility box

clamping screw

A/C

FIGURE 11.6   Modern armoured cable, with ground wire, secured to box by an A/C connector

Steel nails driven into the wooden members can easily puncture the aluminum armour. As with NMSC, where the A/C is closer than 3 cm to the face of the wooden member, a *steel plate* is located in front of the cable. (See Chapter 8, Fig. 8.11) Section 12 of the Canadian Electrical Code lists regulations for armoured cable installations.

Fishing Armoured Cable

When installing armoured cable in a building that is completed, access to existing outlet boxes is gained by lowering the cable down inside the wall from a hole in the attic or ceiling. This is called *fishing*.

The popular practice of finishing off rooms or areas with false ceilings promotes the extensive use of armoured cable, particularly in shopping, industrial and commercial malls. Lighting and power circuits can be run through the "open" spaces above false ceilings with ease, and armoured cable is a natural choice for these areas. When more circuits are added in the years following a building's completion, armoured cable can be easily fished over and around the many beams, girders, support rods and wires in the ceiling area. However, with so many obstacles in the ceiling area, reasonable care must be taken not to damage the outer sheath while doing this.

ACL for Wet Areas

When equipped with a lead sheath between the conductors and the armour (ACL or BXL), armoured cable can be used to supply power to outdoor lights, signs, and related equipment. ACL can also be used for underground supply lines to electrical equipment. Pumps, garden lighting, post lights, and supply lines between buildings often require ACL's water-resistant feature. (See Fig. 11.3)

Water-tight connectors *must* be used to secure this cable to a box or fitting in a wet area. (See Fig. 11.7) There are several types of connectors, all of which compress a moisture-proof lead sleeve around the cable. Figure 8.10 shows this method used to supply an outdoor post light. Figure 11.8 shows a garden receptacle installation. Because it has steel armour, ACL can be buried in masonry or concrete where excessive moisture is present.

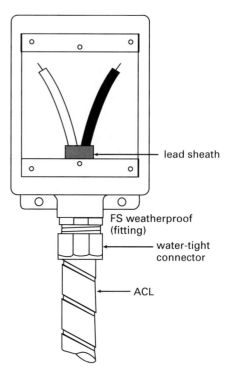

FIGURE 11.7   ACL connected to moisture-proof fitting

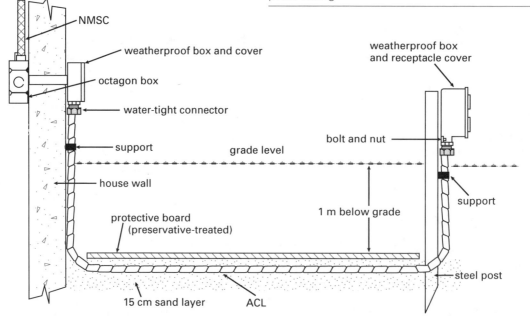

FIGURE 11.8   Typical garden receptacle installation using ACL

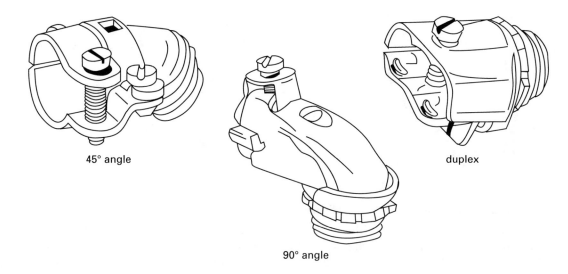

45° angle

duplex

90° angle

FIGURE 11.9 Special A/C connectors

Note: ACL does *not* have a grounding strip. It relies on a secure mechanical connection to the box for its ground. The soft lead sheath is allowed to extend past the connector and in to the box. This eliminates the need for an anti-short bushing.

Armoured Cable Accessories

A/C can be bent only so far before the interlocking armour separates. Cable left in this condition will not receive approval of the electrical inspection department. Special connectors providing 45° and 90° angles are used where a sharper angle than the cable will allow must be made. Duplex connectors allow a pair of cables to enter one knockout in a box. (See Fig. 11.9)

Straps and/or staples of the proper size, which are similar to those used with NMSC, are made for supporting armoured cable. (See Chapter 8, Fig. 8.7)

For Review

1. Why is A/C suitable for both concealed and open wiring systems?
2. Why does A/C have a higher voltage rating than NMSC?
3. List the A/C conductor combinations.
4. Explain in your own words the procedure for terminating A/C at a box.
5. What care must be taken when fishing A/C?
6. How does A/C for wet locations differ from A/C used in dry locations?
7. How is moisture prevented from entering the box or fitting in an outdoor cable installation?
8. How is the ground circuit maintained in A/C installations?
9. Why must care be taken when bending A/C?
10. What devices are used for terminating armoured cable where a sharp bend must be made?

12

Aluminum-Sheathed Cable

An aluminum-sheathed cable, which is a *factory-assembled* wiring system, is a seamless metal or welded wrap-around sheath enclosing a single or multiple conductor assembly. The aluminum sheath is both vapour- and liquid-tight, and the conductors enclosed in the sheath can be either copper or aluminum. Conductors were at one time insulated by rubber, but are now insulated by *Cross-Link polyethlene,* rated at a maximum temperature of 90°C.

Smooth-Sheathed Cable

Aluminum-sheathed cable was basically a *cable-in-a-conduit*, assembled and

tested at the factory. Using it simplified installation and saved considerable time and labour. While much smooth-sheathed cable is still in use (see Fig. 12.1), but manufacturers are now producing corrugated cable in all sizes, instead.

Corrugated Cable

The conductors of this cable are enclosed in an oversize aluminum tube. The tube is then passed through a *revolving die* that compresses two-thirds of the tube on to the insulated conductors. (See Fig. 12.2) The soft, metal *arches* allow the cable to be bent easily while the work-hardened *flat spiral* areas

FIGURE 12.1   Old-style smooth-sheathed aluminum cable

FIGURE 12.2   Corrugated aluminum sheathed cable with a type "W" moisture-proof connector

Courtesy Canada Wire & Cable Ltd.

support the conductors. Large cables can thus be bent without the use of special tools such as those required for a conduit system.

Sheath Covering

A *polyvinyl chloride (PVC)* jacket is extruded over the aluminum sheath. The jacket enables the cable to resist corrosive chemicals and to be used underground. Jackets are made in various colours to aid identification of circuits and voltages.

Electrical Ratings

Aluminum-sheathed cable is made for use on any voltage required. Most of the common sizes of conductor units are readily available with a 600 V rating. Part 1 of the Canadian Electrical Code lists the current-carrying capacity of the cables in its ampacity tables.

The conductors of aluminum-sheathed cable carry a maximum temperature rating of 90°C.

Single Versus Multi-Conductor Cables

Single-conductor cables have a much *higher current rating* than multi-conductor cables with the same gauge of wire. Therefore, it is more economical to use several single-conductor cables, rather than one large multi-conductor cable, for construction of heavy-current circuits.

Sheath Currents

A conductor carrying electrical current produces a magnetic field around itself. When the current is alternating (AC), the magnetic field rises and falls in strength as the current pulses through the conductor. This results in a *moving magnetic field*.

Any metal substance within the range of this moving magnetic field will have a current induced into it by the field. When both ends of the cable are grounded on a single-conductor unit, a circuit is formed and the sheath currents circulate. The sheath currents of cables running side by side will circulate between them, if the cables are connected to a common metal box at each end.

As the sheath currents increase in volume, *heat* is generated in the sheath. This heat can be severe enough to damage the insulation around the conductors within the cable.

Single-conductor cables carrying higher amounts of current, for example, 425 A and up, must receive special care when run side by side. Copper cables of 250 MCM and aluminum cables of 350 MCM and larger must have their sheaths insulated from one another at one end of the cable run and be grounded at the opposite end. These measures prevent sheath currents from circulating within the cable system. (See Figs. 12.3 and 12.4)

Note: Do not use magnetic materials, such as steel, for supporting single-conductor cables. These supports can become magnetized and have currents circulating within them. These eddy currents, as they are called, heat the support and cause damage to the cable insulation.

Connectors used to terminate single-conductor cables must also be made of *nonmagnetic* material. In a three conductor system (3 phase system), all three cables must enter the box or cabinet through one, nonmagnetic *plate*. When this plate is fastened to the box or cabi-

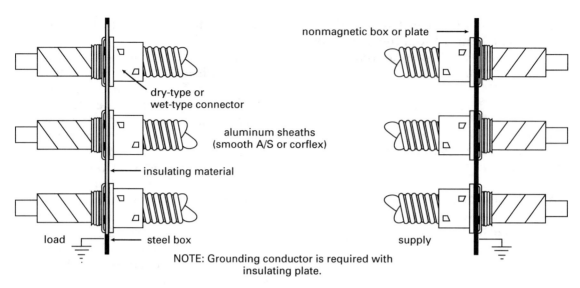

FIGURE 12.3 Typical cable installation (no sheath currents)

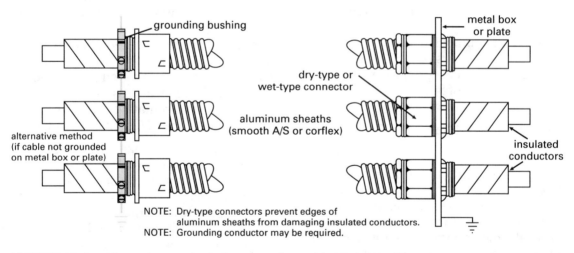

FIGURE 12.4 Alternative methods for grounding cables at boxes (Sheath currents present)

net where the cables terminate and currents are in excess of 200 A, serious eddy currents will not occur at the box.

Multi-conductor cables are more expensive than the single conductor units required to replace them, but they tend to eliminate the problem of sheath currents. When two or more conductors are enclosed by the same sheath, the magnetic fields around each conductor cancel one another. Multi-conductor cables may be encircled or supported by magnetic materials without risk of eddy current damage.

Cable Termination

The old-style *smooth-sheathed* cables in the smaller sizes made use of a connector that was similar to an armoured cable connector. Both dry-type and moisture-proof connectors were used. (See Figs. 12.5 and 12.6)

Modern *corrugated* cable requires connectors of a design different from that used with smooth-sheathed cable. Both dry-type and moisture-proof connectors are available. (See Figs. 12.7 and 12.8)

Corrugated cable with a *PVC jacket* must be terminated with a different type of connector. (See Fig. 12.9 on page 147.)

Preparation of Cable for Termination

Removing the aluminum sheath from the cable is easy to do. Calculate how much free conductor is required in the box, mark the cable, then score the outer sheath with a knife. Take care not to cut through the aluminum: just dent the surface. Bend the end of the sheath to be removed up and down by hand, cracking the sheath at the score line. Pull on the sheath to be removed. It will slip off the conductors easily.

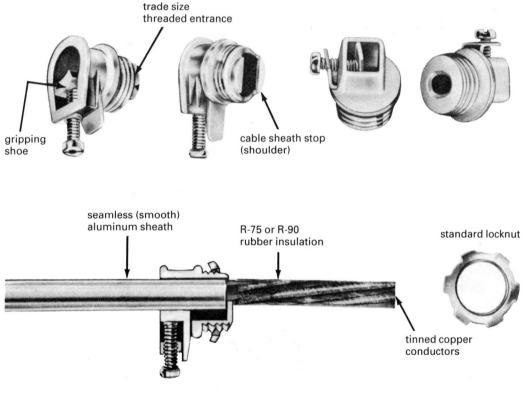

NOTE: X-link insulation is now being used in aluminum-sheathed cables.

FIGURE 12.5   Dry-type connectors used on smooth-sheathed cables (no longer produced)

Courtesy Canada Wire & Cable Ltd.

NOTE: Compresses clamping ring

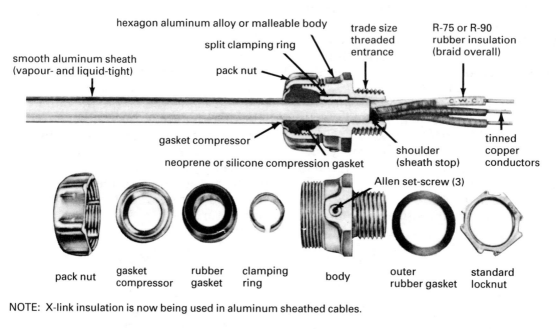

smooth aluminum sheath
(vapour- and liquid-tight)

hexagon aluminum alloy or malleable body

split clamping ring

pack nut

trade size
threaded
entrance

R-75 or R-90
rubber insulation
(braid overall)

gasket compressor

neoprene or silicone compression gasket

shoulder
(sheath stop)

tinned
copper
conductors

Allen set-screw (3)

| pack nut | gasket compressor | rubber gasket | clamping ring | body | outer rubber gasket | standard locknut |

Courtesy Canada Wire & Cable Ltd.

NOTE: X-link insulation is now being used in aluminum sheathed cables.

FIGURE 12.6 Moisture-proof connectors used on smooth-sheathed cables (no longer produced)

NOTE: X-link insulation is now being used in aluminum sheathed cables.

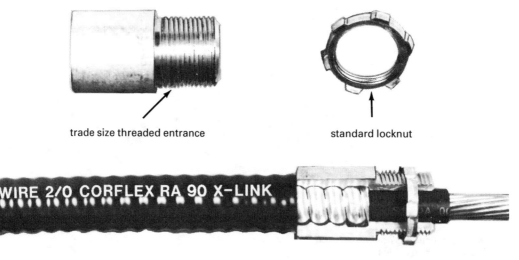

trade size threaded entrance

standard locknut

Courtesy Canada Wire & Cable Ltd.

WIRE 2/0 CORFLEX RA 90 X-LINK

FIGURE 12.7 Type ''D'' dry location connector for corrugated cable (Suitable for cable without sheath covering)

NOTE: X-link insulation is now being used in aluminum sheathed cables.

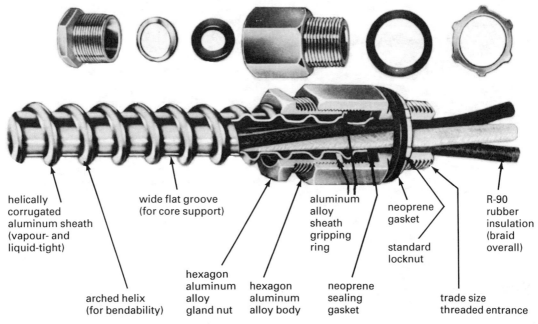

helically
corrugated
aluminum sheath
(vapour- and
liquid-tight)

wide flat groove
(for core support)

aluminum
alloy
sheath
gripping
ring

neoprene
gasket

R-90
rubber
insulation
(braid
overall)

standard
locknut

hexagon
aluminum
alloy
gland nut

hexagon
aluminum
alloy body

neoprene
sealing
gasket

arched helix
(for bendability)

trade size
threaded entrance

Courtesy Canada Wire & Cable Ltd.

FIGURE 12.8   Type ''FM'' moisture-proof or submersible connector for corrugated cable
(Suitable *only* for cable with sheath covering)

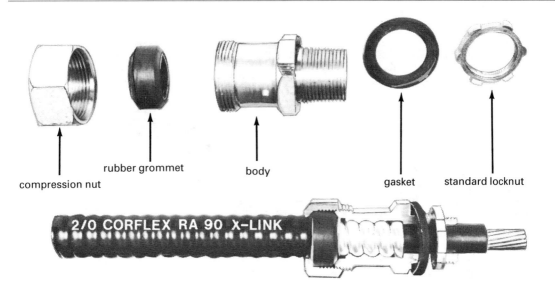

compression nut

rubber grommet

body

gasket

standard locknut

2/0 CORFLEX RA 90 X-LINK

Courtesy Canada Wire & Cable Ltd.

FIGURE 12.9   Type ''W'' submersible connector for corrugated cable with PVC jacket
(Suitable *only* for cable with sheath covering)

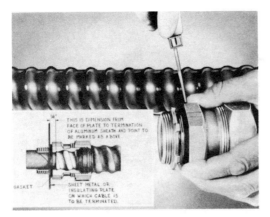

STEP 1. Locate and mark end of aluminum sheath and temporary end of jacket (sheath covering).

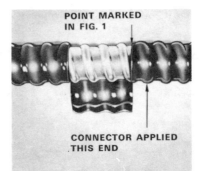

STEP 2. Strip off short length of jacket (sheath covering) from point marked in Step 1 towards end of cable. Cut covering square (see Step 6).

STEP 3. Use fine tooth hacksaw to cut fully through helix of aluminum sheath, using jacket edge as guide. Score flat portion of sheath to about half of sheath thickness. Crack scored sheath by bending cable gently back and forth.

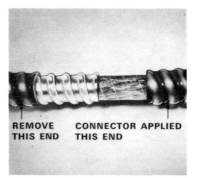

STEP 4. Pull off left-hand end of cable sheath (exposed sheath and jacket), leaving jacketed end (at right) for application of connector. Smooth off burr on aluminum sheath this end.

STEP 5. Use mark on hexagon of connector body to locate point for removal of jacket.

STEP 6. Wrap piece of paper around jacket as guide to ensure jacket cut is square with axis of cable.

STEP 7. Jacket removed and components of connector placed as shown. Wet rubber grommet to facilitate turning it on jacket of sheath.

STEP 8. Body of connector turned firmly by hand onto aluminum sheath. Do not use wrench.

STEP 9. Rubber grommet turned firmly against body of connector.

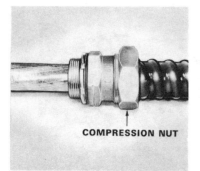

STEP 10. Compression nut threaded onto body of connector. Use wrenche and compress grommet until it bulges slightly from under compression nut.

Courtesy Canada Wire & Cable Ltd.

FIGURE 12.10   Procedure for terminating corrugated cable with moisture-proof connector

Corrugated cable, with its PVC jacket, is a little more difficult to handle. Figure 12.10 shows how to prepare this cable for a connector.

Cable Supports

Aluminum-sheathed cable requires *non-magnetic aluminum* supports. (See Fig. 12.11) The Canadian Electrical Code requires a strap every 2 m, but some sag in the cable may occur in the smaller sizes. One cable manufacturer recom-

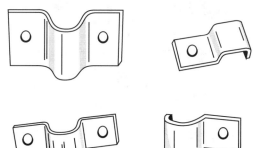

FIGURE 12.11   Aluminum clips for A/S cable

motor wiring

oil refinery

pulp and paper mill

FIGURE 12.12 Cable applications

Courtesy Canada Wire & Cable Ltd.

mends a strap every 1 m to 2 m on cables up to 3 cm in diameter. In many industrial installations, multi-conductor cables are laid in a trough or rack designed for this purpose. Figure 12.12 shows typical cable installations.

For Review

1. What are the two main types of aluminum-sheathed cable?
2. Which aluminum-sheathed cable is easier to bend in large sizes?
3. What jacket material is used for corrosion-resistant cable?
4. What is the most common voltage rating for aluminum-sheathed cable?
5. What is the maximum temperature rating for the conductors in aluminum-sheathed cable?
6. Why is it more economical to run single-conductor cables, rather than one multi-conductor cable?
7. Explain in your own words how sheath currents are produced in aluminum-sheathed cable installations.
8. Why are there no sheath currents in multi-conductor cable systems?
9. Magnetic materials must not be used to support or terminate single-conductor cables. Explain.
10. Describe briefly how to remove the aluminum sheath from a cable that is to be terminated in a distribution panel.
11. What spacing is required by the Canadian Electrical Code for straps supporting aluminum-sheathed cable?

13

Mineral-Insulated Cable

Mineral-insulated (MI) cable was developed to provide the electrical industry with a *fire-resistant* wiring product. This cable cannot cause or contribute to a fire because it contains *no* inflammable materials. It was first introduced in the 1930s. Since then, it has proved to be so versatile that new applications are constantly being devised.

Features of MI Cable

MI cable resists fire up to the point of the copper melting in and around the cable. It is moisture-proof, corrosion resistant, immune to oil product damage, and not prone to aging. It provides a compact, neat, surface-wiring cable system. It does not sag and is easy to install. It can be used indoors or out and is strong enough for direct burial in the earth. It will withstand an almost unbelievable amount of physical abuse before an electrical breakdown occurs.

How MI Cable Is Made

From 1 to 7 high-conductivity *copper rods,* 9 m in length, are inserted into a 5 cm tube of seamless copper. *Magnesium oxide*, which is an excellent electrical insulator and conductor of heat, is packed under pressure around the rods.

The ends of the tube are sealed, and the 9 m section is drawn through a series of *reducing dies*. These dies decrease the diameter of the cable and increase its length.

The magnesium oxide insulation is the same density as the copper used in the tube and rods. This means that as the tube assembly is pulled through the reducing dies, the tube, insulation, and rods are reduced in size simultaneously. The relative spacing between the conductors and the tube is always maintained, because the rods are reduced in direct proportion to the tube itself.

MI cable can be produced in any conductor size by continuing the drawing process until the conductors have been reduced to the desired wire gauge number. (See Fig. 13.1)

Cable Size and Voltage Range

Mineral-insulated cables are produced in the 300 V and 600 V ranges. Figure 13.2, on page 153, illustrates the actual size of each conductor in the cable, as well as typical groupings for multi-conductor cables, in the 300 V range.

Figure 13.3, on page 154, shows 600 V cables and their conductor groupings.

These cables range in size from No. 16 AWG to 250 MCM. Single-conductor, 350 MCM and 500 MCM cables are also available.

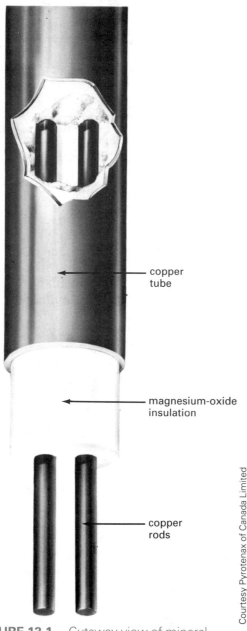

copper tube

magnesium-oxide insulation

copper rods

Courtesy Pyrotenax of Canada Limited

FIGURE 13.1 Cutaway view of mineral-insulated cable

Durability

MI cable can withstand severe physical abuse without electrical failure. Its ability to change the shape of the outer sheath, mineral insulation, and internal conductors at the same time makes this possible. A crushing blow from a heavy object will merely change the cable's shape. (See Fig. 13.4 on page 155.)

MI cable is as resistant to heat as it is to mechanical injury. Figure 13.5 shows how Pyrotenax MI cable compares with other wiring systems under severe heat conditions.

The Underwriters' Laboratories of Canada tested the cable's resistance to heat by mounting several 300 V and 600 V MI cables on the inside of a furnace. The cables were subjected to the furnace's fire for a two-hour period while carrying a full electrical load. Temperatures reached a level of 1010°C. As a result of this test, Pyrotenax cables were given a two-hour fire rating under standard S101. Figure 13.6 illustrates a fire-rated cable and its attached label.

Fire Endurance and Comparison Test

Further testing of wiring products was carried on at the Warnock Hersey Testing Labs in Vancouver, British Columbia. Aluminum-sheathed cable, armoured cable, 90°C X-link wire in conduit, and Pyrotenax cable were subjected to temperatures reaching 939°C. The wiring products were mounted on a double layer of drywall material and exposed to an open furnace. All cables were energized during the one-hour test period.

The conventional wiring products failed within the first 3 min at temperatures no higher than 316°C. The aluminum-sheathed cable failed to ground in

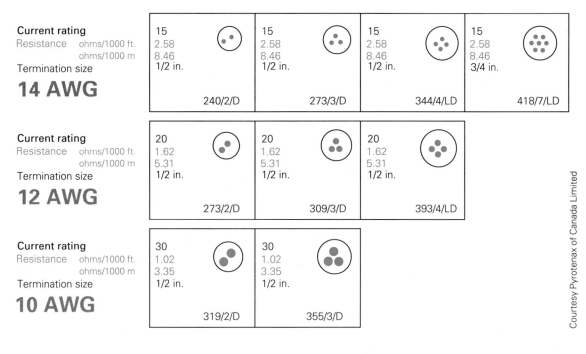

Current rating
Resistance   ohms/1000 ft.
                ohms/1000 m
Termination size

14 AWG

15	15	15	15
2.58	2.58	2.58	2.58
8.46	8.46	8.46	8.46
1/2 in.	1/2 in.	1/2 in.	3/4 in.
240/2/D	273/3/D	344/4/LD	418/7/LD

Current rating
Resistance   ohms/1000 ft.
                ohms/1000 m
Termination size

12 AWG

20	20	20
1.62	1.62	1.62
5.31	5.31	5.31
1/2 in.	1/2 in.	1/2 in.
273/2/D	309/3/D	393/4/LD

Current rating
Resistance   ohms/1000 ft.
                ohms/1000 m
Termination size

10 AWG

30	30
1.02	1.02
3.35	3.35
1/2 in.	1/2 in.
319/2/D	355/3/D

Courtesy Pyrotenax of Canada Limited

FIGURE 13.2 Typical 300 V MI cable configurations and sizes

1 min, 50 s; the armoured cable failed to ground in 3 min; and the X-link wire in conduit failed to ground in 2 min, 25 s. The MI cable, however, withstood the test temperatures for the entire one-hour period and continued to operate satisfactorily after the test was completed.

A major reason for the cable's fireproof characteristic is that the magnesium oxide insulation and the copper sheath tend to conduct heat away from the cable. Fire alarm or pump circuits can use to advantage this cable's ability to operate under high-temperature conditions.

Fireproof Applications

Modern, complex smoke detection systems and signalling methods no longer require a person to "pull-the-alarm" to give warning of danger from fire. As well as notifying a building's inhabitants of fire, ancillary systems such as smoke control dampers, pressurization fans, door closing, and elevator homing are activated immediately. Many of these systems are required by the Building Code. Others are installed by concerned building management companies.

Despite the progress made in alarm and detection systems, little or no progress has been made in the way they are interconnected. The Building Code recognizes that fire protection and alarm circuits require special treatment. If the conductors are installed in a service space containing combustible materials, they must be isolated from these materials by a one-hour, fire-rated separation. Unfortunately, as mentioned previously, raceways and other wiring products cannot withstand fire for more than a couple of minutes.

16 AWG Resistance ohms/1000 ft. ohms/1000 m Termination size	4.094 13.43 1/2 in. 215/1	4.094 13.43 1/2 in. 340/2	4.094 13.43 1/2 in. 355/3	4.094 13.43 3/4 in. 387/4

4.094 13.43 3/4 in. 449/7

14 AWG
Current rating — 20, 15, 15, 15, 12
Resistance 2.58 ohms/1000 ft., 8.46 ohms/1000 m
Termination size: 1/2 in., 1/2 in., 1/2 in., 3/4 in., 3/4 in.
230/1 371/2 387/3 418/4 496/7

12 AWG
Current rating — 25, 20, 20, 20, 16
Resistance 1.62 ohms/1000 ft., 5.31 ohms/1000 m
Termination size: 1/2 in., 1/2 in., 1/2 in., 3/4 in., 3/4 in.
246/1 402/2 434/3 465/4 543/7

10 AWG
Current rating — 40, 30, 30, 30, 24
Resistance 1.02 ohms/1000 ft., 3.35 ohms/1000 m
Termination size: 1/2 in., 3/4 in., 3/4 in., 3/4 in., 1 in.
277/1 449/2 480/3 527/4 621/7

8 AWG
Current rating — 70, 50, 50, 40
Resistance 0.641 ohms/1000 ft., 2.1 ohms/1000 m
Termination size: 1/2 in., 3/4 in., 3/4 in., 3/4 in.
309/1 512/2 543/3 590/4

6 AWG
Current rating — 100, 70, 70, 56
Resistance 0.403 ohms/1000 ft., 1.32 ohms/1000 m
Termination size: 1/2 in., 3/4 in., 3/4 in., 1 in.
340/1 590/2 621/3 684/4

4 AWG
Current rating — 135, 90, 90
Resistance 0.253 ohms/1000 ft., 0.83 ohms/1000 m
Termination size: 1/2 in., 1 in., 1 in.
402/1 684/2 730/3

3 AWG	2 AWG	1 AWG	1/0 AWG	2/0 AWG
155 0.201 0.66 1/2 in. 434/1	180 0.159 0.52 3/4 in. 465/1	210 0.126 0.41 3/4 in. 496/1	245 0.100 0.33 3/4 in. 543/1	285 0.0795 0.26 3/4 in. 590/1

Current rating
Resistance ohms/1000 ft.
ohms/1000 m
Termination size

3/0 AWG	4/0 AWG	250 MCM
330 0.0630 0.21 3/4 in. 637/1	385 0.0500 0.16 1 in. 699/1	425 0.0431 0.14 1 in. 746/1

Current rating
Resistance ohms/1000 ft.
ohms/1000 m
Termination size

(The ratings shown for single conductor cables are those designated by the Canadian Electrical Code, Part 1, for cables in free air.)
Screw-on seals for this cable range.

Courtesy Pyrotenax of Canada Limited

FIGURE 13.3 Typical 600 V MI cable configurations and sizes

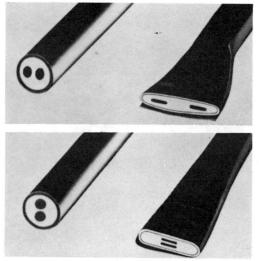

FIGURE 13.4 MI cable continues to function when flattened to one-third its original diameter. Only its shape changes.

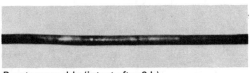

Pyrotenax cable (intact after 2 h)

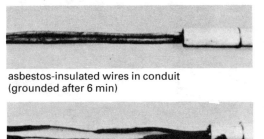

lead-sheathed rubber-insulated cable (destroyed after 25 s)

asbestos-insulated wires in conduit (grounded after 6 min)

rubber-insulated wires in conduit (destroyed after 2.5 min)

FIGURE 13.5 Fire test of 960°C shows MI cable durability under severe heat conditions

FIGURE 13.6 Fire-resistant MI cable with a two-hour fire rating

MI cables, tested and fire rated for two hours, are invaluable in ensuring the continuous operation of elevators, pumps, emergency lighting, sprinkler systems, smoke control systems, communication systems, and fire detection systems. The additional operating time guaranteed by MI cables enables people to leave a building safely, while contributing immensely to the control and eventual put-out of the fire. As commercial high-rise buildings continue to be built in large cities, the ability to successfully evacuate large buildings becomes increasingly important. MI cable is one way of providing the much needed time to do so.

Special Applications

Several special MI cables have been developed for specific areas. For example, in keeping with the cable's fire protection ability, a special instrument cable containing a pair of twisted conductors inside a double sheath provides single point grounding. It also provides a shield from stray signals or electrical interference. Figure 13.7 illustrates this cable.

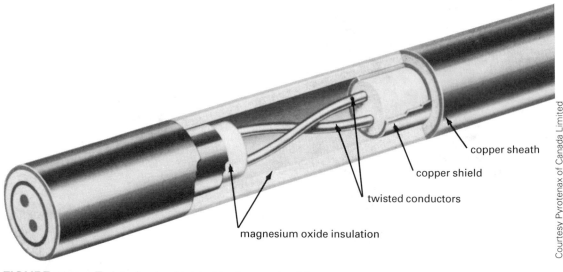

Courtesy Pyrotenax of Canada Limited

copper sheath
copper shield
twisted conductors
magnesium oxide insulation

FIGURE 13.7   Twisted pair, shielded instrument cable

Both round and square cables having solid or hollow conductors can be used in high radiation environments. (See Fig. 13.8) The sturdiness and heat resistance of the MI cable make it most useful around a space shuttle's launch platform.

(See Fig. 13.9) Petrochemical plants continue to develop automated circuits and equipment where danger is

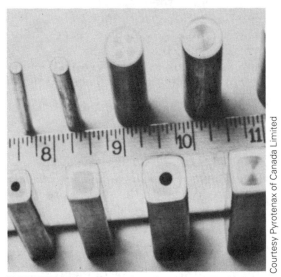

Courtesy Pyrotenax of Canada Limited

FIGURE 13.8   Round and square MI cables suitable for high radiation environments

Courtesy Pyrotenax of Canada Limited

FIGURE 13.9   Pyrotenax MI has been used on the space shuttle's launch platform where its ruggedness and ability to withstand high ambient temperatures are required.

Applications of Electrical Construction

ever present from flash fire. Figure 13.10 shows a typical MI cable installation in this area. Large-scale mining operations also provide a challenge for this versatile product. (See Fig. 13.11) Many other industries such as breweries use MI cable for their control circuits. (See Fig. 13.12) Also, a stainless steel MI cable can be used near food and beverage products or near chemicals that could damage the copper sheath and its conductors.

FIGURE 13.10 MI cable is used in the petrochemical industry.

FIGURE 13.12 This control panel for a brewery bottling plant uses MI cable.

FIGURE 13.11 MI was chosen for its durability and fire resistance in this large-scale mining operation.

Cable Termination

Prepackaged termination kits are used to connect MI cable to boxes, cabinets, and fittings. (See Fig. 13.13) Remove the copper sheath with a stripping tool. Place a *gland connector* on the cable just before threading the self-tapping pot on to the outer sheath. Press plastic sealing compound into the pot. Make sure your hands are clean so that no metal particles are pressed into the sealing compound. A short circuit will likely occur if metal particles of any kind are allowed to enter the pot.

Next, install the preassembled insulating sleeves. Use a *crimping* and

STEP 1. Remove outer sheath with stripping tool.

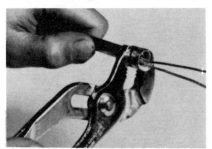

STEP 2. After installing gland connector over cable, thread self-tapping pot onto copper sheath.

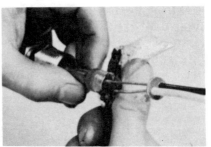

STEP 3. Force plastic sealing compound into pot by hand. Be sure hands are clean.

STEP 4. Crimping and compression tool secures sleeve sub-assembly to pot.

FIGURE 13.13   Procedure for typical MI cable termination

---

compression tool to force the insulating sleeves into place and lock them there.

Figure 13.14 shows a simple screwdriver-operated tool and a more elaborate crimping tool that is also available.

Figure 13.15 shows a cutaway view of the termination and an assembled unit. These terminal fittings are designed for use at temperatures up to 150°C. The gland is equipped with a standard electrical conduit thread that will fasten to electrical boxes with conduit locknuts or thread into pretapped holes in moisture-proof boxes and fittings.

Corrosive-resistant thermoplastic-jacketed cables and terminations are available. (See Fig. 13.16)

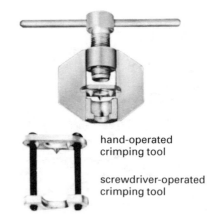

hand-operated crimping tool

screwdriver-operated crimping tool

FIGURE 13.14   Crimping and compression tools

Applications of Electrical Construction

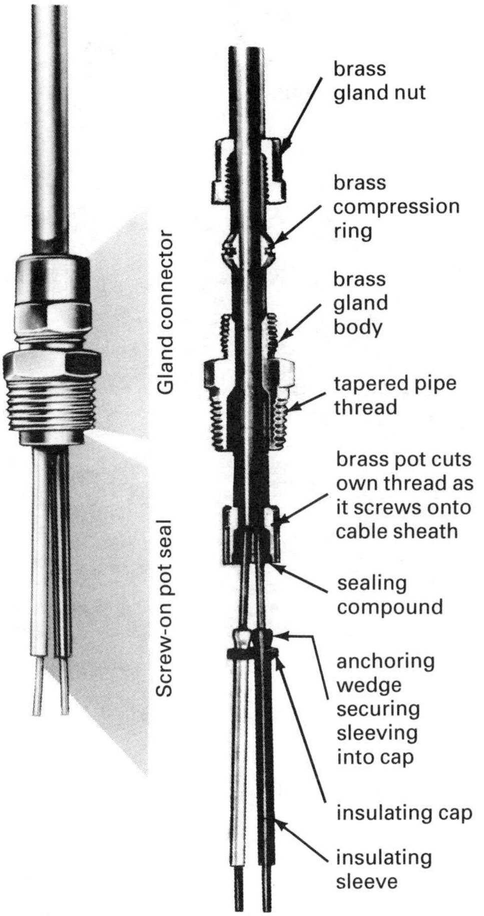
Gland connector
Screw-on pot seal
brass gland nut
brass compression ring
brass gland body
tapered pipe thread
brass pot cuts own thread as it screws onto cable sheath
sealing compound
anchoring wedge securing sleeving into cap
insulating cap
insulating sleeve

Courtesy Pyrotenax of Canada Limited

FIGURE 13.15 An MI cable termination and assembled unit (cutaway view)

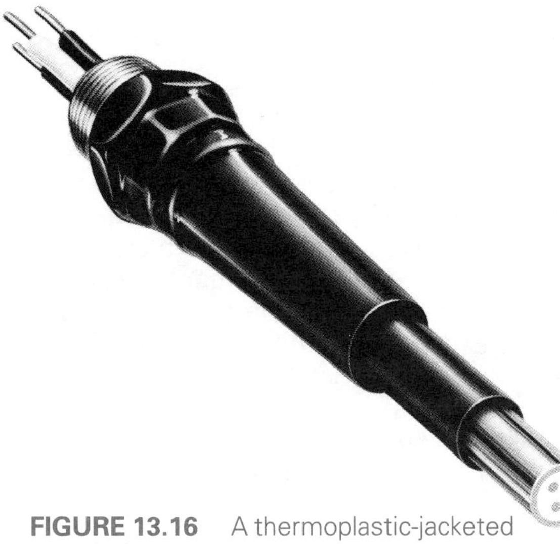

Courtesy Pyrotenax of Canada Limited

FIGURE 13.16 A thermoplastic-jacketed cable termination for corrosive areas

All cable terminations should be checked with an *insulation tester*. This high-resistance measuring instrument will apply approximately 500 V to the termination and indicate whether or not it is safe for use.

Cable Installation

MI cable is supported by *copper straps* located every 1 m to 2 m throughout the run. A wooden block and hammer are used to straighten any irregularities in the cable.

MI cable can be formed into a bend having a radius no less than *six times* the diameter of the cable being bent, without placing undue stress on it.

Multiple runs of MI cable require care and precision in planning. (See Figs. 13.17 and 13.18)

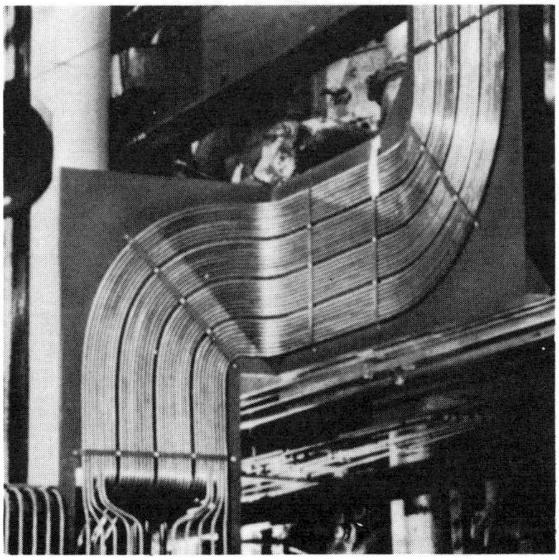

Courtesy Pyrotenax of Canada Limited

FIGURE 13.17 MI cable multiple runs

Sheath Currents

Induced currents can flow in the outer sheaths of single-conductor cables in alternating current circuits. (See Chapter 12) The electromagnetic fields surrounding nearby cables cause them. Since heat generated by sheath currents does not affect MI cables much, these cables do not need to be spaced apart. They can be grouped together under a common strap. (See Fig. 13.18)

Cables carrying in excess of 200 A must be fastened to a nonmagnetic plate at both ends of the cable, prior to securing to a box or panel. Doing so eliminates heat induction which could damage the box and its contents.

General Purpose Applications

The versatile MI cable is used for many purposes and in many situations besides fire protection and warning systems. Commercial buildings, factories, houses, and apartments, as well as processing plants, and railway and subway systems, make use of this cable. It is used for electric heating of driveways and steps to aid in snow removal and of water pipes to provide protection from frost. It can also eliminate solidification of waxes

FIGURE 13.18   Single-conductor MI cables entering distribution panel

Courtesy Pyrotenax of Canada Limited

and other materials in oil refinery pipes and systems.

In addition to these uses, MI cable is found in communication and transmission systems and has particular application in all kinds of marine vessels and hazardous areas that contain dust, explosive vapours, or liquids. Figure 13.19 shows MI cable applications.

to supply motor control equipment

Courtesy Pyrotenax of Canada Limited

to supply electricity to apartment building

Courtesy Pyrotenax of Canada Limited

FIGURE 13.19 Typical MI cable applications

Conduit Wiring

A conduit wiring system offers a mechanical protection to electrical circuits that is rare with other wiring methods. Voltage in a conduit system is limited only by the insulation on the conductors within the system. Metallic conduits provide a high degree of fire protection, as well as the ability to safely contain overloaded or short-circuited conductors that could cause or contribute to a fire.

Unlike any other wiring method, conduit allows the conductors within the system to be removed easily, *without* dismantling the system. A change in circuit design or equipment will often require conductors of a different size, colour, material, or quantity to be installed in the conduit. Consult the Canadian Electrical Code when changing the size or quantity of conductors in a conduit.

Conduit Applications

Conduit wiring systems are used for surface wiring in apartments, factories, garages, warehouses, and public buildings and for service entrance equipment for a house.

Conduit can be buried directly in masonry construction. Commercial and industrial buildings constructed of poured concrete will often have a con-

duit wiring system installed before the concrete is poured. Conductors to be installed under ground can be adequately protected by a conduit system.

When installed properly, conduit is both water- and vapour-tight. Hazardous areas, where explosive liquids, gases, or dusts are present, can be wired safely with conduit and electrical fittings approved for the purpose. Plastic-coated conduits are available for use in areas where corrosive materials are present. Motors and similar equipment subject to vibration or movement can be connected with flexible conduit. Liquid- and dust-tight varieties of flexible conduit are also available.

Conduit Sizes

Conduit is usually produced in 3 m lengths. This length, regardless of diameter, provides an easy-to-estimate, practical unit for installation and allows for ease of bending and handling. Under normal conditions, these lengths can be quickly assembled into a continuous run.

The second important dimension of the conduit is internal diameter. This measurement determines the quantity and size of the conductors that can be safely installed in the conduit. Section 12

Applications of Electrical Construction

of the Canadian Electrical Code lists general installation rules and allowable conduit capacities.

Conduit is available in the following *trade sizes*, as determined by the internal diameter, in millimetres: 13, 20, 25, 32, 38, 51, 64, 76, 89, 102, 127, and 152.

Types of Conduit

There are several types of conduit. This chapter discusses rigid (thickwall), EMT (thinwall), rigid aluminum, rigid PVC, flexible, and PVC flexible (liquid-tight) conduit.

Rigid Conduit

Rigid, or *thickwall*, conduit is produced in aluminum or steel. This thickwall type provides the greatest amount of mechanical protection to conductors. It is available with a choice of external coatings, such as electroplating, baked-on enamel, or polyvinylchloride, to reduce the damaging effect of corrosive chemicals in certain installations. (See Fig. 14.1)

Rigid conduit *must* be supported by approved straps, clips, or hangers at regular intervals. These supports must be located in accordance with Section 12 of the Canadian Electrical Code, Part 1. Section 12 outlines intervals as follows:

13 mm and 20 mm conduit: Not exceeding 1.5 m intervals;

25 mm and 32 mm conduit: Not exceeding 2 m intervals;

38 mm and larger: Not exceeding 3 m intervals.

Normal practice is to locate a support within 75 cm of a box or cabinet.

The inside of steel conduit is often coated with an *insulating paint* or *varnish* to ease the installation of conductors, insulate damaged conductors

rigid steel conduit (electroplated finish)

steel thinwall (EMT) conduit (electroplated finish)

rigid steel conduit (baked enamel finish)

rigid aluminum conduit

rigid steel conduit (PVC-jacketed)

PVC-jacketed thinwall (EMT) conduit

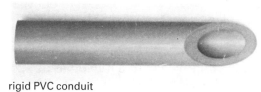

rigid PVC conduit

Courtesy Ron Ridsdill, Toronto

FIGURE 14.1   Electrical conduit raceways

from the metal wall of the conduit, and prevent internal conduit rusting. Lengths of rigid conduit must be *threaded* at each end for connection to couplings, boxes, or fittings.

Threading Rigid Conduit. The best method for holding conduit securely for threading is in a *pipe vise*. (See Fig. 14.2) For the best straight cut, hold the *hacksaw* at an angle of approximately 45° to the horizontal. Figure 14.2 shows the hacksaw being drawn back with one hand, so that the pipe vise would not be hidden. In actual practice, the second hand should grip the front portion of the saw. A more secure grip and a straighter cut would result.

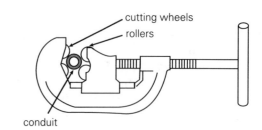

FIGURE 14.3   Pipe cutter

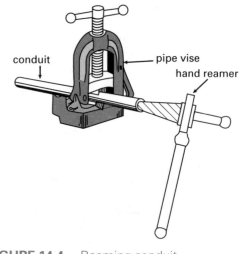

FIGURE 14.4   Reaming conduit

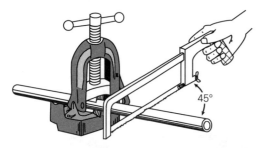

FIGURE 14.2   Cutting conduit with hacksaw (*Note*: Hacksaw should have 24 teeth per 2.5 cm.)

A *pipe cutter* is also excellent for producing neat cuts for threading. (See Fig. 14.3) Using one, however, will necessitate extensive reaming to the inside of the conduit.

Whichever method is used to cut the conduit, a sharp *burr*, capable of damaging the conductor insulation, will be produced on the inside of the conduit. This must be removed with a *round file* or *reamer*. (See Fig. 14.4)

Reaming the pipe before threading is best, because some conduit reamers expand the ends of newly threaded conduits. An expanded edge, flared like a bugle, makes starting the threaded conduit fitting on to the new thread difficult.

The actual thread is cut into the prepared end of the conduit by a *stock and die set*. (See Fig. 14.5) At one time, electrical stocks and dies cut parallel threads. Modern threading tools produce tapered threads similar to those used with water pressure systems. (See Figs. 14.6 and 14.7) These tools allow conduit fittings to be installed securely on the new thread.

Cut sufficient thread into the conduit to completely engage the threads available on the conduit fitting. Doing so

Applications of Electrical Construction

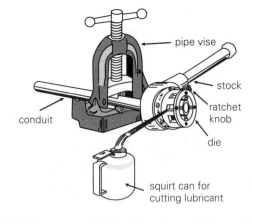

FIGURE 14.5 Cutting thread on rigid conduit

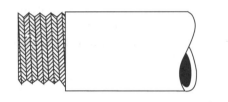

FIGURE 14.6 Parallel thread

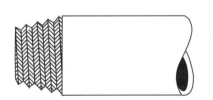

FIGURE 14.7 Tapered thread

provides the greatest mechanical security and ground continuity.

Application of a *cutting liquid-lubricant* will ease the physical effort required to cut a thread and also extend the life of the die's cutting edges. There are several excellent threading lubricants, but oil or liquid soap can be used when the proper lubricant is not at hand. The simplest method of applying the cutting lubricant is with an oil squirt can.

Bending Rigid Conduit. Installers need many hours of practice to master the art of bending and forming conduit systems.

Conduits up to 25 mm internal diameter are usually formed manually with the help of a bending tool called a *hickey*. (See Fig. 14.8) Hickeys are available in standard conduit sizes; for best results, one of the proper size should be used. Conduit of a larger trade size than the hickey usually does not fit into the bending tool. Smaller conduit than the tool is designed for often slips in the tool, producing kinks, inaccurate bends, and/or bruises on the operator.

Bends must be made *without* reducing the internal diameter of the conduit. A kink in the bend will make installation of conductors much more difficult and

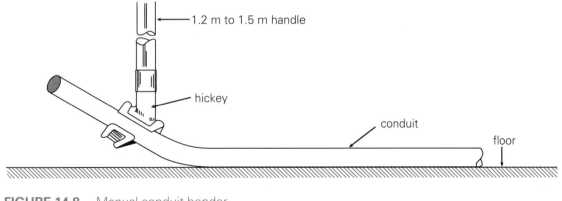

FIGURE 14.8 Manual conduit bender

might even damage the insulation. (See Fig. 14.9) Figure 14.10 shows two safe methods for using the manually operated hickey.

Conduit is bent to go around corners, pass over obstacles, or enter a box at the proper angle. Figure 14.11 shows five bends and their uses.

The *offset* bends are used to enter a box or cabinet and can be also used to bring the conduit from one level to another on any surface, such as a wall or ceiling.

The *square saddle* is used to by-pass protrusions in a wall or to pass over several conduits located side by side.

The 90° bend is used to go around inside corners. (An outside corner requires a conduit fitting of the type shown in Figure 14.27.)

Note: The saddle bends take the most time to form. Care and practice, however, will result in smooth, well-aligned bends.

The *45° offset* bends are often used to overcome obstructions in the path of the conduit.

The operator who uses body weight rather than arm strength can greatly reduce the strain of bending conduit. A 25 mm diameter conduit can be very resilient at times—a challenge to the operator. Good technique and practice are essential. The emphasis is on matching the bender to the conduit. Installers, however, often have to use a building support beam, a hole in a wall, or some other means when a bend must be made and no hickey is available. Take great

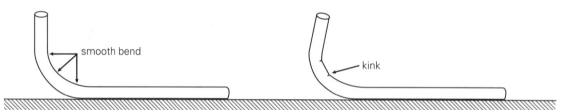

FIGURE 14.9 Smooth conduit bends are important. Conduit bends with kinks are not suitable for conductor installation.

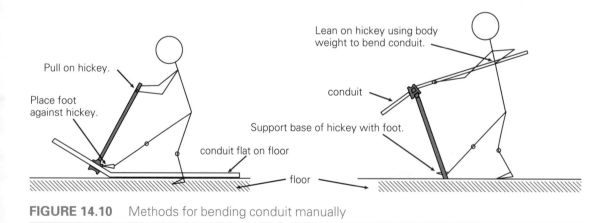

FIGURE 14.10 Methods for bending conduit manually

Applications of Electrical Construction

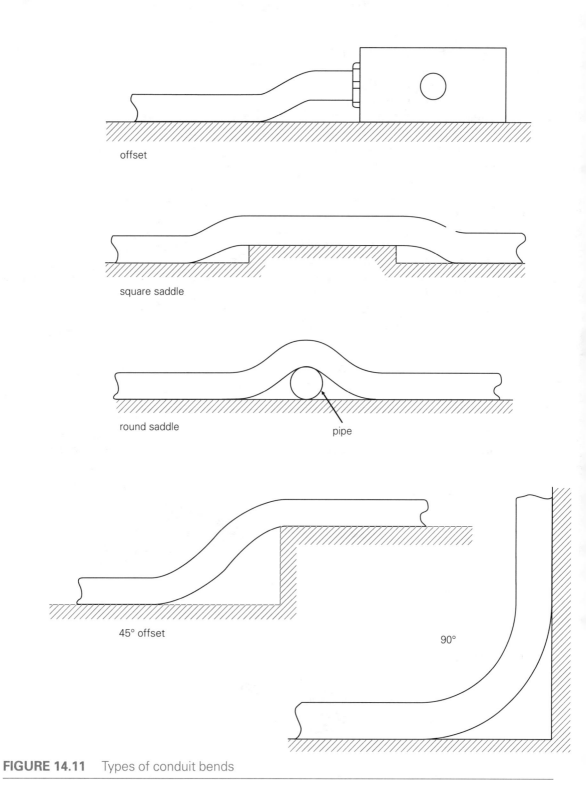

offset

square saddle

round saddle

pipe

45° offset

90°

FIGURE 14.11 Types of conduit bends

care not to kink the conduit in such conditions.

When bending conduit, the operator should not try to make a full bend (90°) with a single grip of the bender. The head of the hickey should be moved away from or towards the operator approximately 25 mm per grip.

Doing so makes the completed bend a series of small, continuous curves (see Fig. 14.12) and creates a symmetrical bend with less chance of the conduit being kinked. The accuracy of the bend can be easily checked by placing the vertical part of the conduit against a wall.

Conduits larger than 25 mm in diameter are usually formed with a *hydraulic conduit bender*. These units are available with hand- or motor-driven pump attachments. (See Figs 14.13 and 14.14) Take care to set the conduit in the bender carefully, because enormous pressure is exerted by the hydraulic system.

As in the case of the hand-operated hickey, bends should be made up of a series of small curves. When using the

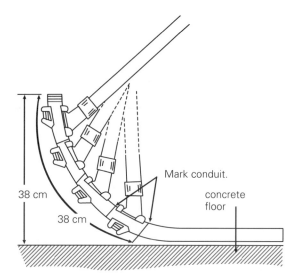

FIGURE 14.12 Typical 90° bend (*Note*: Many small curves are required for a smooth bend.)

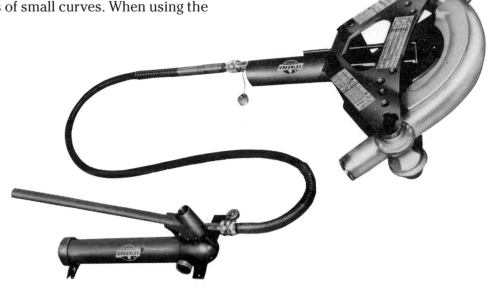

Courtesy Greenlee Tool Company

FIGURE 14.13 A hand-operated hydraulic bender simplifies forming large diameter conduits.

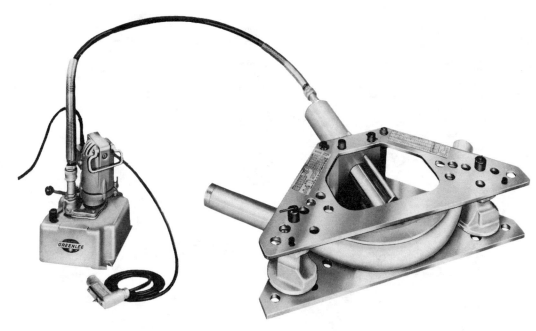

Courtesy Greenlee Tool Company

FIGURE 14.14 A motor-driven hydraulic bender speeds forming large diameter conduits.

hydraulic bender, release the pressure and reposition the conduit for each curve. Place a series of equally spaced pencil marks on the conduit to assist in locating the pressure points needed to produce a smooth, symmetrical bend. Since removing an excess of bend from large conduits is difficult, take care not to overbend. It is easier to place the conduit back in the bender and add a few degrees of bend than to take out bend. If, however, a small amount of bend must be taken out, drop the conduit onto the base of the bend from about 60 cm off the floor. The weight of the conduit and the force of hitting the floor will straighten the bend slightly. Take care not to flatten the bottom of the bend or damage the floor. (See Fig. 14.15)

Conduit systems are often installed in large commercial buildings before the concrete is poured. Figure 14.16 shows a rigid-steel conduit system for a high-rise office building.

There are mechanical conduit benders to help form medium size conduits, up to 38 mm in diameter. (See Fig. 14.17)

Electric-powered benders, as shown in Figure 14.18, can speed up conduit bending. These newly developed benders, which can be equipped with digital display read-outs on their control boxes, can be preset to quickly make several identical bends or offsets with a minimum of set-up time. They are sold with

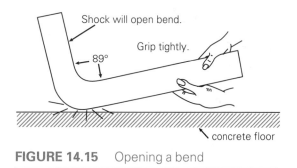

FIGURE 14.15 Opening a bend

FIGURE 14.16 Installing conduit before pouring concrete

FIGURE 14.18 An electric bending machine

all forms of roller support and shoe attachments included. They control the accuracy and consistency of their bends electronically.

Thinwall Conduit

The proper name for thinwall conduit is *electrical metallic tubing (EMT)*. This conduit does not provide the same degree of mechanical protection as rigid conduit, but it has several important advantages.

EMT is made of *lightweight, steel tubing*, and so does not require as much physical strength on the part of the installer as does rigid conduit, when assembling the conduit system. (See Fig. 14.19) Also, because of its lightweight construction it is not practical to thread EMT. This feature alone saves much time

FIGURE 14.17 Typical mechanical conduit bender

Applications of Electrical Construction

Courtesy Ron Ridsdill, Toronto

FIGURE 14.19   Steel thinwall conduit (EMT). Electroplated finish

compression type   set-screw type

Courtesy Elliott Electric Mfg. Corp.

FIGURE 14.20   EMT connectors

compression type   set-screw type

Courtesy Elliott Electric Mfg. Corp.

FIGURE 14.21   EMT couplings

and work when installing EMT. Because EMT is not threaded into a fitting or box, the job does not depend on the installer's ability to rotate a length of conduit, which is full of bends and in an awkward location, until the threaded end is secure in the box or fitting. Mechanical connectors eliminate this chore.

Section 12 of the Canadian Electrical Code lists the conditions under which EMT can be substituted for rigid conduit.

Fittings for EMT.   There are *mechanical fittings* to couple lengths of EMT and/or connect the conduit to boxes and cabinets. These fittings are made of steel or zinc alloy (die-cast).

> **Note:** Take care not to subject the die-cast zinc fittings to great pressure or stress, as they break easily. Zinc alloy is used because it is easily die-cast and moderately priced, but a person stepping on a run of conduit coupled with die-cast fittings is likely to crack a coupling. For this reason, die-cast zinc fittings should not be used in poured concrete buildings where they can be damaged (and go unnoticed) before the pouring of concrete.

There are two main types of EMT fittings: the *set-screw* and the *compression*. (See Figs. 14.20 and 14.21) The *set-screw* type is used in *dry* locations; only a screwdriver is needed to secure the connector to the conduit. Glandular connectors are fastened to the conduit by using

Courtesy Ron Ridsdill, Toronto

FIGURE 14.22   PVC-jacketed thinwall conduit (EMT)

adjustable pliers or a wrench. These *rain-tight connectors* are used outdoors, in poured concrete installations, or in other damp areas.

PVC-jacketed EMT.   Some EMT is jacketed with *polyvinylchloride (PVC)*, which resists corrosive chemicals and vapours. (See Fig. 14.22) Take care not to cut or damage this outer jacket when forming bends. *Liquid* PVC is applied to the connectors and couplings of such installations to give complete protection in corrosive areas.

Bending EMT.   A special conduit bender (the *hickey*) is used to form offsets and bends in EMT. (See Fig. 14.23) The hickey forms the conduit with a radius that will tend not to kink the conduit where it has been bent. Also, it supports the sides of the conduit as it

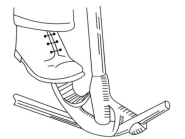

FIGURE 14.23　An EMT bender (hickey)

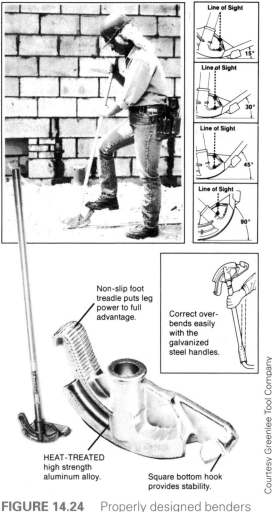

Non-slip foot treadle puts leg power to full advantage.

Correct over-bends easily with the galvanized steel handles.

HEAT-TREATED high strength aluminum alloy.

Square bottom hook provides stability.

Courtesy Greenlee Tool Company

FIGURE 14.24　Properly designed benders and handles allow ease of bending and personal safety.

bends, further reducing the possibility of kinking. The operator places a foot on the hickey to make sure that the conduit remains within the hickey during bending. This action reduces the tendency of the EMT to kink. Properly designed handles permit the operator to bend the conduit with a minimum of strain, while maintaining perfect balance for personal safety. (See Fig. 14.24)

Because the bender fits the conduit closely, a separate hickey is needed for each size of conduit from 13 mm to 32 mm. EMT 32 mm in diameter and larger is usually formed in a *hydraulic bending unit*. (See Fig. 14.25) The hickey completely encloses the EMT at the point of pressure, virtually eliminating kinks in the conduit.

Both manual and hydraulic benders can form a 90° bend in one continuous movement of the bender. However, unlike rigid conduit, EMT does not have to be bent in a series of small curves. EMT benders can be very accurate: a skilled operator, following the measurements provided on the unit, can produce a 90° bend that is within 2 mm of the required dimension. Such accuracy is necessary because EMT is difficult to straighten and rebend without kinking.

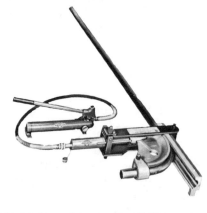

Courtesy Greenlee Tool Company

FIGURE 14.25　A hydraulic EMT bender

Applications of Electrical Construction

Even though EMT does not require as much physical strength to bend as does rigid conduit, considerable practice is needed before truly professional results will be obtained. Many people who install conduit take great pride in their ability. Figure 14.26 illustrates a high-quality installation in a commercial/industrial area.

FIGURE 14.26 A conduit installation requiring a high degree of skill and craftsmanship

Threaded for Rigid Conduit

type C

type LR

type LB

type LL

type E

type T

type X

Courtesy Greenlee Tool Company

Set-screw type for EMT

type C

type LB

type LL

type LR

type T

Courtesy Elliott Electric Mfg. Corp.

FIGURE 14.27 Conduit fittings (condulets)

Conduit Fittings

Often, during a conduit installation, a sharper bend than the conduit will allow must be made. A *conduit fitting* is then used. (See Fig. 14.27 on page 173.) These fittings, which are often called *condulets*, assist the conduit installer to go around corners and provide access to the conduit for the installation and removal of conductors.

Some fittings are designed to provide a branch path from the main conduit run. (See Fig. 14.28)

The Canadian Electrical Code requires that access to the conduit be provided every 30 m of conduit or every 360° of accumulated bends, whichever occurs first. There are special fittings for this purpose. (See Fig. 14.29)

Threadless, set-screw fittings are available for direct application to EMT. The standard fitting with its internal conduit thread is suitable for use with rigid conduit or connector-equipped EMT.

There are many forms of *electrical boxes* for conduit systems. Manufacturers produce such a variety that complete catalogues are necessary to list them. Figure 14.30 shows a number of them.

Some fittings are listed with the prefix *L* in combination with another letter, for example, LB, LL, or LR.

The first letter of the prefix indicates the shape of the fitting. For example, the *L* of *LB* indicates that the fitting is shaped like the capital letter *L*. (See Fig. 14.30: FSC with switch; FSX with switch; FS with explosion-proof outlet and cord connector.)

FIGURE 14.29 A "C" conduit fitting provides access to a conduit for installing or removing conductors.

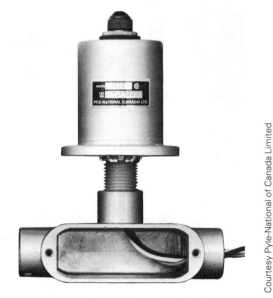

FIGURE 14.28 A "T" conduit fitting provides a branch path from the main run.

explosion-proof motor starting units

FS fitting with explosion-proof outlet and cord connector

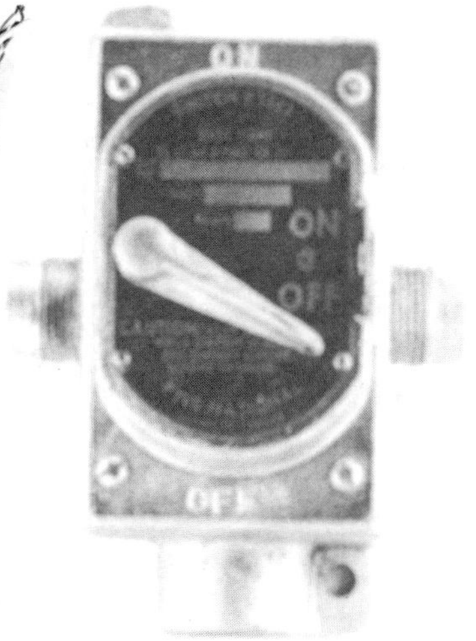

FSX fitting with switch (4 conduit openings)

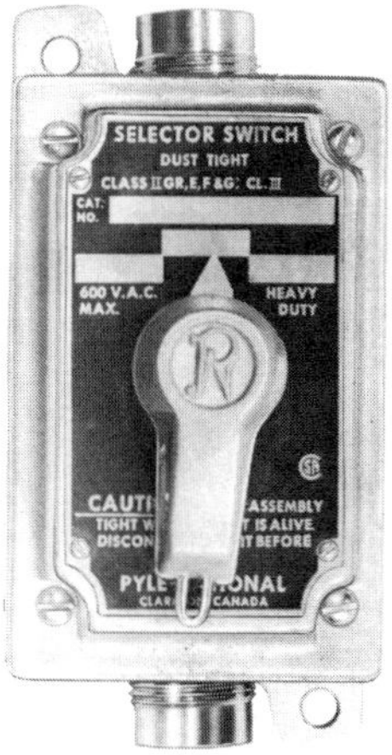

FSC fitting with switch (2 conduit openings)

FS fitting (single conduit opening)

FS fitting with motor starting switch and pilot light (2 gang)

10 cm diameter round box

FIGURE 14.30 Typical electrical boxes for conduit fittings

Courtesy Pyle-National of Canada Limited

FIGURE 14.30 **(continued)**

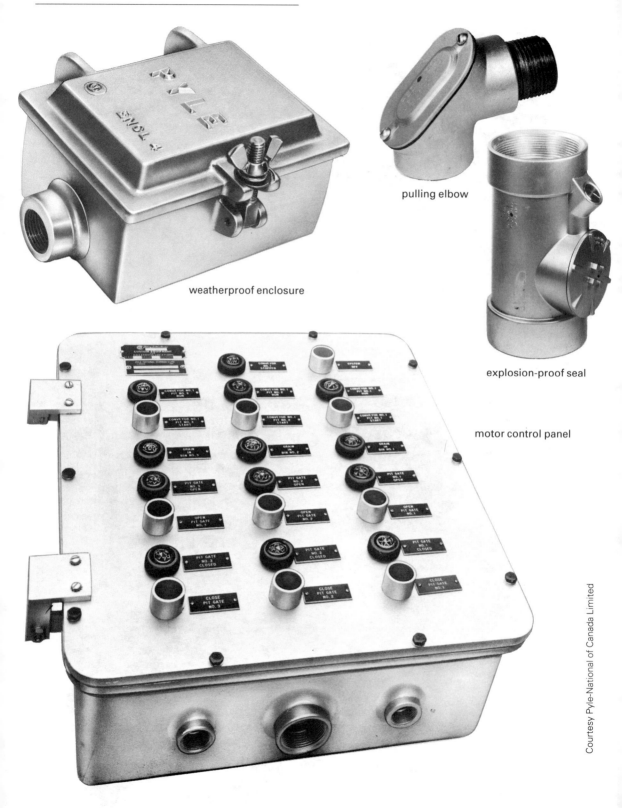

weatherproof enclosure

pulling elbow

explosion-proof seal

motor control panel

The second letter of the prefix indicates the location of the *conductor access* opening in relation to the shorter conduit opening; that is, the direction the short arm of the fitting is pointing when the fitting is held with the conductor access opening facing the viewer and the longer arm pointing upwards. For example, when an LR fitting is held with the conductor access opening facing the viewer and the longer arm is pointing upwards, the short arm of the fitting will be pointing to the right. An LL fitting held the same way will have its short arm pointing to the left. An LB fitting will have its short arm pointing backwards.

Condulets can be capped with covers made from aluminum, steel, porcelain, or a fibre composition. When excessive moisture is present, a *cork* or *rubber gasket* must be used between cover and fitting. (See Fig. 14.31)

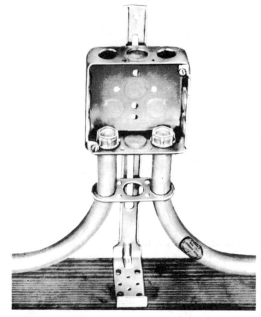

FIGURE 14.32   Conduit connected to outlet box

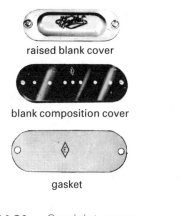

raised blank cover

blank composition cover

gasket

FIGURE 14.31   Condulet covers

Terminating Conduit

When rigid conduit is brought into boxes through knockout holes, *locknuts* and *bushings* are used to secure it to the box. (See Fig. 14.32) These fasteners must be installed tightly enough to provide both mechanical security

(strength) for the installation and ground circuit continuity.

While it is always wise to use a tool for its intended purpose, common practice among electricians is to tighten the locknut and bushing by placing the blade of a *screwdriver* against the unit, and then striking the handle of the screwdriver with the flat side of a pair of *pliers*. (See Step 4, Fig. 11.4) A hammer and cold chisel are often needed with very large conduits.

Locknuts are usually made of steel and have sharp teeth on one side to bite in to the box and establish the ground circuit.

Bushings, which provide a smooth opening through which the conductors can enter the box, are made of die-cast aluminum, steel, or nylon. Steel bushings with a nylon insert are also available. (See Fig. 14.33)

locknut   insulated bushing

bushing

FIGURE 14.33   Locknut and bushing terminating devices

Courtesy Elliott Electric Mfg. Corp.

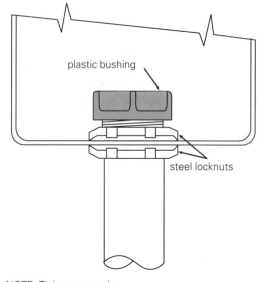

plastic bushing

steel locknuts

NOTE: Tighten securely.

FIGURE 14.35   Double locknut and bushing
installation

Figure 14.34 shows how conduits of less than 32 mm in diameter are secured to a box. When securing conduits of 32 mm to 150 mm in diameter, a system of two locknuts and a plastic bushing is used. This is because insulation on large and heavy cables can be punctured or

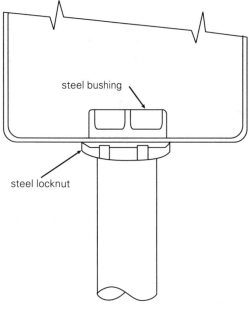

steel bushing

steel locknut

NOTE: Tighten securely.

FIGURE 14.34   Single locknut and bushing
installation

pierced by the weight of the cable pressing on a solid metal bushing. A metal bushing with a plastic insert is also available. This *double-locknut* method is also used on conduit systems that have an applied voltage of more than 250 V. (See Fig. 14.35)

Rigid Aluminum Conduit

This lightweight, easily handled conduit is often used to enclose residential service entrance conductors. Aluminum is rustproof and will not stain or streak the surface on which it is mounted. The high-electrical conductivity of aluminum provides a safe ground circuit for the system. As a nonsparking metal, it is safe for use near explosive gases or vapours. It is also ideal for alternating-current systems that require a nonmagnetic substance to enclose the conductors. (See Fig. 14.36)

Applications of Electrical Construction

FIGURE 14.36   Rigid aluminum conduit

FIGURE 14.37   Rigid PVC conduit

Bending Rigid Aluminum Conduit.

Bending aluminum conduit does not require much physical effort. It does however require the use of a hickey or some other *forming* device that will not kink the bend. Standard rigid conduit hickeys can be used if extreme care is taken not to flatten the bend with too small a radius. EMT benders one size larger than the conduit are often used successfully. Hydraulic benders also usually produce satisfactory bends.

Threading Rigid Aluminum Conduit.

Dies used for threading aluminum conduit should be sharp and well-lubricated while cutting the thread. Chips of metal should be cleared from the die occasionally. Otherwise, they will lodge between the die and the conduit and tear the newly cut threads. When reaming this conduit, take care not to flare and expand the end because bugling makes it difficult to attach conduit fittings.

Installing Rigid Aluminum Conduit.

Avoid *cross-threading* aluminum conduit when attaching couplings, locknuts, bushings, condulets, or similar devices. The soft aluminum will quickly lose its threads if an attempt is made to remove a cross-threaded fitting. Putting lubricating oil on the threads before assembly will help eliminate this problem.

Aluminum conduit should be supported by *straps* placed at regular intervals, as is rigid steel conduit. The material's light weight is appreciated, particularly during conduit installations from a ladder or some other precarious position.

If aluminum conduit is to be embedded in concrete below ground where it may be wetted every so often, a *bituminous base paint* or *pitch* should be applied to the conduit. This substance will protect the conduit from any moisture in the concrete that may attack the metal and corrode it.

Rigid PVC Conduit

Rigid PVC (Polyvinylchloride) conduit protects conductors in the worst corrosive locations. Moisture or condensation has no effect on this durable product. Since PVC will not conduct electricity and is nonmagnetic, it is not affected by sheath currents. It has no voltage limitations. It resists aging, exposure to ozone, sunlight, and underground environments. It is immune to electrolytic action. Heavy blows do not cause permanent damage to it. (See Fig. 14.37)

Weight Advantage of Rigid PVC Conduit.

This conduit is approximately *five times* lighter than steel conduit and *twice* as light as aluminum conduit of the same size. As a result, the installer finds it much easier to handle, especially when working from ladders or scaffolds.

Applications of Rigid PVC Conduit.

This conduit is used in areas where corrosion is a problem, for example, in paper mills, meat-packing plants, chemical and electroplating plants, barns and animal shelters, hospitals, and food preparation plants. It is also being used more and more for residential service entrance conduit, because it is easily handled, nonrusting, and nonstaining.

Joining and Terminating Rigid PVC Conduit.

A metal-cutting *hacksaw* or *carpenter's wood saw* can be used to cut the conduit to length. A *pocketknife* will quickly remove any burr on the inside of the cut end. Lengths of PVC and fittings are assembled by using a form of *solvent welding*. No threads are required. Dirt and grease from handling are removed with a *PVC cleaner*. *Solvent cement* is brushed on to the prepared end of the conduit and the inside of the fitting to be joined. The conduit is then pushed into the fitting, given a quarter turn, and set aside. After a few minutes the connection will withstand the strain of use. (See Fig. 14.38)

Bending Rigid PVC Conduit.

This conduit is a thermoplastic material that will soften at temperatures between 115°C and 130°C. If proper care is taken, the conduit will not flatten and/or kink while being bent. All bends should have a radius of at least ten times the diameter of the conduit. Commercially produced bends are available from the conduit's manufacturer.

There is an approved *PVC heater* that directs a heated-air stream at the conduit. This air stream is capable of raising the conduit's temperature to the desired level. Open flame will damage the conduit and should *not* be used. Cold water or natural cooling will maintain the shape of the bend. The conduit should be *overbent* slightly to allow for *springback* as the conduit is cooling. (See Figs. 14.39 and 14.40) Figures 14.41 and 14.42 illustrate several other forms of heating equipment for bending rigid PVC conduit. The *heating blanket* is used on all sizes of conduit and is thermostatically controlled for uniform bending. The motor-driven electric PVC heater is capable of automatically rotating the conduit for even heating. Controls include a heating chart and timer to bring the PVC up to bending temperature.

Rigid PVC Conduit Expansion.

Whenever the temperature variation exceeds a range of 15°C, PVC conduit will expand and contract enough to warrant the use of expansion joints. An *expansion joint* is simply one PVC tube telescoping within another. A 30 m length of conduit will expand approximately 9 cm. Expansion joints are available from the manufacturers of PVC conduit.

Rigid PVC Conductor Installation.

Less physical effort is needed to pull conductors into this conduit than into other types. The conduit's smooth, friction-free interior is the major reason.

Rigid PVC Conduit Supports.

Supports for this conduit do not need to be as sturdy as those for other forms of conduit. They should be spaced at regular intervals and take into account the possibility of snow, ice, and wind loads. For example, adequate support would be one support every 80 cm for conduit 13 mm, 20 mm, and 25 mm in diameter.

Metallic Flexible Conduit

Flexible conduit (*flex*) combines mechanical protection with maximum flexibility for nonhazardous locations. This versatile raceway is made of an interlocking steel or aluminum strip.

Applications of Electrical Construction

Courtesy Scepter Manufacturing Co. Ltd.

FIGURE 14.38   Apply cement to pipe and fitting. Insert pipe into fitting and give it a quarter turn.

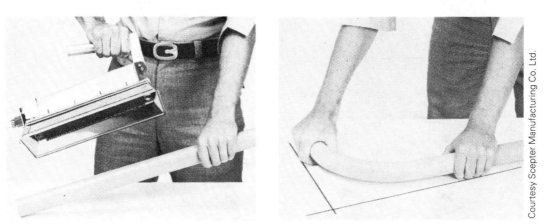

Courtesy Scepter Manufacturing Co. Ltd.

FIGURE 14.39   Use a PVC heater to soften pipe for bending. Use guidelines to establish proper bend angle.

Courtesy Scepter Manufacturing Co. Ltd.

FIGURE 14.40   Using a PVC heater before bending rigid PVC conduit

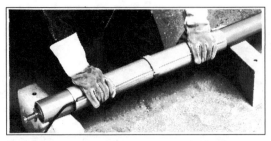

thermostatically controlled, even heating for
uniform bending

Wrap heating blanket around PVC and heat.

Remove blanket and bend PVC.

FIGURE 14.41 Using a heating blanket for the bending of PVC conduit

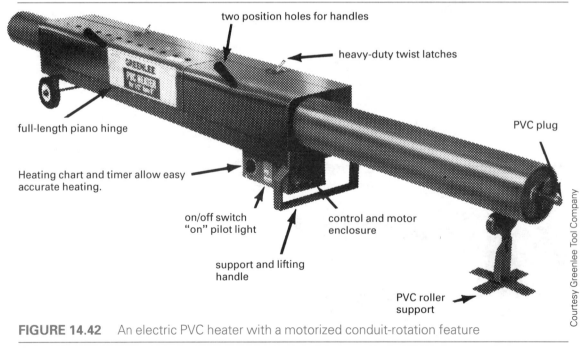

two position holes for handles

heavy-duty twist latches

full-length piano hinge

PVC plug

Heating chart and timer allow easy
accurate heating.

on/off switch
"on" pilot light

control and motor
enclosure

support and lifting
handle

PVC roller
support

FIGURE 14.42 An electric PVC heater with a motorized conduit-rotation feature

Applications of Electrical Construction

TABLE 14.1	Metallic Flexible Conduit	
Internal Diameter		
8 mm	9.5 mm	11 mm
13 mm	20 mm	25 mm
32 mm	38 mm	51 mm
64 mm	75 mm	100 mm
Standard Length of Coil			
15 m	7.5 m	30 m	75 m

(See Fig. 14.43) Flex is available in the lengths and internal diameters shown in Table 14.1.

Metallic Flex Applications. Conduit protection is sometimes needed in a location where rigid conduit cannot be formed to the contours required because of close-working conditions. An example is a wiring system that must pass over and around steel girders, existing machinery and equipment, or be fished through masonry walls and ceilings. Using flex greatly simplifies the installation work. Another example is motors and/or machines with vibrating moving parts. These parts must be provided with a supply system that will not allow metal fatigue to set in and fracture the raceway. If the metallic raceway is broken or separated, ground continuity can be lost. To accommodate sound mechanical and electrical installation methods, approved flex connectors must be used. (See Fig. 14.44)

Supports for Metallic Flex. Approved flex supports are similar to those used with armoured cable. They must be installed within 30 cm of each box or cabinet and at regular intervals of not more than 1.4 m throughout the run. (See Fig. 14.45) Where flex is fished or used in lengths of up to 1 m requiring flexibility, supports are not needed.

FIGURE 14.43   Metallic flexible conduit

straight connectors (squeeze type)

90° angle connector (squeeze type)

straight connector

90° angle connector (3 screw type)

FIGURE 14.44   Flexible conduit connectors

Nonmetallic Flexible Conduit

A new form of extruded, flexible tubing, which is non-corrosive, non-conductive,

FIGURE 14.45 Flexible conduit strap (2-hole)

Courtesy Elliott Electric Mfg. Corp.

and moisture resistant, is now available. This lightweight tubing can be obtained in coils up to approximately 100 m in length.

Special connectors and couplings exist for use with this product only. With their fast-acting, *snap-on* feature, they allow the installer to insert the flex into the connector/coupling without the use of any tool or piece of equipment. It should be noted, however, that sometimes the solvent-welding technique (as used with PVC conduit) may be required by local wiring codes. Whichever method is used, the flex is not easily removed from the connector/coupling once installed. Figure 14.46 illustrates the snap-on connector. Figure 14.47 illustrates a snap-on coupling.

Nonmetallic Flex Applications.
Nonmetallic flexible conduit is often used in metal-stud partitions (see Fig. 14.48) and poured concrete work. (See Fig. 14.49) Care must be taken not to use this product in hazardous locations (as described in the Canadian Electrical Code), however. It should also not be buried in the earth, exposed to mechanical injury, or enclosed in thermal insulation materials.

Supports for Nonmetallic Flex.
Fishing of wires into the flex is eased by the corrugated design on the interior: the corrugation results in less friction when the conductors are pulled into the tubing. (See Fig. 14.50) Flex should have a support within 1 m of a junction box, coupling, or fitting, plus supports no

more than 1 m apart on a run. It is available in 13 mm, 20 mm, and 25 mm internal diameters.

FIGURE 14.46 A PVC flex, snap-on connector

Courtesy Scepter Manufacturing Co. Ltd.

FIGURE 14.47 A PVC flex, snap-on coupling

Courtesy Scepter Manufacturing Co. Ltd.

FIGURE 14.48 Typical use of PVC flexible conduit in metal stud partitions

Courtesy Scepter Manufacturing Co. Ltd.

FIGURE 14.49 This PVC flexible conduit is being prepared for use in a concrete slab.

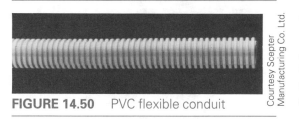

Courtesy Scepter Manufacturing Co. Ltd.

FIGURE 14.50 PVC flexible conduit

Liquid-Tight Flexible Conduit

This *PVC-jacketed flex* is excellent for use in damp locations, corrosive areas, or around machines where coolants, cutting, and/or lubricating liquids are likely to splash on to the flex. (See Fig. 14.51) *Moisture-proof connectors* are used to terminate it. (See Fig. 14.52 on page 186.) These connectors make use of a *nylon compression ring* to grip the outer jacket; regular flex connectors have a metallic clamping device.

FIGURE 14.51 Liquid-tight, flexible metal conduit with a PVC jacket

Courtesy Canada Wire and Cable Limited

Take care not to puncture the PVC jacket either during installation or later when it is in use.

The Canadian Electrical Code requires that this conduit not be used where temperatures are higher than 60°C. These temperatures could damage the PVC jacket. Also, take care not to use the conduit in temperatures that are low enough to cause injury to the jacket when it flexes.

Section 12 of the Canadian Electrical Code lists guidelines for the type and size of conductors allowed in liquid-tight flexible conduit.

Conduit Grounding and Termination

According to the Canadian Electrical Code, neither metallic nor nonmetallic conduit can be relied upon for ground continuity.

A separate conductor *(green insulation)* with the sole purpose of grounding the equipment *must* be installed together with the other current-carrying conductors in the system.

Cutting the Flex. Cut flex to length with a *hacksaw* in the same manner as for armoured cable. (See Fig. 11.4)

Terminating the Flex. Flex connectors are used to secure the flex to boxes, cabinets, or condulets, and to couple the flex with other forms of conduit. (See Fig. 14.44 on page 183.) An *insulating sleeve*, similar to the anti-short bushing used with armoured cable, must be inserted between the conductors and the armour to prevent chafing of the conductors. Some modern flex connectors have a *plastic insert* built into the threaded end of the unit.

Flex connectors should be fastened

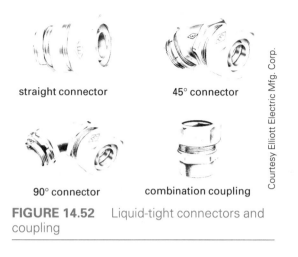

straight connector

45° connector

90° connector

combination coupling

Courtesy Elliott Electric Mfg. Corp.

FIGURE 14.52 Liquid-tight connectors and coupling

securely to the box or cabinet with a conduit locknut to provide mechanical security and assist in ground continuity.

Knockout Cutters

It is often necessary to increase the size of an available knockout in a box or cabinet. Using a reamer or file tends to be clumsy, time-consuming, and damaging to the worker's hands. Instead a *knockout cutter* could be used.

These cutters are made in all standard conduit sizes. Using, for example, a 20 mm knockout cutter will produce an opening that will accept a 20 mm conduit or connector. The actual diameter of the hole will be approximately 27 mm.

The *hand-operated* cutter unit has a hardened-steel cutter, which is drawn through the metal box by a wrench-tightened bolt. (See Fig. 14.53)

Once the knockout cutter has removed a ring-shaped section of metal from the box, the installer may have some difficulty in removing the metal piece from the cutter. He or she can easily waste valuable time before making the next hole in the box or panel. However, a recently designed type of cutter

Courtesy Greenlee Tool Company

FIGURE 14.53 Hand-operated knockout punches

Applications of Electrical Construction

- Punch up to 10 gauge mild steel
- Splits slug for easy removal

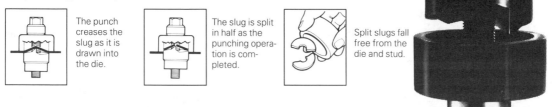

The punch creases the slug as it is drawn into the die.

The slug is split in half as the punching operation is completed.

Split slugs fall free from the die and stud.

Courtesy Greenlee Tool Company

FIGURE 14.54 Newly designed knockout punches require less time to remove slugs than traditional cutters.

breaks each metal slug into two separate pieces. The installer can thereby remove the metal pieces quickly. This cutter is available in traditional knockout cutter sizes. (See Fig. 14.54)

Hydraulic-powered units use the same type of hardened-steel cutter, but require much less physical effort on the part of the operator. (See Fig. 14.55)

Smaller, compact, lightweight *hydraulic punch drivers* are available to installers. These units can be stored and their parts organized in custom-fitted cases. (See Figs. 14.56 and 14.57)

Courtesy Greenlee Tool Company

FIGURE 14.56 Compact hydraulic punch systems ease hole-making operations in metal boxes and cabinets.

Courtesy Greenlee Tool Company

FIGURE 14.55 Hydraulic-powered knockout punch

Courtesy Greenlee Tool Company

FIGURE 14.57 Hydraulic punch driver set

Installing the Conductors

The Canadian Electrical Code requires that conductors be drawn into the conduit system after the system has been completely assembled. Otherwise, conductors could be damaged while bending, forming, or fastening the conduit to boxes.

A flattened, tempered-steel wire, which is produced in 7.5 m, 15 m, 30 m, and 60 m lengths, is pushed into the conduit between boxes and fittings. This steel wire, called a *fish tape*, is available in 3 mm, 5 mm, and 6 mm widths to suit light and heavy conductor installations.

For ease of handling and storage, the tape is wound onto a metal or plastic/nylon reel. Figure 14.58 illustrates the older style metal reel. Figure 14.59 displays the most common sizes of plastic/nylon reels available to the installer.

The tape is usually inserted into the conduit by means of a series of short frequent pushes. It is important to keep the tape in motion, because constant pulsing and vibration allow the tape to be eased around bends and offsets in the conduit. Figure 14.60 illustrates a typical use of the fish tape/reel for the insertion of conductors into conduit wiring systems.

FIGURE 14.59 Numerous sizes of nylon/plastic reels are available for use with fish tapes.

FIGURE 14.60 Typical use of fish tape and reel with a conduit installation

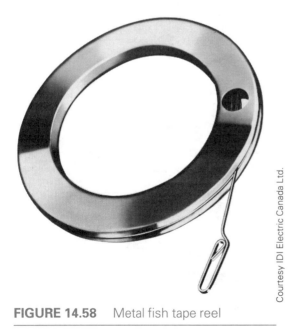

FIGURE 14.58 Metal fish tape reel

Applications of Electrical Construction

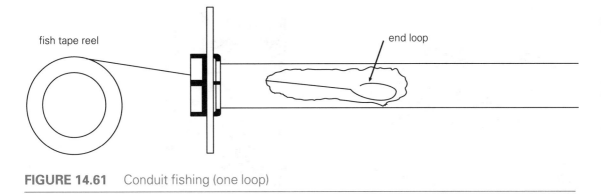

FIGURE 14.61   Conduit fishing (one loop)

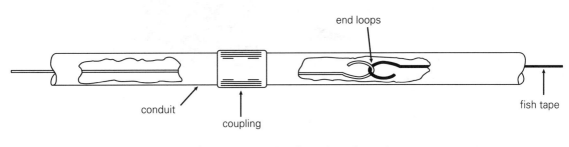

FIGURE 14.62   Conduit fishing from both ends of run (two loops)

A *loop* that is almost closed helps guide the tape around bends, couplings, and fittings. (See Fig. 14.61) Often a second fish tape is pushed into the conduit from the opposite end. It is hooked onto the first tape by rotating the first tape several times to engage the end loops, then used to draw the first tape through the conduit. (See Fig. 14.62) This method is especially useful on conduit runs that have a number of 90° bends.

When pulling large conductors or many conductors into a large conduit, the fish tape is often used to draw a *rope* or *cable* into the conduit. The rope is then used to pull the conductors through.

Larger cables are extremely hard to pull into a conduit system when the conduit run is long or built with several bends. For this reason, installers may rely on heavy-duty electric winch pull-ers. (See Fig. 14.63) These units require the use of special minimum-stretch rope which removes some of the danger associated with this system. A tremendous amount of energy can be stored in a tightly drawn rope, and if the rope breaks or releases, the installer can be seriously hurt by it or objects fastened to its ends. Care must be taken to stand clear and out of the way of the rope whenever possible. Figures 14.63 and 14.64 illustrate two winch pullers.

Wire-pulling *lubricants* are available. If a generous amount of lubricant is applied to the conductors, the physical strain of installation is greatly reduced. In fact, there are many other ways to reduce the physical strain. Every experienced electrician has devised some unusual apparatus for pulling conductors into a conduit system. (See Figs. 14.65, 14.66 and 14.67)

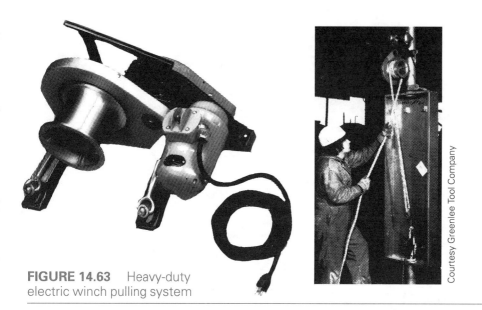

FIGURE 14.63 Heavy-duty electric winch pulling system

Courtesy Greenlee Tool Company

FIGURE 14.64 Electric winch pulling system

Courtesy Greenlee Tool Company

FIGURE 14.65 Wire-pulling lubricant is available in a variety of easy-to-use containers.

Courtesy IDI Electric Canada Ltd.

Applications of Electrical Construction

FIGURE 14.66 Wire-pulling lubricant is being applied to conductors.

FIGURE 14.67 The pulling of conductors is eased by the application of wire-pulling lubricant.

Securing Conductors to Fish Tape.
When two or more conductors are to be pulled through a conduit at the same time, take care to fasten them *securely* to the fish tape. Much time and effort will be wasted if any of the conductors break away from the tape. Figure 14.68 shows two methods for securing the conductors to a tape.

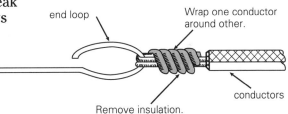

NOTE: It is good practice to cover a connection with tape.

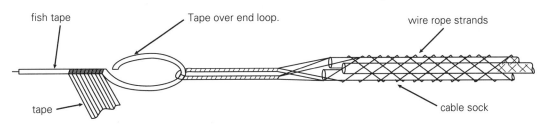

NOTE: The greater the pull on the fish tape, the tighter the sock grips.

FIGURE 14.68 Methods for securing conductors to fish tape

Compressed-Air Fishing

Compressed air can be used to force a lightweight ball through a conduit. The ball can be almost as large as the diameter of the inside of the conduit. A length of strong string is attached to the ball. Once the string reaches the opposite end of the run, it can be used to draw in a heavier fish tape or rope. The compressed air can be obtained from a *tank* or *compressor*. A *hose* and *nozzle* is used to control the rate of air flow. (See Fig. 14.69) This method is simple and saves much time and effort.

One manufacturer produces a series of *coiled-string, plastic-coated projectiles* that let out a fine, extremely strong nylon string when forced through the conduit by compressed air. A heavier string or line can then be pulled into the conduit by the nylon string. These projectiles are produced in sizes to match the conduit being fished.

Vacuum/Blower Fishing

More recent developments in the conduit tool and accessory industry have made possible *power fishing* with either a *blowing* or *vacuum/suction* system. A series of solid foam pistons, sized to match the conduit in use, are connected to a length of strong nylon line. Air pressure or suction from a vacuum/blower can then force or draw the pistons along the conduit system. Figure 14.70 illustrates a vacuum/blower in action. Once lightweight nylon string is inside the conduit, it can pull in either a heavier, stronger line or a metal tape, then the conductors themselves. Figure 14.71 shows the components of a vacuum/ blower power fishing system.

Conduit Fill

As mentioned in Chapter 7, conductors require air space for cooling. For the

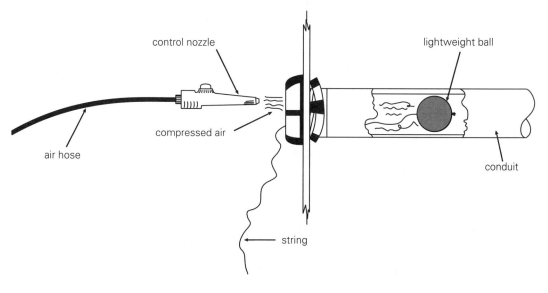

FIGURE 14.69 Conduit fishing using compressed air

FIGURE 14.70   A vacuum/blower power fishing system

FIGURE 14.71   Components of a vacuum/blower power fishing system

same reason, the number of conductors in any given size of conduit or tubing must be limited. The Canadian Electrical Code specifies the number allowed. This number depends on the conductor size and type of insulation. (See Table 14.2)

To determine the number of conductors allowed, Table 14.2 is used with Table 5.1. (See Chapter 5) For example, a No. 14 gauge stranded conductor has a TW-75 insulation thickness of 0.8 mm or 1.6 mm when you consider both sides of the conductor. When 1.6 mm is added to the conductor's approximate diameter of 3.4 mm, a fairly accurate diameter of 5.0 mm is established. (See Table 5.1) To determine how many of these conductors are allowed in 13 mm tubing, look for the diameter entry equal to (and, if necessary, larger than) 5.0 mm on Table 14.2. In this case, the nominal overall diameter is 5.1 mm, which allows three conductors in the tube.

Since editions of the Code book continue to be in the imperial system of measurement, the following conduit fill method is included.

For simplified conduit fill calculations, refer to Table 14.3 on page 195. This table lists the common trade sizes and capacities of conduit to be installed with RW-75 or R-90 insulated conductors. Its information is most useful when the conductors to be installed in the conduit are all of one type and gauge size. A more detailed method of arriving at conduit fill is recommended when conductors of various size and gauge number are to be installed. (See Tables 14.4 and 14.5 on pages 196 and 197.)

TABLE 14.2 Maximum Number of Conductors of One Size in Trade Sizes of Conduit or Tubing

Nominal* Overall Diameter of Condr.**	Maximum Number of Conductors in Conduit or Tubing												
	Size of Conduit or Tubing—Millimetres												
millimetres	13	20	25	30	40	50	65	75	90	100	115	130	150
2.5	15	27	44	76	101	171	—	—	—	—	—	—	—
2.8	12	22	36	63	85	141	—	—	—	—	—	—	—
3.0	10	18	30	53	72	119	169	—	—	—	—	—	—
3.3	9	15	26	45	61	105	143	—	—	—	—	—	—
3.6	7	13	22	39	53	87	124	192	—	—	—	—	—
3.8	6	11	19	33	46	76	108	163	—	—	—	—	—
4.1	6	10	17	29	40	67	95	146	197	—	—	—	—
4.3	5	9	15	26	35	59	84	130	174	—	—	—	—
4.6	4	8	13	23	32	53	75	116	155	199	—	—	—
4.8	4	7	12	21	28	47	67	104	139	178	—	—	—
5.1	3	6	10	19	26	42	60	94	126	162	—	—	—
5.7	3	5	8	15	20	33	48	74	99	127	161	—	—
6.4	1	4	7	12	16	27	38	60	80	103	130	162	—
7.0	1	3	5	10	13	22	32	49	66	85	108	134	194
7.6	1	3	4	8	11	19	27	41	56	71	91	113	164
8.3	1	1	4	7	9	16	23	35	47	61	77	96	139
8.9	1	1	3	6	8	13	19	30	41	52	66	83	120
9.5	1	1	3	5	7	12	17	26	35	46	57	72	104
10.0	1	1	2	4	6	10	15	23	31	40	51	63	92
10.8	1	1	1	4	5	9	13	20	27	35	45	56	81
11.0	1	1	1	3	5	8	12	18	24	31	40	50	72
12.0	—	1	1	3	4	7	10	16	22	28	36	45	65
13.0	—	1	1	3	4	6	9	15	20	25	32	40	58
14.0	—	1	1	1	3	5	8	12	16	21	27	33	48
15.0	—	1	1	1	2	4	6	10	14	18	22	28	40
16.5	—	—	1	1	1	4	5	8	11	15	19	24	34
18.0	—	—	1	1	1	3	4	7	10	13	16	20	30
19.0	—	—	1	1	1	3	4	6	8	11	14	18	26
20.0	—	—	—	1	1	2	3	5	7	10	12	15	23
22.0	—	—	—	1	1	1	3	5	6	8	11	14	20
23.0	—	—	—	1	1	1	3	4	6	7	10	12	18
24.0	—	—	—	1	1	1	2	4	5	7	9	11	16
25.0	—	—	—	1	1	1	1	3	5	6	8	10	14
28.0	—	—	—	—	1	1	1	3	4	5	6	8	12
30.0	—	—	—	—	—	1	1	2	3	4	5	7	10
33.0	—	—	—	—	—	1	1	1	2	3	4	6	8
36.0	—	—	—	—	—	1	1	1	1	3	4	5	7
38.0	—	—	—	—	—	1	1	1	1	2	3	4	6
41.0	—	—	—	—	—	—	1	1	1	1	3	3	5
43.0	—	—	—	—	—	—	1	1	1	1	3	3	5
46.0	—	—	—	—	—	—	1	1	1	1	1	3	4
48.0	—	—	—	—	—	—	1	1	1	1	1	2	4
50.0	—	—	—	—	—	—	1	1	1	1	1	1	3
64.0	—	—	—	—	—	—	—	1	1	1	1	1	1
***													

* For intermediate sizes, use the next larger dimension (e.g., for conductor with diameter 5.3 mm, use fill for 5.7 mm).

** For the purpose of conduit fill, 'conductor' means either insulated conductor, single or multiconductor cable.

Based on the Canadian Electrical Code

If, for example, a No. 8 gauge conductor with R-90 insulation was to be installed, you could refer to Table 14.4 and find that the conductor has an area of 0.076 0 in.^{2}. If 1 in. conduit is to be installed, you could refer to Table 14.5 which lists the cross-sectional areas of the conduits and their allowable percentages of fill. From Table 14.6 you can find that three or more conductors in a given conduit must not occupy more than 40% of the available space in the conduit.

The 40% column of Table 14.5 shows that 1 in. conduit has a cross-sectional area of 0.344 in.^{2}. If you divide 0.344 in.^{2} by the conductor's area of 0.076 0, you will arrive at an answer of 4.52 conductors. Therefore, you can determine that four conductors could be installed in the 1 in. conduit, as shown in Table 14.3.

TABLE 14.3 Maximum Number of Conductors of One Size in Trade Sizes of Conduit or Tubing

NOTE: For ampacity derating factors for more than three conductors in raceways, see Rule 4-004 in the *Canadian Electrical Code*.

Size of Conduit or Tubing—Inches	½	¾	1	1¼	1½	2	2½	3	3½	4	4½	5	6
Conductor													
Type / Size AWG, MCM													
14	3	6	10	18	25	41	58	90	121	155	195	200	200
12	3	5	9	15	21	35	49	77	103	132	166	200	200
10	2	4	7	13	17	29	41	64	86	110	138	174	200
8	1	2	4	8	10	17	25	39	52	67	84	105	152
6	1	1	2	5	6	11	15	24	32	41	51	64	93
4	0	1	1	3	5	8	12	18	24	31	39	50	72
3	0	1	1	3	4	7	10	16	21	28	35	44	63
2	0	1	1	3	4	6	9	14	19	24	31	38	56
1	0	1	1	1	3	5	7	11	14	18	23	29	42
RW-75 R-90													
0	0	0	1	1	2	4	6	9	12	16	20	25	37
00	0	0	1	1	1	3	5	8	11	14	18	22	32
000	0	0	1	1	1	3	4	7	9	12	15	19	28
0 000	0	0	0	1	1	2	4	6	8	10	13	16	24
250	0	0	0	1	1	1	3	5	6	8	10	13	19
300	0	0	0	1	1	1	3	4	5	7	9	11	17
350	0	0	0	1	1	1	1	3	5	6	8	10	15
400	0	0	0	0	1	1	1	3	4	6	7	9	14
500	0	0	0	0	1	1	1	3	4	5	6	8	11
600	0	0	0	0	0	1	1	2	3	4	5	6	9
700	0	0	0	0	0	1	1	1	3	4	4	6	8
750	0	0	0	0	0	1	1	1	3	3	4	5	8
800	0	0	0	0	0	1	1	1	2	3	4	5	8
900	0	0	0	0	0	1	1	1	2	3	4	5	7
RW-75 R-90													
1 000	0	0	0	0	0	1	1	1	1	2	3	4	6
1 250	0	0	0	0	0	0	1	1	1	1	3	3	5
1 500	0	0	0	0	0	0	0	1	1	1	2	3	4
1 750	0	0	0	0	0	0	0	1	1	1	1	2	4
2 000	0	0	0	0	0	0	0	1	1	1	1	2	3

TABLE 14.4 Dimensions of Insulated Conductors for Calculating Conduit and Tubing Fill

Size AWG MCM	Rubber (Thermoset)- and Thermoplastic-Insulated Conductors (0 V—600 V)							
	Types RW–75 and R–90		Types T, TW, TWH, THHN‡, RW–75 (XLPE) §, RW–90 (XLPE)§, R–90 Silicone, R–90 (XLPE) §		Types TWU, RWU–75 (XLPE) §, RWU–90 (XLPE) §		Types RWU–75 EP, RWU–90 EP	
	Diameter Inches	Area Square Inches	Diameter Inches	Area Square Inches	Diameter Inches	Area Square Inches	Diameter Inches	Area Square Inches
14	(2/64) 0.171	0.023 0	0.131	0.013 5	—	—	—	—
14	(3/64) 0.204*	0.032 7*	0.166	0.021 6	—	—	—	—
14	—	—	—	—	0.193	0.029 3	0.231	0.041 9
12	(2/64) 0.188	0.027 8	0.148	0.017 2	—	—	—	—
12	(3/64) 0.221*	0.038 4*	0.183	0.026 3	—	—	—	—
12	—	—	—	—	0.209	0.034 3	0.247	0.047 9
10	0.242	0.046 0	0.168	0.022 4	—	—	—	—
10	—	—	0.204	0.032 7	—	—	—	—
10	—	—	—	—	0.230	0.041 5	0.268	0.056 4
8	0.311	0.076 0	0.248	0.047 5	0.324	0.082 4	0.345	0.093 5
6	0.397	0.123 8	0.323	0.081 9	0.363	0.103 5	0.456	0.163 3
4	0.452	0.160 5	0.372	0.108 7	0.412	0.133 3	0.505	0.200 3
3	0.481	0.181 7	0.401	0.126 3	0.440	0.152 1	0.533	0.223 1
2	0.513	0.206 7	0.433	0.147 3	0.473	0.175 7	0.566	0.251 6
1	0.588	0.271 5	0.508	0.202 7	0.544	0.232 4	0.649	0.330 8
0	0.629	0.310 7	0.549	0.236 7	0.585	0.268 8	0.690	0.373 9
00	0.675	0.357 8	0.595	0.278 1	0.632	0.313 7	0.737	0.426 6
000	0.727	0.415 1	0.647	0.328 8	0.684	0.367 5	0.789	0.488 9
0 000	0.785	0.484 0	0.705	0.390 4	0.744	0.434 7	0.849	0.566 1
250	0.868	0.591 7	0.788	0.487 7	0.822	0.530 7	0.977	0.749 7
300	0.933	0.683 7	0.843	0.558 1	0.878	0.605 5	1.033	0.838 1
350	0.985	0.762 0	0.895	0.629 1	0.930	0.679 3	1.085	0.924 6
400	1.032	0.836 5	0.942	0.696 9	0.978	0.751 2	1.133	1.008 2
500	1.119	0.983 4	1.029	0.831 6	1.064	0.889 1	1.219	1.167 1
600	1.233	1.194 0	1.143	1.026 1	1.180	1.093 6	1.301	1.329 4
700	1.304	1.335 5	1.214	1.157 5	1.252	1.231 1	1.373	1.480 6
750	1.339	1.408 2	1.249	1.225 2	1.287	1.300 9	1.408	1.557 0
800	1.372	1.478 4	1.282	1.290 8	1.321	1.370 6	1.442	1.633 1
900	1.435	1.617 3	1.345	1.420 8	1.385	1.506 6	1.506	1.781 3
1 000	1.494	1.753 1	1.404	1.548 2	1.444	1.637 7	1.565	1.923 6
1 250	1.676	2.206 2	1.577	1.953 2	1.616	2.051 0	1.809	2.570 2
1 500	1.801	2.547 5	1.702	2.274 8	1.741	2.380 6	1.934	2.937 7
1 750	1.916	2.889 5	1.817	2.593 0	1.858	2.711 3	2.051	3.303 9
2 000	2.021	3.207 9	1.922	2.901 3	1.966	3.035 7	2.159	3.661 0

*These are the dimensions for types RW–75 and R–90.
NOTE: To calculate conduit and tubing fill using metric measurements, refer to page 193 of the text for guidance.

Applications of Electrical Construction

TABLE 14.5 Cross-Sectional Areas of Conduit and Tubing

Trade Size Inches	Internal Diameter Inches	Per Cent Cross-Sectional Area of Conduit and Tubing—Square Inches							
		100%	55%	53%	40%	38%	35%	31%	30%
1/2	0.622	0.30	0.165	0.159	0.120	0.114	0.105	0.09	0.090
3/4	0.824	0.53	0.292	0.281	0.212	0.202	0.185	0.16	0.159
1	1.049	0.86	0.473	0.456	0.344	0.327	0.301	0.27	0.258
1 1/4	1.380	1.50	0.825	0.795	0.600	0.570	0.525	0.47	0.450
1 1/2	1.610	2.04	1.122	1.081	0.816	0.776	0.714	0.63	0.612
2	2.067	3.36	1.848	1.780	1.344	1.277	1.176	1.04	1.008
2 1/2	2.469	4.79	2.635	2.540	1.916	1.820	1.677	1.48	1.437
3	3.068	7.38	4.060	3.910	2.952	2.805	2.585	2.29	2.214
3 1/2	3.548	9.90	5.450	5.250	3.960	3.765	3.465	3.07	2.970
4	4.026	12.72	7.000	6.745	5.088	4.840	4.450	3.94	3.820
4 1/2	4.506	15.94	8.771	8.452	6.378	6.060	5.581	4.94	4.784
5	5.047	20.00	11.000	10.600	8.000	7.600	7.000	6.20	6.000
6	6.065	28.89	15.900	15.320	11.556	10.980	10.120	8.96	8.670

NOTE: Metric equivalents are not provided here, because they have no application in the electrical industry.

TABLE 14.6 Maximum Allowable Per Cent Conduit and Tubing Fill

	Maximum Conduit and Tubing Fill Per Cent				
	Number of Conductors or Multi-conductor Cables				
	1	2	3	4	Over 4
Conductors or multi-conductor cables (not lead-sheathed)	53	31	40	40	40
Lead-sheathed conductors or multi-conductor cables	55	30	40	38	35

For Review

1. List three main advantages that a conduit wiring system has over other wiring systems.
2. List three areas where a conduit wiring system is used to protect the power supply.
3. In what length is conduit usually produced? in what diameters?
4. How is conduit measured to determine its trade size?
5. List five main types of conduit.
6. Why is the inside of rigid conduit coated with insulating paint or varnish?
7. Explain in your own words the steps for cutting and threading rigid conduit.
8. What is the main disadvantage of using a pipe cutter to prepare an end of rigid conduit for threading?
9. What can be used as a substitute for thread-cutting lubricant?
10. What are the two types of conduit bender used for rigid conduit? Which is used for the smaller sizes of conduit? Which is used for larger sizes?
11. Why must care be taken when forming a bend in rigid conduit?
12. For what three purposes are bends in rigid conduit made?
13. Explain in your own words how to make a 90° bend in a length of 20 mm rigid conduit.
14. Why are EMT mechanical fittings made of zinc-alloy? What is the main disadvantage of using zinc-alloy?
15. What are the two main types of EMT fittings? Where is each used?
16. What protection does a PVC jacket give to EMT conduit?
17. What is the main difference between a rigid conduit bender and one used for thinwall conduit?
18. List three uses for condulets.
19. List five different shapes of condulet fittings.
20. Of what materials are condulet covers made?
21. Describe two methods for securing rigid conduit to boxes. Why are plastic bushings sometimes used?
22. List four advantages of rigid aluminum conduit.
23. What is the main advantage of flexible conduit?
24. List three examples of how flex is useful.
25. How is ground continuity maintained in a flex system?
26. How is flex terminated? How is PVC-jacketed flex terminated?
27. Why must care be taken when installing PVC-jacketed flex?
28. What is a *knockout cutter*? How and why is it used?
29. Why is it important to keep the fish tape in motion when inserting it into a run of conduit?
30. What method is used to help draw a fish tape in to a run of conduit with many bends?
31. What are the advantages of compressed-air fishing?
32. Explain how power fishing is done.
33. Why is it necessary to limit the number of conductors in conduit or tubing?
34. How many No. 12 gauge conductors with 1.2 mm RW-90 insulation can be placed in 20 mm conduit?
35. What precautions should be taken when using a power winch to pull conductors into a conduit?

Residential Service Wiring

A 3 wire distribution system is used to supply residences with 120 V and 240 V. The two main methods for bringing the 3 wire system into a house are by *overhead* wiring from a pole transformer and by *underground* wiring from a distribution transformer mounted below or above grade-level. (See Chapter 1) This chapter discusses both methods in greater detail.

Supply Authority

The conductors carrying current from the transformer—whether it is pole-mounted, at grade level, or below grade in a transformer vault—belong to the *local supply authority* (the hydro utility). It is the supply authority's responsibility to install, maintain, and service the conductors, which are known as *service supply conductors* (hydro supply lines).

The cost of installing service supply conductors is assumed by the local supply authority, providing the residence is within 30 m of a distribution pole or transformer. Home–owners beyond the 30 m distance often have to pay the cost of placing a pole or running conductors from their houses to the 30 m margin.

Consumer's Service

The consumer's service includes all service boxes and related equipment, up to and including the point at which the supply authority makes its connections.

Figure 15.1 shows a typical residential service with overhead supply lines. Section 6 of the Canadian Electrical Code provides up-to-date regulations for the installation of a consumer's service.

Service Size and Capacity

A residence used to be considered well equipped if it had a 60 A main switch, with an *8 circuit* plus range (stove fuses) distribution panel. The meter was usually located inside the house between the service boxes. (See Fig. 15.2)

As the demand for electrical appliances increased, the 60 A service installation became inadequate. Also, local hydro employees, who had to enter homes to read meters so that customers could be billed, often found no one there to admit them.

The need for more circuits in the home and, therefore, more fuses in the distribution panel made the 100 A, *20*

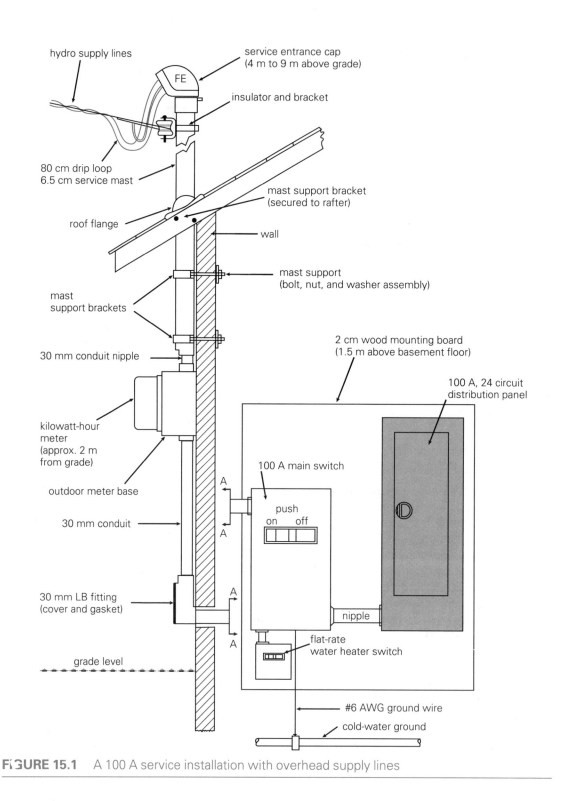

hydro supply lines

service entrance cap
(4 m to 9 m above grade)

FE

insulator and bracket

80 cm drip loop
6.5 cm service mast

mast support bracket
(secured to rafter)

roof flange

wall

mast support
(bolt, nut, and washer assembly)

mast
support brackets

2 cm wood mounting board
(1.5 m above basement floor)

100 A, 24 circuit
distribution panel

30 mm conduit nipple

kilowatt-hour
meter
(approx. 2 m
from grade)

outdoor meter base

100 A main switch

A
A

push
on   off

30 mm conduit

A
A

30 mm LB fitting
(cover and gasket)

A

nipple

flat-rate
water heater switch

grade level

#6 AWG ground wire

cold-water ground

FIGURE 15.1   A 100 A service installation with overhead supply lines

Applications of Electrical Construction

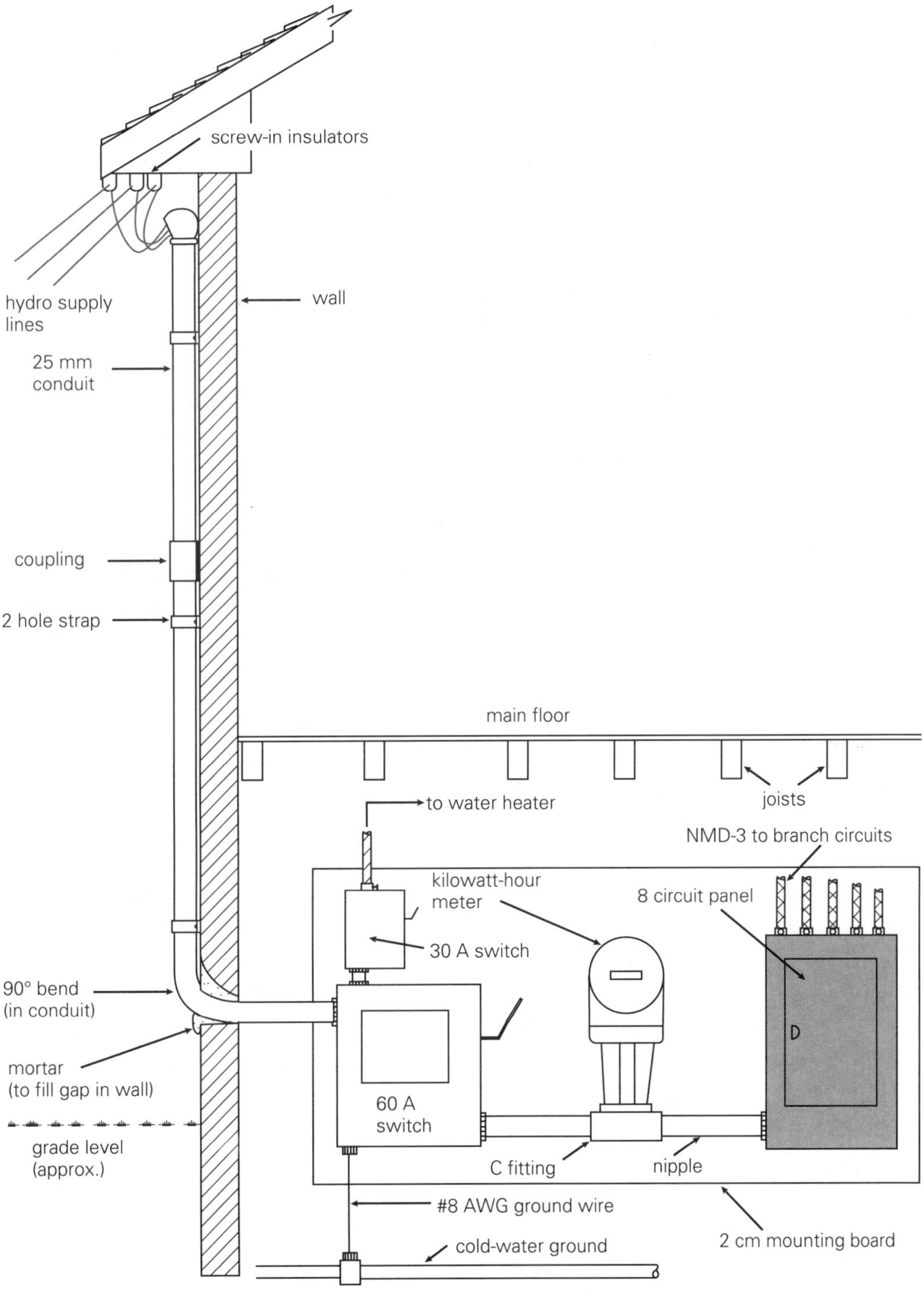
screw-in insulators
hydro supply lines
wall
25 mm conduit
coupling
2 hole strap
main floor
joists
to water heater
NMD-3 to branch circuits
kilowatt-hour meter
8 circuit panel
30 A switch
90° bend (in conduit)
mortar (to fill gap in wall)
60 A switch
grade level (approx.)
C fitting
nipple
#8 AWG ground wire
cold-water ground
2 cm mounting board

FIGURE 15.2 A 60 A service installation with overhead supply lines (indoor method)

circuit panel a logical choice. The extra circuits (fuses) available allowed appliance receptacles in the kitchen and laundry areas to be fused separately. The high current demands of modern appliances in these areas made this feature necessary. At the same time, the meters were moved outdoors, eliminating the access problem for meter readers.

The 100 A service can be installed with a *separate* main switch and distribution panel or in the new *combination* unit. (See Figs. 15.1 and 15.3)

The switch and panel *combination unit* reduces installation time and costs by eliminating the nipple, locknuts, and bushings required when the units are installed separately. Since most combination units are factory-prewired (between main switch and distribution panel), even more time is saved. There is, however, a metal *dividing wall* between the two sections, so that a short circuit in the distribution panel cannot flash over to the main switch (and vice versa). This dividing wall also isolates the non-fused service entrance conductors. As a result, when a person attempts to change a circuit or fuse, the

conductors are protected from damage. Despite the advantages of such a combination unit, before long manufacturers were encouraged to produce *24 circuit* panels. The necessity of separate circuits for clothes driers, electric water heaters and other modern appliances led to this.

Many installers have discovered the versatility and convenience of circuit breaker combination panels for houses. These panels are available from several companies specializing in the manufacture of service equipment. Figure 15.4 shows a typical 24 circuit combination panel.

Circuit Breakers

A more extensive look will be taken at circuit breakers in Chapter 16. There are, however, several basic features that make a circuit breaker panel attractive for residential installation. Such a panel is normally smaller than a fuse panel with the same number of circuits. It is thus ideal for installation in crowded areas when upgrading the service on an older house. Additional features such as tamper-proof current ratings, ease of resetting (instead of replacing), and absence of exposed live parts further enhance this panel's practicality. (See Figs. 15.5 and 15.6)

For a time, the 100 A service was considered the ultimate in residential equipment. But as larger, more spacious houses were designed and often equipped with electric heating, the 100 A service went the way of the 60 A installation. The 200 A, *40 circuit* service took its place. (See Figs. 15.5 and 15.7)

The 200 A unit is designed to provide 100 A and 20 circuits for *heating*, with the other 100 A and 20 circuits for *lighting, receptacles*, and *heavy*

FIGURE 15.3 Combination service entrance panel with circuit breaker main disconnect and pull-out units for branch circuit fuses

Applications of Electrical Construction

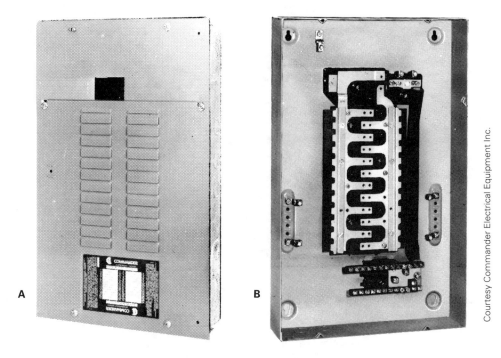

A   B

Courtesy Commander Electrical Equipment Inc.

FIGURES 15.4A and B   Combination service entrance load centre with 100 A main breaker and provision for up to 24 individual circuit breakers. With the cover removed, the breaker connection points can be easily seen.

appliances. With this system, the electric heating part of some service boxes can be turned off during the summer months.

As with 100 A panels, circuit breakers are part of larger service panels. Figure 15.5 illustrates a 200 A circuit breaker panel.

Even 400 A, 600 A, or 800 A services can be found in some private residences. That is because larger modern homes demand more circuits and have more area to heat. The installation of more and more electrical conveniences requires a corresponding increase in service size.

Installation Techniques

Service entrance equipment must be installed in a location suitable to both the local supply authority and the inspection department. Supply authorities are usually quite willing to visit homes on request and advise on the location of service equipment. Consultation with them can prevent installation of service equipment in an unacceptable manner; equipment installed contrary to supply authority standards must be removed and reinstalled to their satisfaction. Mast height, meter height and location, point of entry into the building, and box location can be selected quickly by an inspector on a service location visit.

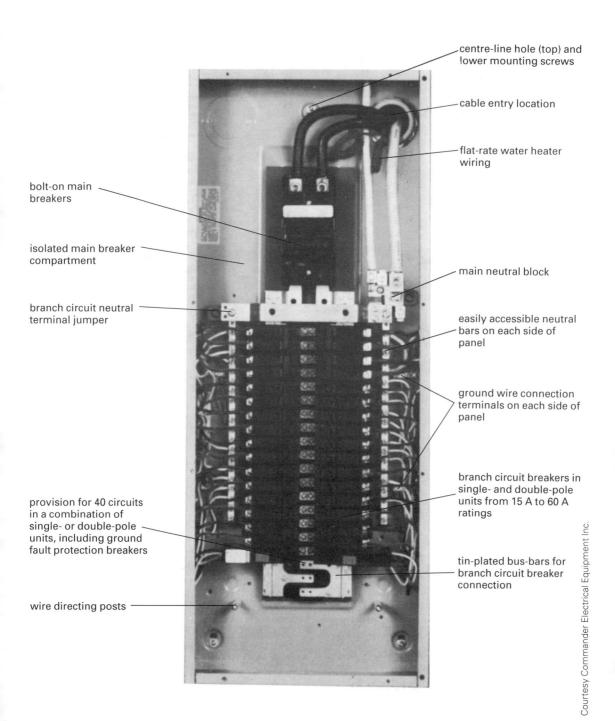

centre-line hole (top) and
lower mounting screws

cable entry location

flat-rate water heater
wiring

bolt-on main
breakers

isolated main breaker
compartment

main neutral block

branch circuit neutral
terminal jumper

easily accessible neutral
bars on each side of
panel

ground wire connection
terminals on each side of
panel

branch circuit breakers in
single- and double-pole
units from 15 A to 60 A
ratings

provision for 40 circuits
in a combination of
single- or double-pole
units, including ground
fault protection breakers

tin-plated bus-bars for
branch circuit breaker
connection

wire directing posts

Courtesy Commander Electrical Equipment Inc.

FIGURE 15.5 Internal features of a combination service entrance load centre, 200 A rating with
40 circuit capacity, in a total circuit breaker unit

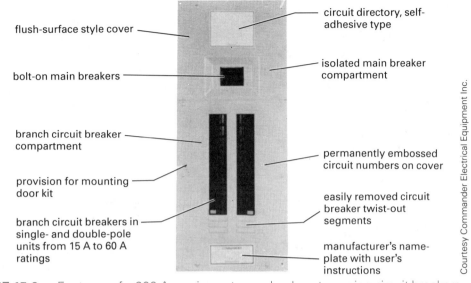

flush-surface style cover

circuit directory, self-adhesive type

bolt-on main breakers

isolated main breaker compartment

branch circuit breaker compartment

permanently embossed circuit numbers on cover

provision for mounting door kit

easily removed circuit breaker twist-out segments

branch circuit breakers in single- and double-pole units from 15 A to 60 A ratings

manufacturer's name-plate with user's instructions

Courtesy Commander Electrical Equipment Inc.

FIGURE 15.6   Features of a 200 A service entrance load centre, using circuit breakers throughout

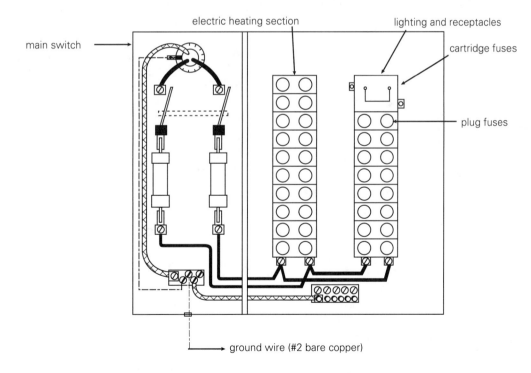

main switch

electric heating section

lighting and receptacles

cartridge fuses

plug fuses

ground wire (#2 bare copper)

FIGURE 15.7   A 200 A combination service entrance panel

Service Mast

The Canadian Electrical Code requires that the *service entrance cap* be located a minimum of 4.6 m above grade. Following this regulation helps ensure hydro supply lines are not damaged by large vehicles and reduces the possibility of a person moving a ladder, for example, coming in contact with the wires. The entrance cap should not be located more than 9 m above grade on high buildings: this would make it difficult for supply authority personnel to reach the cap. Figure 15.8 show a typical mast unit and its individual components, which are available with 100 A and 200 A service fittings.

The service mast, which is sometimes called the *standpipe*, must be securely fastened to the building to prevent hydro supply lines from pulling it loose. Wind, rain, snow, ice, and contraction of the wire in cold weather often exert tremendous pressures on the mast.

There are several methods for installing the mast. (See Figs. 15.8, 15.9, 15.10; see also Fig. 15.17) One common method for securing it is to use long *bolts* that pass completely through the wall and fasten over a wooden member on the inside of the house. Large square *washers* and *nuts* are used to complete the system. (See Fig. 15.1 on page 200.)

The mast is usually located about 1 m from the corner of the building and in such a way that no supply lines will pass within 1 m of a window or any other point that might provide access to the wires. Thus, if a person shook a mop from a window, there should be no likelihood of touching the wires accidentally.

An 80 cm length of free conductor must be left at the entrance cap. Functioning as a *drip loop*, it ensures that rain will not run back into the mast. It also gives the supply authority enough conductor for the connection.

Meter Socket

The local hydro authority determines where the meter socket is installed, usually at about 1.8 m above grade. The meter socket, a metal enclosure with threaded hubs for conduit, is designed so that the service conductors to be metered are attached to *solderless lugs* within the enclosure. (See Fig. 15.11) The *meter* is then plugged into the enclosure in much the same way as a cord cap is plugged into a receptacle. In fact, this is where the name *meter socket* came from. (See Fig. 15.13)

Meter sockets, which are often called *meter bases*, are available in several styles. (See Figs. 15.12, 15.14, 15.15, and 15.16) *Threaded hubs* are sized to accept the rigid conduit required for the service installation, for example, 30 mm for a 100 A service and 50 mm for the 200 A unit. Some meter sockets have hubs that bolt onto the actual metal box of the socket.

One type of meter socket is equipped with a pair of special solderless lugs on the *line* side of the socket. These lugs allow a pair of No. 10 gauge, red-insulated conductors to be brought in to the main switch to supply the flat-rate water heater. The smallest conductor allowed in a service conduit is a No. 10 gauge wire. Since these conductors are attached to the line side of the meter, any current passing through them will not register on the meter.

Flat-Rate Water Heater System

Some local utilities operate on a *flat-rate* water heater supply system. The consumer pays a *set amount* each month

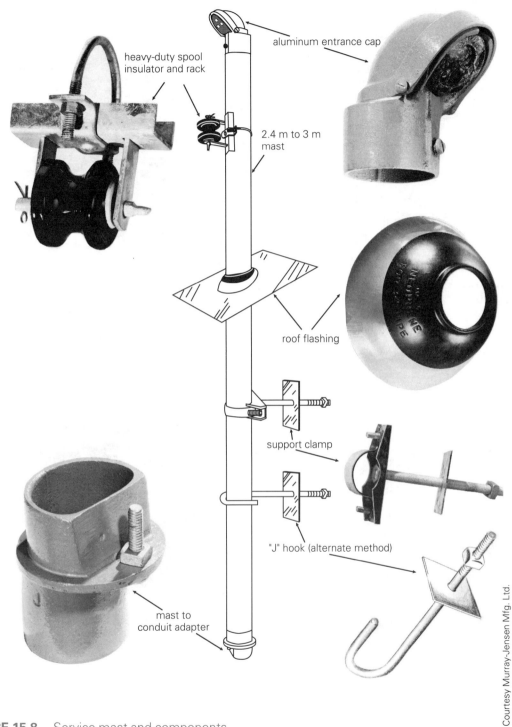

heavy-duty spool insulator and rack

aluminum entrance cap

2.4 m to 3 m mast

roof flashing

support clamp

"J" hook (alternate method)

mast to conduit adapter

FIGURE 15.8   Service mast and components

Courtesy Murray-Jensen Mfg. Ltd.

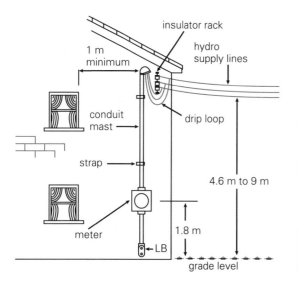

FIGURE 15.9 Mast installation on a 2-storey house

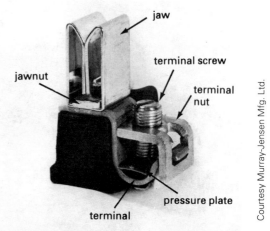

FIGURE 15.11 Standard terminal block for 100 A and 200 A meter sockets

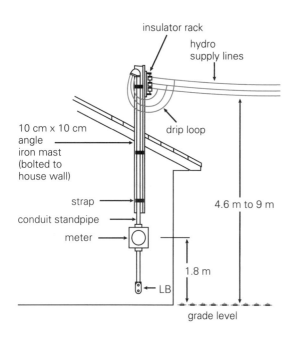

FIGURE 15.10 Mast assembly on a 1-storey house

FIGURE 15.12 A modern, 100 A meter socket is rectangular in shape to provide more space inside the socket when connecting service conductors.

Applications of Electrical Construction

FIGURE 15.13 A plug-in style of residential kilowatt hour meter for use with 100 A and 200 A services

Courtesy Sangamo Company Ltd.

FIGURE 15.14 A 200 A service meter socket

Courtesy Murray-Jensen Mfg. Ltd.

and may use as much power for heating water as required. Large families, which need a lot of hot water, find this an economical system. Small families, however, are usually better off if the water heater, together with the lights and other appliances, operate through the *meter*. With this system, they pay only for the amount of power used.

The flat-rate water heater system is good for both the consumer and the local utility. The demand for power is greatest during the hours of 07:00 to 09:00 and 15:00 to 18:00. During these times, the utility can turn off all the flat-rate water heaters in the area. Power from the water heater units can then be directed to industrial customers. A *high-frequency signal* is sent through the supply lines, then picked up by the *relay* in the basement of the house. The relay then turns off the power to the heater. The same process is used to restore the power to the water heater.

Some communities require the two red-insulated conductors to be enclosed in the service conduit and mast all the way up to the weather head fitting. An 80 cm length of drip loop is left for connection to a *fourth* conductor, which is brought to the home from the closest hydro pole (on overhead service instal-

FIGURE 15.15 A meter socket assembly for use on a duplex or semi-detached house

FIGURE 15.16 An old-style, 60 A/100 A service meter socket in the original, round configuration

lations). This fourth conductor carries the high-frequency signal to one red wire, while the second red conductor is connected to one of the incoming supply lines for the house. Underground service installations normally use the meter socket with the special lugs.

Service Entrance Elbows

A 90° *corner* is used to bring the *standpipe* from the meter socket through the wall and into the service boxes. A 90° bend can be made in the service conduit, but a *groove* must be chiselled in the wall for approximately 30 cm above the hole to allow the conduit to fit tightly

Applications of Electrical Construction

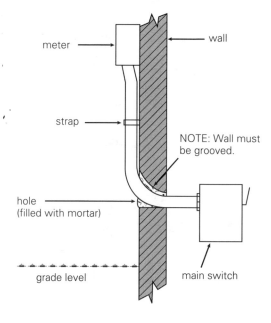

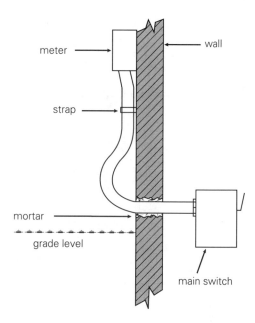

FIGURE 15.17 Method for driveway installation of standpipe (90° bend)

FIGURE 15.18 Alternative method for driveway installation of standpipe (goose-neck bend)

against the wall. (See Fig. 15.17) Doing this is worthwhile when the service conduit is in a driveway or similar area, where the conduit may be abused by vehicles.

There is another way of handling the 90° bend. (See Fig. 15.18) It does away with the need for grooving the wall, but the standpipe protrudes from the wall a few extra inches (several centimetres). This *goose-neck* bend method of entering the building was once used widely. With the increase in conduit size and the development of condulet fittings, however, this difficult bend has become less popular.

A *condulet* fitting, such as the LB, is approved for use where there is little chance of mechanical injury cracking the casting. (See Fig. 15.19)

The hole in the wall must be made slightly larger to allow the short arm of

the condulet to enter the wall. A weatherproof *gasket* and *cover* fastened to the LB seal out any moisture. The LB fitting provides access to the conduit and makes it easier to pull in the large conductors required for modern houses.

A special fitting called the *service ell* can also be used for entering a building. (See Figs. 15.20 and 15.21) Its main advantage is that the short arm does *not* extend in to the wall. Because the hole in the wall does not need to be made any larger than the conduit nipple, this type of installation is more weatherproof than other types.

No matter which method is used to enter the building, the Canadian Electrical Code requires that all openings around the area where the conduit enters the wall be filled. Otherwise, there may be water damage to the house

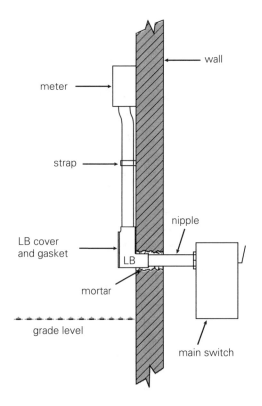

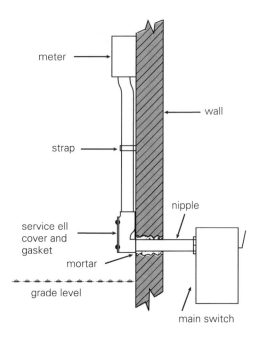

FIGURE 15.21   A service ell installation

FIGURE 15.19   LB condulet method for entering building

FIGURE 15.20   Service ell fitting with cover and gasket

Courtesy Pyle-National of Canada Limited

and/or the service equipment. *Mortar*, or a similar material, trowelled into place will stop moisture or cold air from coming in.

Underground Service

Most new subdivisions have underground service installations to help their streets look neat and uncluttered. The distribution transformer can be located under ground in a transformer vault or on a concrete pad. The vault or pad can be out near the street or behind the house, with the supply lines coming under the backyard. (See Fig. 15.22)

Location of Service Boxes

Section 6 of the Canadian Electrical Code lists the places where service boxes must *not* be located. In general, service boxes should be mounted on a wall about 1.5 m above the floor and as close as possible to the point where the service conduit enters the building. If there is any doubt about where to install a box, the local inspection authority should be asked for advice.

Applications of Electrical Construction

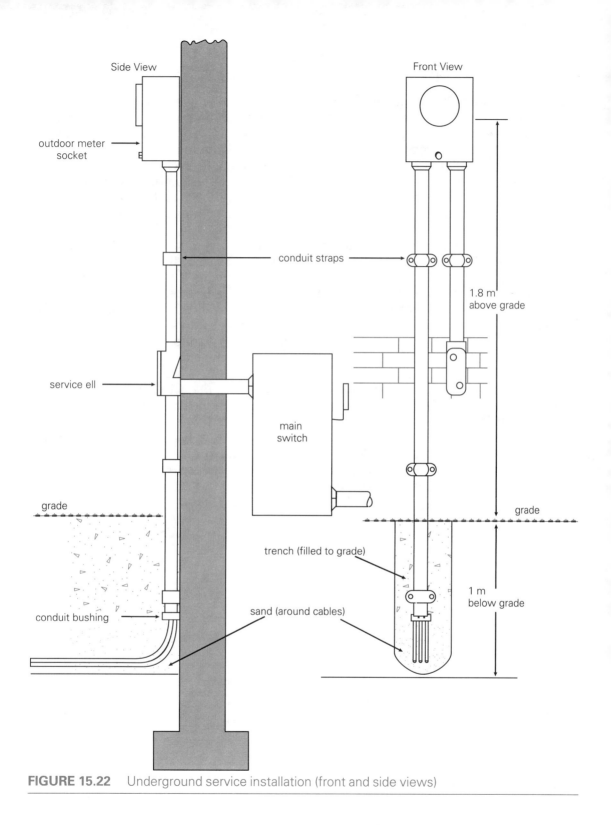

Side View

outdoor meter
socket

conduit straps

service ell

main
switch

grade

conduit bushing

sand (around cables)

Front View

conduit straps

1.8 m
above grade

grade

trench (filled to grade)

1 m
below grade

FIGURE 15.22   Underground service installation (front and side views)

Residential Service Wiring   213

Service Mounting Board

Service boxes must be supported on a *wooden mounting board* approximately 2 cm thick. Plywood does an excellent job, but regular 2 cm lumber can be used. A 100 A service requires a 1 m² panel. A 200 A service will likely need a 1.2 m square board, and larger services require a full 120 cm x 240 cm plywood sheet. The board should be fastened securely to the wall with such devices as concrete nails or masonry plugs and screws.

The wooden panel allows branch circuit conductors to be supported close to the distribution panel. Future equipment can also easily be secured to it.

Service Boxes

Service boxes used to be arranged as two separate units. (See Figs. 15.1 and 15.2 on pages 200 and 201.) Modern service equipment, in 100 A and 200 A ranges, is usually arranged in the form of *combination* units. (See Figs. 15.3, 15.4, and 15.5) Larger service boxes are usually arranged as shown in Figure 15.23.

Service Wiring

Table 5.5, on page 61, lists the size of copper conductors to be used with their various ampacities. Table 5.6, on page 62, covers the requirements of aluminum conductors. Once conductor size has been determined, these tables and Tables 14.2, 14.3, 14.4, 14.5 and 14.6 can be used to establish conduit size for the installation.

When an *outdoor meter* is used, the conductors must first pass through the *meter terminals* (lugs). The neutral wire must not be broken in the meter socket. It must pass directly to the *neutral block* in the main switch. A neutral wire that is not making a good contact will cause uneven voltage distribution in an unbalanced system, which can be dangerous.

The two *live* wires in the service conduit are connected directly to the *line* terminals in the main switch. These terminals are nearly always located at the top of the switch and are designed so that when the switch is *off*, only these two terminals are live. Some manufacturers design their equipment so that the line terminals are covered by an *insulating barrier* to prevent a person from

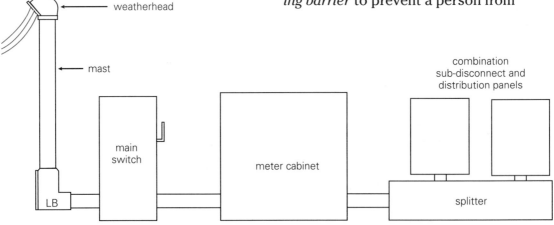

FIGURE 15.23   A large service layout

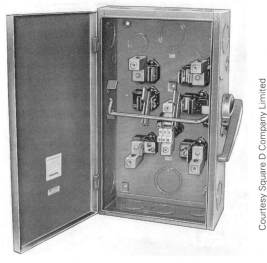

FIGURE 15.24 Typical main switch with line terminals at top and load terminals at bottom

accidentally touching them. (See Fig. 15.24) The *load* side of the switch (bottom terminals) is then connected to the distribution panel.

Figure 15.25 shows the wiring of a typical 100 A service. The 200 A service uses larger conductors, but is wired in the same way.

Figure 15.26, on page 217, shows the box arrangement and wiring for a semi-detached house.

Figure 15.27, on page 218, shows the box arrangement and wiring for a duplex.

Apartment buildings, where the apartments are metered separately, must be handled differently. Figure 15.28 shows a typical service installation for a four-suite apartment building. Large services of this type have indoor metering facilities.

When a *single-meter* system is needed for a service with a capacity larger than 200 A, the current flow must be metered *without* actually passing full line current through the meter. Since most meters have a maximum current

capacity of 200 A, a system of *current-reducing transformers* is used. These current transformers carry the full line current in a main (*primary*) winding, which usually consists of a solid bar of copper. They reduce the line current to a safe level through a secondary winding. A 5 A current is then passed on to a specially designed meter. (See Fig. 15.29)

The current transformers and meter are housed in a metal *meter cabinet*. There are several cabinet sizes: the smallest measures about 1 m square and is about 30 cm deep.

Figure 15.30, on page 221, shows an indoor meter for use in a meter cabinet.

Figures 15.31, 15.32 and 15.33A and B, on page 221, show current transformers. Transformers are available in a variety of sizes and ampere capacities.

Service Grounding

Electrical services must be grounded for two reasons. The first is that the *steel mast*, which rises 4.6 m to 9 m, is an attractive target for lightning. Grounding the mast reduces the chance of lightning striking the house. If lightning does strike, grounding provides a direct path to the earth.

The second reason is that grounding of the *boxes* and *cabinets* gives another, equally important, form of protection. If one of the live wires in the system comes in contact with any of the metal enclosures, there will be a short circuit. The excessive short-circuit current will flow along the ground conductor and pass harmlessly into the earth. The usual result is a blown fuse. Once the fault has been located and repaired, the fuse can be replaced easily. If the boxes were *not* grounded, all of the conduit, boxes, and cabinets would become *alive* and dangerous. A person standing on

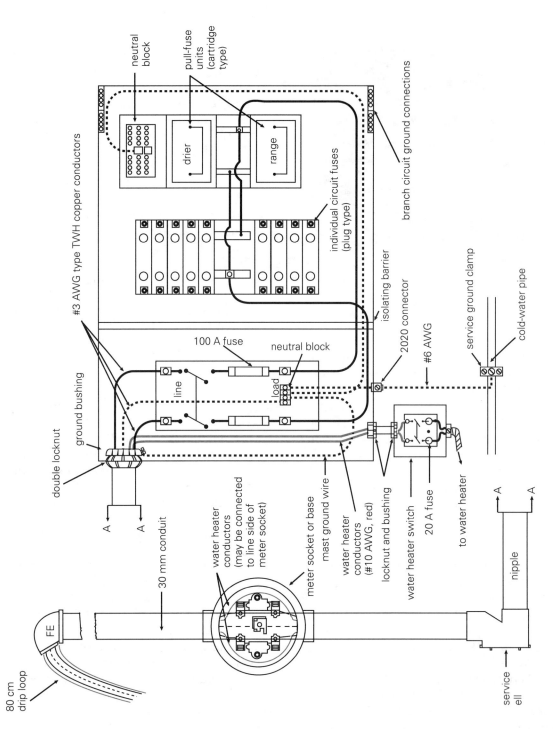

FIGURE 15.25 The once popular 100 A combination main switch and fuse panel, with which a round meter socket (covers removed) is used

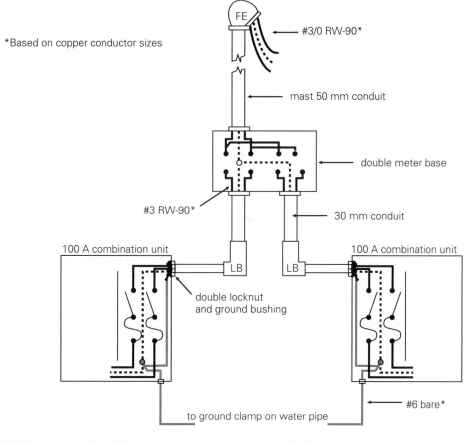

*Based on copper conductor sizes

#3/0 RW-90*

mast 50 mm conduit

double meter base

#3 RW-90*

30 mm conduit

100 A combination unit

100 A combination unit

double locknut
and ground bushing

LB

LB

#6 bare*

to ground clamp on water pipe

FIGURE 15.26   Semi-detached house service installation

moist earth or a damp floor and touching part of the metallic system would receive a 120 V shock. In a large building, with no blown fuse on an ungrounded system, the fault may not show up for some time and then be almost impossible to locate.

The most common method for providing a ground for a residential service is to connect a *conductor* from the *neutral* block in the main switch to the *cold-water supply pipe* as it comes out from the basement floor. The neutral block is connected to the switch box by a *brass bolt*. Thus, both the box and neutral conductors are grounded. The

ground clamp must be fastened on to the water pipe *ahead* of the meter, because a *leak-proof compound* (a nonconductor of current) is applied to the plumbing threads *before* assembly. (See Fig. 15.34 on page 222.) The ground wire from the main switch box is usually bare copper or covered with a white insulation. Table 16.1, on page 231, lists the conductor size to be used.

When a house has *copper* plumbing, which is properly soldered at the fittings, the ground wire can be run from the main switch to the closest cold-water pipe. The hot-water tank interferes with the electrical continuity of the hot-

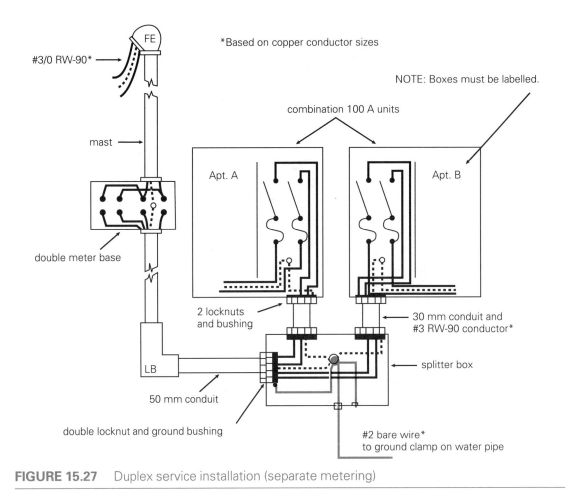

#3/0 RW-90*

*Based on copper conductor sizes

NOTE: Boxes must be labelled.

mast

combination 100 A units

Apt. A

Apt. B

double meter base

2 locknuts
and bushing

30 mm conduit and
#3 RW-90 conductor*

LB

splitter box

50 mm conduit

double locknut and ground bushing

#2 bare wire*
to ground clamp on water pipe

FIGURE 15.27   Duplex service installation (separate metering)

water pipe and should not be used. Since the water meter with its water-proof connections is still a problem, a *jumper* of the same gauge conductor as the ground wire must be installed to by-pass the meter. Then, if the water meter is removed at any time, ground continuity is maintained. (See Fig. 15.35 on page 222.)

In some areas, houses do not use a water meter. In these cases, the ground conductor should be connected by a ground clamp to the cold-water pipe where it first enters the basement. (See Fig. 15.36 on page 222.)

Two precautions must be taken to make sure that the mast is grounded effectively. First, two securely tightened *locknuts*, one inside and one outside the box, are used. Second, a special *ground bushing* with provision for a grounding conductor is threaded on to the conduit. This conductor, which should be about No. 6 gauge, joins the ground bushing to the neutral block. Figure 15.37, on page 222, shows a ground bushing.

Rural communities and cottage areas usually do not have the cold-water supply pipe used for grounding services in urban communities. There are several different methods for grounding available for these areas. The most common

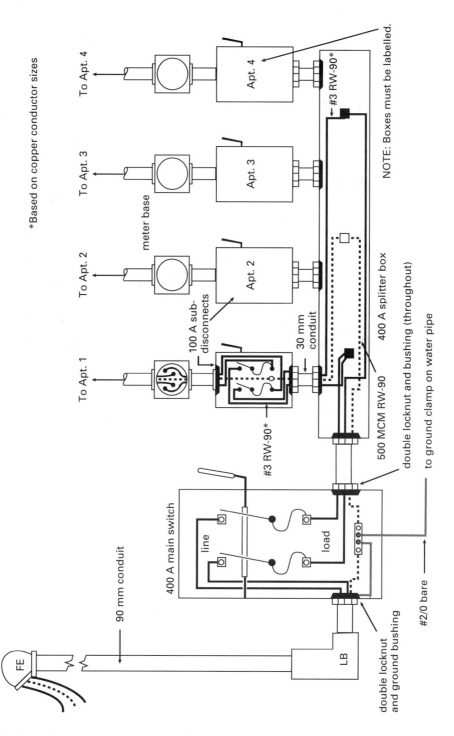

*Based on copper conductor sizes

To Apt. 4

To Apt. 3

To Apt. 2

To Apt. 1

meter base

Apt. 4

Apt. 3

Apt. 2

Apt. 1

#3 RW-90*

NOTE: Boxes must be labelled.

100 A sub-
disconnects

30 mm
conduit

400 A splitter box

#3 RW-90*

500 MCM RW-90

double locknut and bushing (throughout)

to ground clamp on water pipe

400 A main switch

line

load

#2/0 bare

500 MCM RW-90

90 mm conduit

FE

LB

double locknut
and ground bushing

FIGURE 15.28   Apartment service installation (separate metering)

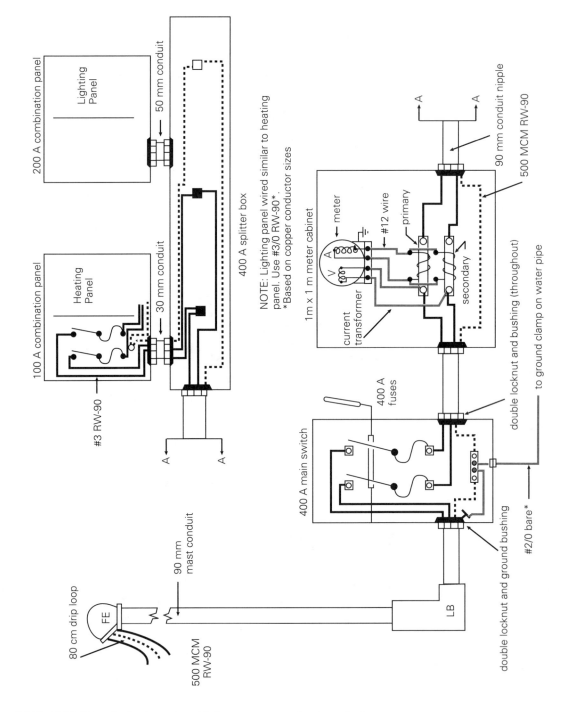

FIGURE 15.29 A single-meter, high-ampere service installation

Applications of Electrical Construction

FIGURE 15.30 An indoor meter for use in a meter cabinet

FIGURE 15.31 A 200 A capacity current transformer

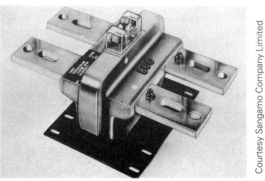

FIGURE 15.32 Typical 800 A capacity current transformer designed to operate with both live wires of the system

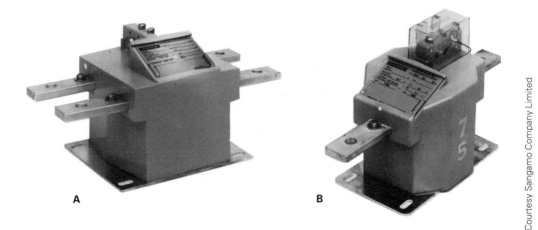

A   B

FIGURES 15.33A AND B Modern current transformers are more compact in size due to a redesigning of the primary conductors' enclosure.

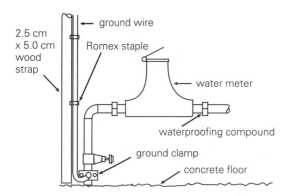

2.5 cm
x 5.0 cm
wood
strap

ground wire

Romex staple

water meter

waterproofing compound

ground clamp

concrete floor

FIGURE 15.34 Ground clamp connected to water pipe ahead of water meter

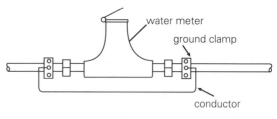

water meter

ground clamp

conductor

FIGURE 15.35 A water meter by-pass

FIGURE 15.36 A water ground clamp for use in a residential service

Courtesy Elliott Electric Mfg. Corp.

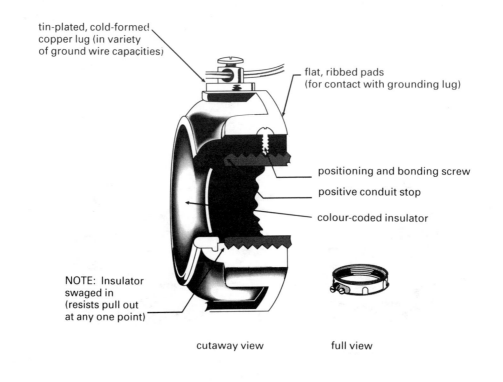

tin-plated, cold-formed copper lug (in variety of ground wire capacities)

flat, ribbed pads (for contact with grounding lug)

positioning and bonding screw

positive conduit stop

colour-coded insulator

NOTE: Insulator swaged in (resists pull out at any one point)

cutaway view

full view

Courtesy Elliott Electric Mfg. Corp.

FIGURE 15.37 Full and cutaway views of a ground bushing

one is to drive two 2 cm diameter, galvanized steel *rods* (ground electrodes) into the earth. They must be spaced approximately 2.5 m apart and driven in a full 3 m. The 3 m depth ensures a high-quality ground contact with permanently moist earth. (See Fig. 15.38)

Commercial *ground rod drivers* are available, but most often a *sledge hammer* is used to drive rods into the earth. Layers of clay and stones or rocky terrain are problems to the installer. Sometimes the rods, striking a layer of rock, gradually turn and move back to the surface, several metres from the starting point. Inspection authorities allow rods to be driven in on an angle if the terrain is hard and the installation difficult.

A grounding conductor *must* run in a *continuous length* from the neutral block in the main switch to the first rod. Here it passes unbroken through a *clamp* on the rod, and then travels to the second ground electrode some 2.5 m away. Take care to check with the inspection authority for up-to-date information about the type of clamp allowed on the ground rod. The clamping device must make a secure electrical connection to both rod and conductor. Once the rods and ground conductor are covered with earth, it may be many years before they are checked again.

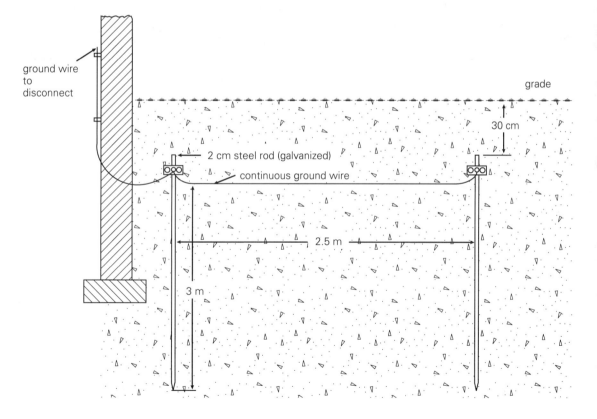

FIGURE 15.38 Ground rod installation

In rocky soil, it may be impossible to drive in ground rods. In these cases, a *steel plate*, 0.2 m² in area and a minimum of 6 mm thick, can be buried in the earth. The plate should be installed as deeply as possible to ensure a good ground contact. (See Fig. 15.39)

There are also other methods of providing a ground circuit under rugged terrain conditions. A pair of ground rods can be laid in a trench, dug as deeply as the rocky terrain will permit, and then covered with top soil. Still another method requires a ground rod or copper conductor to be buried or encased within the concrete footings supporting a building's walls. Sufficient conductor size and length should be provided to make a connection to the main neutral block within the main service disconnect. Consult Section 10.700 of the Electrical Code for confirmation of ground-

ing techniques in difficult areas. (See Fig. 15.40)

Often, when ground rods (artificial grounding) are used, inspection authorities require a *grounding conductor* of at least No. 8 gauge to be run from the neutral block of the main switch to a clamp on the copper plumbing. This happens most often in rural communities to prevent the chance of a voltage difference existing between the earth and the plumbing system in the house. If a plastic (PVC) water system is used throughout the house, this extra ground conductor is *not* required.

Meters

The meter records the number of kilowatt hours (1 kW is equal to 1000 W) of power consumed. Modern meters give a *direct reading* the way odometers do on

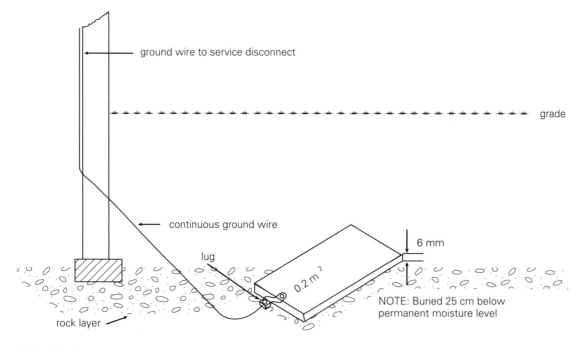

ground wire to service disconnect

grade

continuous ground wire

6 mm

lug

0.2 m²

NOTE: Buried 25 cm below permanent moisture level

rock layer

FIGURE 15.39   Plate electrode grounding method

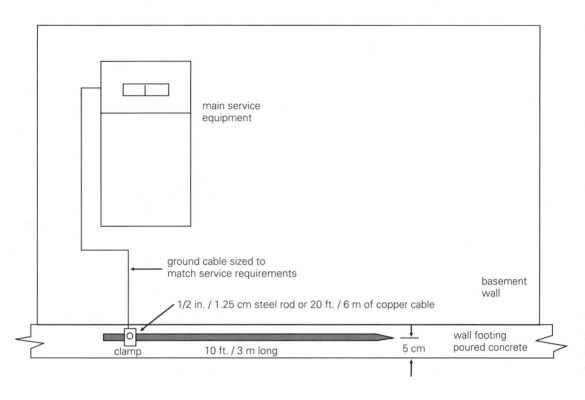

ground cable sized to
match service requirements

main service
equipment

basement
wall

1/2 in. / 1.25 cm steel rod or 20 ft. / 6 m of copper cable

wall footing
poured concrete

clamp   10 ft. / 3 m long   5 cm

FIGURE 15.40   Artificial ground electrode system

cars. (See Fig. 15.30) There are still many old–fashioned meters in use, however. These meters have *five small dials* to indicate the consumption. (See Fig. 15.41) Moving from right to left across the dials, the *first* dial shows single units, the *second* tens, the *third* hundreds, the *fourth* thousands, and the *fifth* tens of thousands. Figure 15.42 shows a five-dial meter reading.

The consumer pays the local utility for power at a rate of so many cents *per kilowatt hour*. The number of kilowatt hours of power used during a given period is read from the meter. The number is multiplied by the rate per kilowatt hour, and the customer is billed.

Note: Many utilities have a *rate structure* whereby the price charged a

FIGURE 15.41   Five-dial meter

FIGURE 15.42   A five-dial meter reading (26 422 kW•h)

customer varies according to the amount of power consumed. For example, the first 50 kW·h would be at a base price, while the next 200 kW·h would be charged at a lower rate per kilowatt hour. This system rewards the consumer using a lot of electrical energy, such as an electric heating customer. It makes electricity more competitive with other fuel types.

Temporary Service

On construction sites power is needed for portable drills, saws, concrete mixers, and other equipment. A *temporary service* is usually erected on a pole. (See Fig. 15.43) This unit is designed with a main switch, a small distribution panel, several duplex receptacles, and a weather-resistant cabinet to protect it. Mast height can be as low as 4.6 m. A single ground rod is usually enough for this type of installation. Often, the local utility charges a flat rate or fee for service, and a meter is not installed.

Inspection Permit

An application for inspection *must* be filed with the local inspection authority for *all* new installations of electrical equipment. A residential service is considered an important installation and receives careful attention from the inspection authority. Power will not usually be allowed on until the inspector is satisfied that every detail is complete and safely installed. This inspection is important, because it guarantees to the owner and the insurance company that every precaution has been taken to ensure a safe and high-quality installation.

For Review

1. What are the two main methods for bringing the 3 wire distribution system to a house?
2. List the main parts of a consumer's service.
3. Why did the 60 A service become inadequate for modern houses? Give two reasons.
4. Why are combination service panels often used in modern houses?
5. What are the minimum and maximum heights for service entrance fittings? Why?
6. Describe a *drip loop*, and explain its purpose.
7. What is the *meter base*? How high above grade is it mounted?
8. Why is the flat-rate water heater system good for both the house owner and the public utility?
9. Describe the two different methods for supplying current to a flat-rate water-heater system.

Applications of Electrical Construction

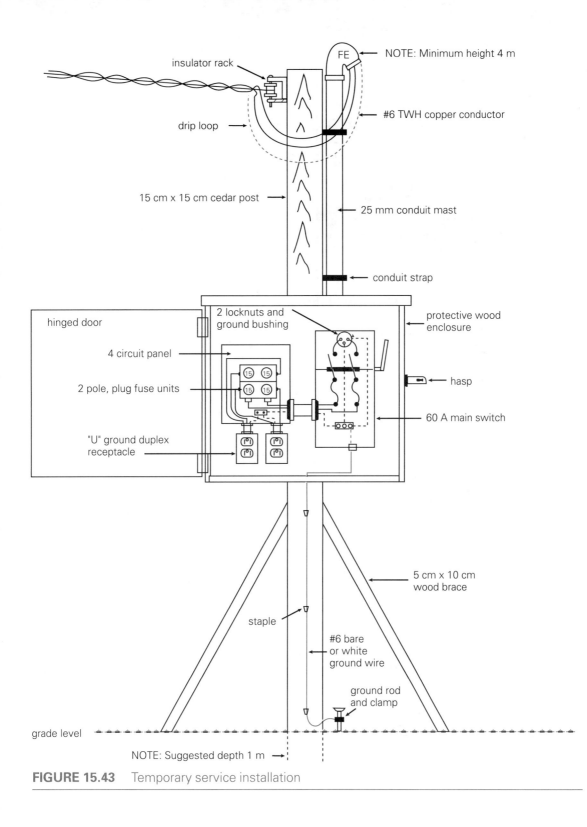

FIGURE 15.43 Temporary service installation

10. Where are cast aluminum condulets approved for use? Why?
11. How is moisture stopped from seeping in where a service conduit enters a wall?
12. Where in the Canadian Electrical Code is information about the size of service conductors and conduit?
13. What danger exists if the neutral wire of the 3 wire system is accidentally broken or loosely connected?
14. Explain why current transformers are used with a service larger than 200 A.
15. List two reasons for grounding an electrical service.
16. Why are services that have water meters grounded ahead of the water meter?
17. How is a service grounded when a house has copper plumbing throughout?
18. Where in the Canadian Electrical Code is information about the size of ground wire to be used on a residential service?
19. What two precautions must be taken when grounding a service mast?
20. Describe in your own words two different methods for grounding a service in a rural community where there is no municipal water supply system.
21. Why are plastic bushings or metal bushings with plastic liners used on service conduits?
22. Why is electrical inspection important to the home-owner?

16

Industrial Services

Small factory and commercial buildings are sometimes equipped with the 120 V/240 V service used in residential installations. The larger capacity 120 V/240 V services (400 A, 600 A, and 800 A) have more than enough circuits for the lights, receptacles, and motors these small, limited production facilities need.

Industrial buildings usually have *motor-driven* equipment. These motors often need far more current than lighting circuits. When being started, a motor draws three to five times' its normal operating current, which places an even greater load on the service equipment. A motor designed to operate on a higher voltage will produce the same amount of power but with a decrease in line current. For example, a 750 W motor operating at 120 V will require approximately 16 A. The same motor, internally connected to operate on 240 V, will require about 8 A.

Motors and equipment designed to operate on even *higher* voltages need still *less* current. When the current is reduced, the *wire size* in the motor windings can also be reduced. This allows, for example, a 550 V motor to be much smaller in size than a 750 W, 120 V motor.

Special voltage systems are available for commercial and industrial use. These are known as *polyphase (3 phase)* systems. Three-phase motors and equipment are more efficient and smaller in size and usually need less current than equivalent units designed for use with a single-phase supply system.

Three-phase voltages are produced by the *alternators* (AC generators) at the power station and can be transformed by the consumer to any one of three common voltage levels. The most popular 3 phase voltage levels are 550 V, 440 V, and 208 V.

The 550 V and 440 V systems use 3 live wires *without* a neutral conductor. The 208 V system is available in 3 wire (3 live conductors) and 4 wire (3 live and 1 neutral conductor) combinations. The advantage of the 4 wire, 208 V system is that there are 120 V between the neutral conductor and any one of the three live conductors for lighting and receptacles. (See Figs. 16.1 and 16.2)

Another advantage of the 3 phase system is that motors can be *reversed* easily. A polyphase motor will rotate in the opposite direction if any of the three live wires are interchanged. To reverse a single-phase motor, the motor often has to be dismantled and the internal connection of the windings changed.

When working on a 3 phase service, take care not to interchange any of the live wires. If the live wires are interchanged,

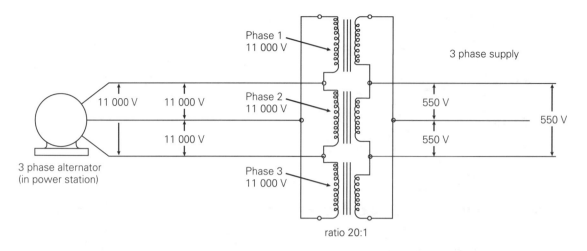

NOTE: Consumer's 3 phase transformer, connected with Delta primary, Delta secondary

FIGURE 16.1 Typical 3 phase 550 V or 440 V supply system

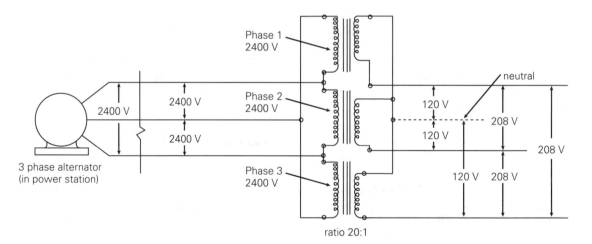

NOTE: Consumer's 3 phase transformer, connected with Delta primary, Wye secondary

FIGURE 16.2 Typical 3 phase 120 V/208 V supply system

Applications of Electrical Construction

every motor in the building will have its direction of rotation reversed. For this reason, high-voltage, 3 phase conductors are often identified with red, white, and blue tags.

550 V and 440 V Systems

The 550 V and 440 V 3 wire services are used primarily for motors and their controls. The *meter* is located within the building in a metal *meter cabinet*. These 3 phase services are available in sizes ranging from 100 A up. The switch boxes and distribution panels have terminals for the three live wires, but not for a neutral conductor, since there is no neutral conductor.

Conductor and conduit sizes are chosen in the same way as for the residential service. Also, mast height, box height, and location are the same as for a residential service. Figure 16.3 shows a typical 550 V and/or 440 V service instal-

lation. The three live conductors are often all the same colour, usually black, red, or blue.

Grounding is limited to the boxes, cabinets, panels, and conduits of the system. A *ground conductor* is placed between the grounding lug in the main switch box and the cold-water supply pipe. Table 16.1 lists ground conductor sizes.

600 V/347 V System

Fairly recent in origin, the *600 V/347 V* system is being used increasingly in commercial lighting and for motor and equipment circuits. A *Wye connected secondary*, formed by the joining of one end of each of the three secondary coils into a common point known as *centre tap*, provides 600 V between any two of the three live wires. The common or centre tap conductor provides a voltage of approximately 347 V between itself and any one of the three live wires.

Figure 16.4 shows internal connections of the meter cabinet.

TABLE 16.1 Minimum Size of Grounding Conductor for AC Systems or Common Grounding Conductor	
Ampacity of Largest Service Conductor or Equivalent for Multiple Conductors	**Size of Copper Grounding Conductor AWG**
100 or less	8
101 to 125	6
126 to 165	4
166 to 200	3
201 to 260	2
261 to 355	0
356 to 475	00
Over 475	000

NOTE: The ampacity of the largest service conductor, or equivalent if multiple conductors are used, is to be determined from the appropriate Code Table taking into consideration the number of conductors and the type of insulation.

Based on the Canadian Electrical Code

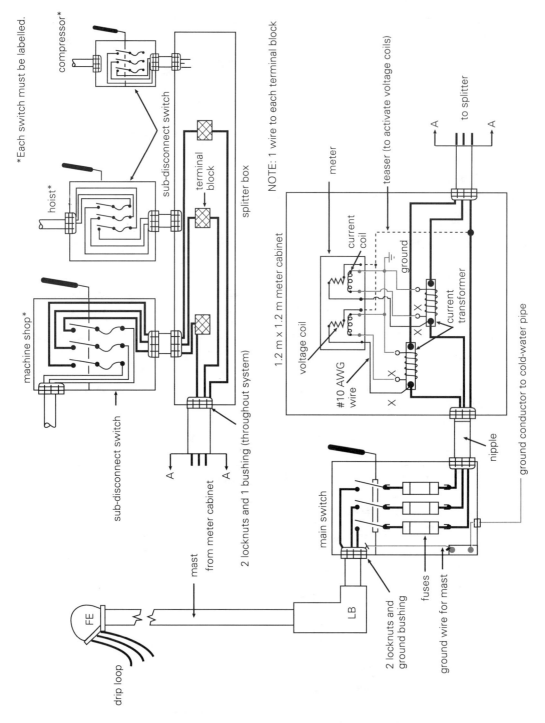

FIGURE 16.3   Typical 3 phase 550 V or 440 V service installation

Metering

The 3 phase system needs a special meter, containing two voltage and two current coils, for recording power. This unit is like two meters in one. The total amount of power used is equal to the sum of both meter sections. (This total is shown as a single reading on the instrument's dials.) (See Fig. 16.5) *Current transformers* are used with the 3 phase meter. (See Figs. 16.6 and 16.7) A special 3 phase "hybrid" meter combining mechanical and electronic features is used, on occasion, for multifunction metering of large industrial loads. (See Fig. 16.8)

Method of Distribution. As Figure 16.3 shows, the service conductors enter a *splitter box* after leaving the meter cabinet. *Sub-disconnect switches* are then joined to the splitter. Conductors of the same ampacity rating as the disconnect switch then distribute 3 phase supply to various parts of the building. Each sub-disconnect must be labelled correctly to show exactly which area or what equipment it controls.

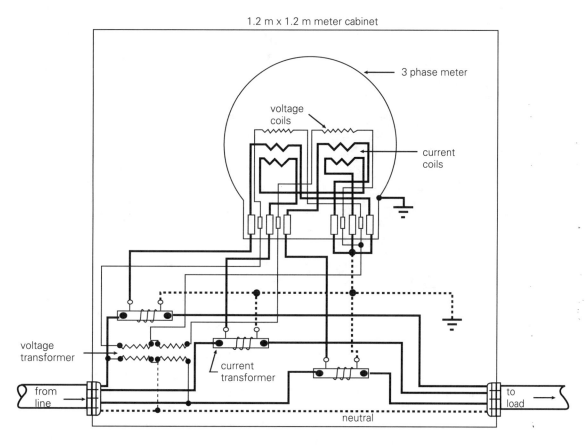

FIGURE 16.4 Current transformers and meter connections for a 3 phase, 600 V/ 347 V service installation

FIGURE 16.5   A 3 phase meter

Courtesy Ron Ridsdill, Toronto

FIGURE 16.7   Current transformer

Courtesy Sangamo Company Limited

FIGURE 16.6   A 3 phase kilowatt-hour meter for use with current transformers

Courtesy Sangamo Company Limited

A *second* disconnect switch is usually mounted on or near the equipment. Designed so that *padlocks* can be placed on it, this switch serves as a safety measure. The padlocks keep the switch out of service until personnel are clear of the machine. Each padlock is removed as each person completes service. When the last padlock has been removed, service may be restored safely.

Grounding.   550 V will both severely shock a person and cause serious burns. As a rule, none of the 3 phase 550 V or 440 V conductors are grounded. This isolated system prevents a person in contact with grounded equipment from being injured if and when a live conductor is also touched. Many services are equipped with three indicating lights *(ground detectors)* to warn that a live conductor has come in contact with a metal box or associated piece of equipment. All metal boxes, panels, conduits, and fittings, however, are grounded.

Take great care when working with a 550 V or 440 V system, because *all* three conductors are alive and dangerous. Sometimes current leaking from faulty equipment establishes a partial ground on one conductor. In these cases, any protection given by the system being isolated is gone.

Applications of Electrical Construction

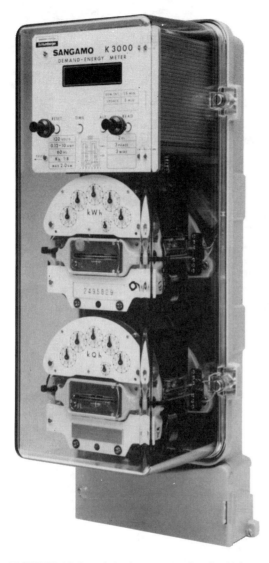

FIGURE 16.8 A 3 phase, mechanical/electronic multifunction meter for large industrial loads

120 V/208 V System

550 V and 440 V systems are excellent sources of power for motors and related equipment, but they are not ideal for general lighting and power requirements. Often the 550 V system is reduced through a transformer to 120 V/240 V,

and then distributed to lighting and general power circuits. This can be expensive, because of the cost of the equipment required. The 3 phase, 4 wire, 120 V/208 V supply system combines the best qualities of both 3 phase and single-phase systems and is ideal for lighting and general power. (See Fig. 16.9)

The 3 phase, 208 V system can be used for motor-driven equipment. The fourth (neutral) conductor in the system provides 120 V between itself and any of the three live conductors for lighting and general power requirements. Figure 16.2 shows how the two voltages are obtained.

Service entrance equipment for the 120 V/ 208 V system looks very much like the equipment for 550 V/ 440 V units. The difference is that a neutral conductor is used in the 120 V/ 208 V system, and it is treated in much the same way as the neutral wire of a single-phase, residential system.

Since 120 V are available between the neutral and live conductors, a group of three *current transformers* must be used to record the power accurately. Figure 16.10 shows a typical 120 V/208 V service installation. These services are usually available in sizes ranging from 200 A up.

Figure 16.12 shows a 3 phase transformer with the three sets of windings for reducing the voltage in each phase of the system.

Large industrial service installations often come in the form of a *cabinet* or *switchboard* system and are ordered from the manufacturer for each individual job or installation. (See Fig. 16.11) Figure 16.13, on page 239, shows a small unit. Figure 16.15, on page 240, shows a larger unit with room for *step-down transformers* in the screened cabinets (on the right-hand side of the photograph).

Courtesy Sangamo Company Limited

550 V main switch   550 V/208 V step-down transformer   splitter trough

sub-disconnects   splitter trough   circuit breaker panel

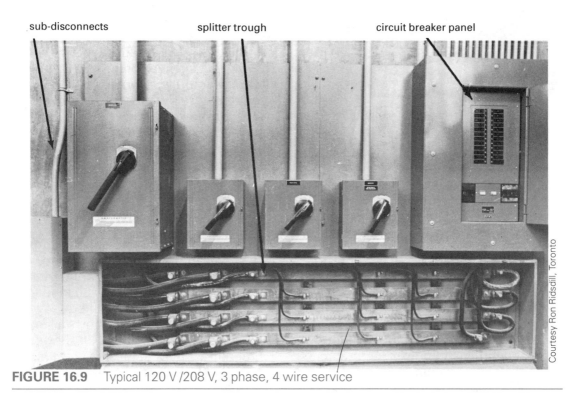

FIGURE 16.9   Typical 120 V /208 V, 3 phase, 4 wire service

Courtesy Ron Ridsdill, Toronto

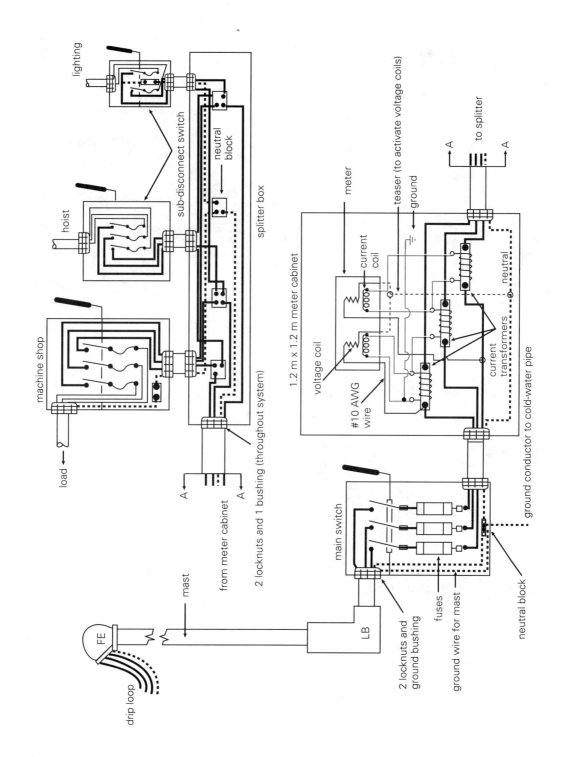

FIGURE 16.10 Typical 3 phase 120 V/ 208 V service installation

FIGURE 16.11   A main switchboard room installation

FIGURE 16.12   A dry-type, core and coil assembly transformer used in a single-ended unit sub-station

Conductors are often run to these switchboard systems from a *vault* or *tunnel* built under the unit. Instead of using one large, hard-to-handle conductor, several smaller wires or cables with a combined ampacity to match the larger cable are installed. Figure 16.16 shows this type of installation. Note the *crimp-on solderless lugs* being used to terminate the cables.

Grounding.   The *neutral* wire of the 120 V/208 V system is *grounded* to the water supply pipe in much the same way as the residential service. This connection gives 120 V between any one of the live conductors and ground. Check Table 16.1 for sizes of the service ground conductors.

Conductor and Conduit Size

Table 5.5, on page 61, lists the current-carrying capacity of the various copper

FIGURE 16.13 A CGE fusible-type distribution switchboard. Rated 347 V/600 V, 3 phase, 4 wire, with a 1200 A main 3-pole fusible QMR switch feeding through the hydro-metering compartment to the branch PDQ fusible-type switches, all with provision for HRC fusing

FIGURE 16.14 A 3 phase combination kilowatt-hour and demand meter

FIGURE 16.15 A CGE main switchboard, double-ended design. Rated 2 kA, for use on 575 V, 3 phase, 3 wire service incoming. Reduced to operating system of 120 V/208 V, 3 phase, 4 wire by internally mounted, dry-type core and coil transformer

FIGURE 16.16 A CGE "hydro collector box" that feeds a main switchboard. Rated 12 kA

conductors, when no more than three are enclosed in one conduit. The 120 V/208 V system requires *four* conductors. Section 4 of the Canadian Electrical Code states that when four conductors are used in a conduit, the *ampacity* of the conductors listed in Table 5.5 must be reduced to 80% of the current value listed in the table. *Conduit size* for this or any service is determined by using Tables 5.1 and 14.2.

Demand Meters

Industrial users of electricity are assessed for power used in two ways.

Applications of Electrical Construction

The first is the total power consumption, calculated in much the same way as for residential customers.

The second way is a charge based on the *demand factor*, the maximum amount of power drawn from the utility at any given time.

A special indicator on the meter records the demand factor by staying at the highest reading reached during a given period of time. The meter reader then takes the reading and turns the indicator back to zero. The demand factor charge helps ensure that an industry will not place unreasonable loads on the utility's supply system for short periods of time. Such loads force the utility to install large, expensive pieces of equipment just for that industry.

Figure 16.14 illustrates a typical 3 phase combination meter which includes kilowatt-hours as well as a thermal-type demand measurement system. This meter is designed to monitor small industrial loads.

Circuit Breakers

Both circuit breakers and fuses have advantages. This section covers only the advantages of circuit breaker protection.

One of the greatest advantages of the *circuit breaker* is that it can be a *manually operated* switch. If an over-current condition causes the breaker to open the circuit, it can be reset by hand and put in operation again without replacing any parts. The built-in switch mechanism allows the breaker to be used as a control switch for a circuit. Often these breakers are used to control lighting or motor circuits, as well as to provide over-current protection. Circuit breakers are more expensive, but save on material and labour that are needed if separate control switches are installed.

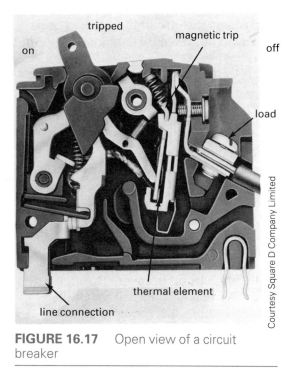

FIGURE 16.17 Open view of a circuit breaker

Courtesy Square D Company Limited

Figure 16.17 shows a circuit breaker.

A circuit breaker has an advantage when a *short circuit* occurs. Since a short circuit reduces the electrical resistance in a circuit to a low value, the current flow in the circuit rises to a high level in microseconds, that is, millionths of a second. As the alternating current rises to a maximum value in its cycle (1/240 s), the circuit breaker senses this drastic increase and opens the circuit safely.

The circuit is opened in two ways. A *heat-sensitive thermal element* (bimetal strip) rises in temperature and triggers the circuit breaker when excessive current flows. Also, a prolonged overload condition will heat up and operate the thermal element. In the sudden overload situation, however, a *latch-on, magnetic trip assembly* quickly operates the circuit breaker. Under short-circuit conditions, the sudden excess in current flow activates this magnetic trip assembly and opens the circuit in microseconds. Fuses do not have this magnetic tripping device.

A third advantage of circuit breakers is that they are much *smaller* in size than fused disconnect switches. Figure 16.18 shows a 30 A, 2 pole fused disconnect. Figure 16.19 shows the much smaller 30 A, 2 pole circuit breaker, which will be mounted in the *distribution centre* shown in the inset. The distribution centre is then mounted in a wall and fitted with a finishing cover once the wall has been covered in. (See Figs.16.20 and 16.21)

Large industrial load centres are built with circuit breakers throughout, and the breaker units are assembled into switchboards. (See Fig. 16.22)

Industrial circuit breakers are made in single-pole, double-pole, and 3 pole units, with voltage and current ratings to match the circuit conductors. Some of the larger units are tripped by *compressed air* to speed up tripping time and help reduce arc damage to the breaker's contacts.

Ground Fault Circuit Interrupters

The ground fault circuit interrupter (*GFI*) is a relatively new device adapted from the basic circuit breaker. Providing what one manufacturer calls *people protection*, it is designed both to prevent dangerous electrical shocks and provide over-current protection. (See Fig. 16.23)

Courtesy Square D Company Limited

FIGURE 16.18   A 30 A fused disconnect switch

Courtesy Square D Company Limited

FIGURE 16.19   A compact circuit breaker for a distribution centre installation

Applications of Electrical Construction

Sometimes a piece of electrical equipment develops a *ground fault*, which is a slight leakage of current from any live conductor to the metal frame or housing of a tool or appliance. A wire pinched between two parts, insulation breakdown, moisture in the equipment, or metal shavings in the unit are a few of the common causes of a ground fault. The higher the voltage of the system, the greater the possibility of a ground fault. The leakage current, however, is often so

FIGURE 16.20 Distribution centre mounted flush in wall

FIGURE 16.21 Distribution centre with finishing cover

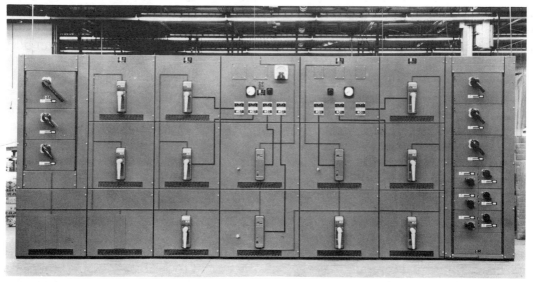

FIGURE 16.22 An industrial switchboard circuit breaker installation

FIGURE 16.23   A ground fault circuit interrupter

TABLE 16.2	Physiological Effects of Electrical Currents
0.001 A	Threshold of sensation
0.005 A	Discomfort and pain
Up to 0.010 A	Severe pain and shock
0.010 A-0.015 A	Local muscle contraction, possible "freezing-to-the-circuit," or being thrown back
0.015 A-0.030 A	Breathing becomes difficult, loss of consciousness possibly resulting
0.050 A-0.100 A	Possible rapid, unco-ordinated contractions of the heart, resulting in loss of synchronism between heart and pulse (ventricular fibrillation)
0.100 A-0.200 A	Ventricular fibrillation of the heart
Over 0.200 A	Severe burns and muscular contractions. Heart is more apt to *stop* than fibrillate.
Over 1 A	Irreparable damage to body tissues

low that it will *not* trip the normal circuit breaker (or blow a fuse). But it is high enough to electrocute, cause serious harm, or give a painful shock to anyone who comes in contact with the faulty equipment. Portable tools often have ground faults and cause many electrical shocks.

Body Resistance

The human body is not a good conductor of electrical current under normal conditions. The amount of moisture present (sweat) on the skin and the muscle structure of the body help determine the resistance to current flow at any given time. Scientific tests have indicated body resistance to be between 1000 Ω and 4000 Ω. The amount of voltage present in the circuit will then determine how well the current will penetrate the skin and flow through the body. The higher the voltage present, the greater the current flow through the body and the greater the effect on vital organs.

Table 16.2 indicates how various amounts of current affect a human body.

Time/Current Factor

The age, general health, and amount of current flow will have an effect on *how long* a person can sustain a shock without serious or permanent damage to the body. Figure 16.24 illustrates how much time and what current combinations are built into a modern GFI unit for the protection of persons using the circuit.

The GFI unit detects leakage currents as low as 2 mA (0.002 A). It then opens the circuit to protect the operator of the equipment.

The average person who receives a

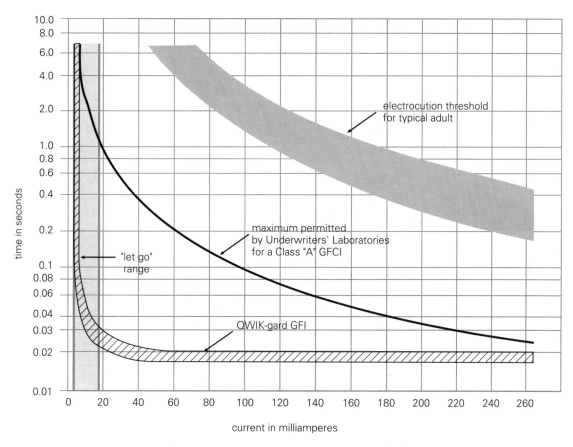

FIGURE 16.24   Time/Current Chart for Ground Fault Circuit Interrupter Devices

shock of 20 mA suffers great pain and loss of muscular control. There is loss of life at approximately 300 mA.

Obviously, the current required to trip a normal 15 A circuit breaker can seriously injure a person. The average power tool equipped with a 3 prong plug and operating on a 15 A circuit is potentially dangerous if the ground prong is removed for any reason.

Operation of the GFI.   The GFI is a self-contained unit that may fit directly in to a distribution centre. (See Figs. 16.21 and 16.23) Other types may require special enclosures.

The GFI operates on the principle

that the current *leaving* a circuit is equal to the current *entering* that circuit (Kirchhoff's Current Law).

Both supply conductors of the circuit pass through a highly developed transformer. When there is *no* leakage current, the magnetic fields around the supply conductors cancel one another. No voltage is produced in the transformer. (See Fig. 16.25)

If a leakage current develops, *more* current is entering the circuit on one supply conductor than is leaving on the other. This *magnetic imbalance* causes a voltage to be induced in to the transformer coils. An amplifier increases the strength of this voltage and uses it to

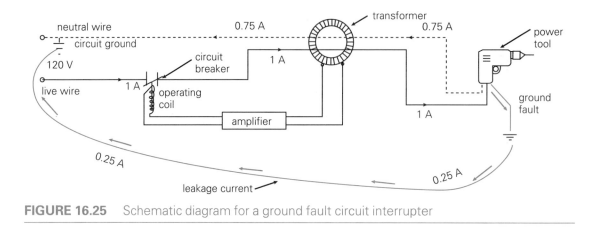

FIGURE 16.25   Schematic diagram for a ground fault circuit interrupter

trip the circuit breaker. Leakage currents as low as 2 mA will trip the GFI breaker.

GFI units are made in a variety of current ratings to suit most circuits. They are recommended for both home and industrial protection. A special (and more sensitive) unit is being made for hospitals. The Canadian Electrical Code

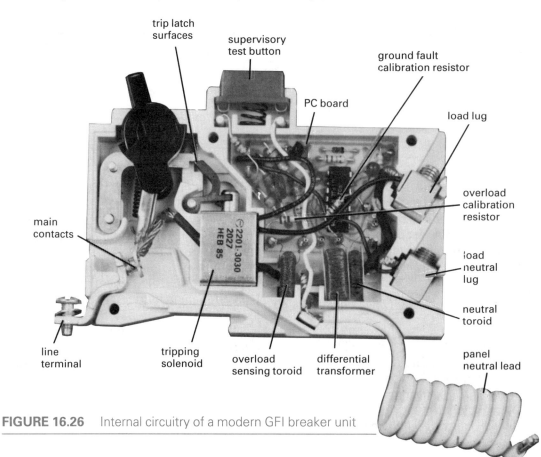

FIGURE 16.26   Internal circuitry of a modern GFI breaker unit

Applications of Electrical Construction

requires that swimming, decorative, or other pools with lighting units be protected with GFIs. Modern residential installations must also use GFIs on all outdoor receptacles, as well as for personal protection in washroom/bathroom areas.

The internal workings of a GFI breaker are not accessible to the installer or breaker user. They are composed of a sensitive electronic circuit which should not be tampered with or adjusted in any way. Figure 16.26 illustrates a typical GFI breaker sensing and monitoring circuit. Figure 16.27 illustrates this GFI unit in the form of a single-pole circuit breaker. Manufacturers frequently test their products to maintain quality and safety. Users of GFI-protected circuits should continue to test their GFI units after installation. Figure 16.28 illustrates a typical test procedure recording chart.

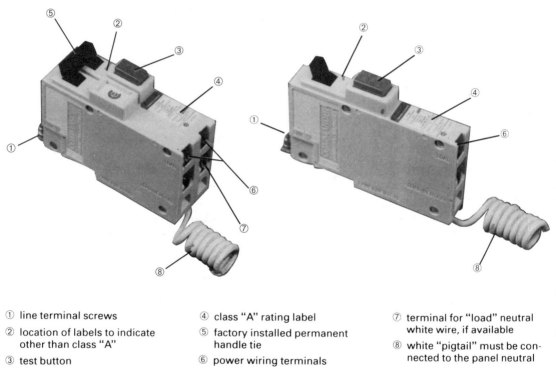

① line terminal screws
② location of labels to indicate other than class "A"
③ test button
④ class "A" rating label
⑤ factory installed permanent handle tie
⑥ power wiring terminals
⑦ terminal for "load" neutral white wire, if available
⑧ white "pigtail" must be connected to the panel neutral

Courtesy Commander Electrical Equipment Inc.

FIGURE 16.27   External view of single- and double-pole GFI breakers

TEST REMINDER

For maximum protection against electrical shock hazard, test your ground fault circuit interrupter at least once a month.

TEST PROCEDURE

1. *Push yellow TEST button. The red RESET button will pop out exposing the word TRIP. Power is now off at all outlets protected by the INTERRUPTER, indicating that the device is functioning properly.*
2. *If TRIP does not appear when testing, do not use any outlets on this circuit. Protection is lost. Call a qualified electrician.*
3. *To restore power, push RESET button. Enter data on record below.*

Month	Jan	Feb	Mar	Apr	May	Jun	Jul	Aug	Sep	Oct	Nov	Dec
19												
19												
19												
19												
19												
19												
19												
19												
19												
19												
19												
19												
19												
19												

FIGURE 16.28 Typical GFI Test Procedure Recording Chart

F o r   R e v i e w

1. What are three advantages of the polyphase, or 3 phase, voltage system in industrial or manufacturing buildings?
2. How is a 3 phase service metered?
3. What is the size of a grounding conductor for a 200 A service? a 400 A service? a 600 A service?
4. List the boxes or enclosures used in a 3 phase service.
5. How is a 3 phase disconnect switch different from a single-phase disconnect?
6. How is the grounding of a 3 phase, 550 V service different from that of a single-phase service?
7. Explain why current transformers are used.
8. What advantage has the 4 wire, 120 V/208 V service over the 3 wire, 550 V or 440 V installation?
9. When installing a 3 phase, 4 wire service, how is the ampacity of the conductors determined?
10. What is a *demand meter*? Why and how is it used?
11. List three main advantages a circuit breaker has over a fuse.
12. List two types of circuit protection provided by circuit breakers.
13. In which two ways does a circuit breaker trip a circuit when under short-circuit conditions?
14. Why is it important to trip a circuit breaker in microseconds when a short circuit occurs?
15. What method is used to trip large industrial circuit breakers? Why?
16. What is a *ground fault*?
17. List three ways in which ground faults often occur.
18. Why is a standard circuit breaker (or fuse) sometimes useless when a ground fault occurs in a circuit?
19. What is the lowest current level at which a GFI in the circuit can operate?
20. Explain in your own words how a GFI unit operates.

Applications of Electrical Construction

17 Fuses

When an electrical current passes through a conductor, some heat is generated. The more current that passes through a conductor, the more heat is generated within the conductor.

The amount of heat produced is proportional to the square of the current. That is, if the amount of current in a conductor is doubled, the heat will be four times as great. If the amount of current in the same conductor is tripled, the heat will be nine times as great.

There are several other factors that affect the amount of heat produced in any given conductor. One is the type of metal used in the conductor. Copper is a better conductor than aluminum, and so will carry more current. Because it can carry more current, a copper conductor of a given size and current load will generate less heat than an aluminum conductor under the same conditions.

The physical size—the *gauge number*—of the conductor also helps determine the current-carrying capacity of the conductor and, therefore, the amount of heat that will be generated under certain loads. Small conductors do not carry as much current as large conductors. Therefore, a small conductor trying to carry too large a current load will generate more heat than a conductor of the right size would.

A third factor is the air surrounding a conductor. Air space has a great deal to do with the conductor's ability to cool or give off its heat. The temperature of the air surrounding the conductor is called the *ambient temperature*. If conductors are crowded into a conduit or box where there is little air circulation, cooling will be difficult, and a higher ambient temperature will result. This ambient temperature will therefore raise the temperature of the conductors. Areas such as boiler rooms and foundries often have problems with conductors because of high ambient temperatures.

A conductor's *insulation* can also affect a conductor's ability to give off heat. Some modern plastic insulations tend to retain the heat in the conductor in the same way that insulation in a building's walls holds in heat. Therefore, the current ratings for these conductors must not be exceeded. If they are, sufficient heat will be generated and retained to melt or damage the insulation. Tables 5.5 and 5.6 list the current-carrying capacities of copper and aluminum conductors. Heat rises to a significant level in these conductors when about 80% of the listed current is passed through the conductor.

Note: Some electrical equipment, such as special circuit breakers and pressure-type switches designed for use with HRC-L fuses, is rated for 100% capacity.

Damage from Overheating

Overheating of conductors results in several problems. One is caused when a conductor reaches a temperature beyond its normal operating range. There is a softening, called *annealing*, of the conductor which removes any resilience (spring-like action) at the terminal and may loosen the connection. (This is much like the *cold flow* of metal experienced with aluminum conductors.)

An increase in temperature also brings about a more rapid *rate of oxidation* on a conductor. Copper is a good conductor, but copper oxide is not. In fact, copper oxide tends to weaken the electrical security of a terminal connection. Aluminum oxide causes an even greater problem. Very close to being an insulator, it will cause further heating at the terminals. The oxide establishes an electrical resistance, which results in a loss of voltage at the terminal. Often, enough heat is generated to completely destroy the terminal connection.

Overheating also causes *insulation* to *dry out* and become hard and brittle. In fact, movement of the conductor will likely cause the insulation to crack and fall off. The exposed conductors may well start a fire if they touch each other or a grounded box. Although it usually takes many months of circuit use to dry out insulation, the problem can easily be overlooked: conductors are seldom checked after installation.

If there is an overload or short-circuit current, conductors can heat up until they actually glow red. When this overheating occurs, the insulation can melt, causing arcing between conductors or between the conductors and ground. A fire can be ignited within the conduit, box, or building wall. Even if the actual damage from the fire is small, the burning insulation gives off an offensive odour that will persist for a long time.

Obviously, the overheating of conductors can have serious effects. The Canadian Electrical Code recognizes this by requiring over-current protection to be provided. One effective way to protect a circuit is with a *fuse*.

Weakest Thermal Link

A *fuse* is a simple, current-sensitive device designed to limit current flow and protect the conductors of a circuit. Protecting the conductors will prevent serious damage to equipment from overload and fire.

A fuse contains a strip of *current-sensitive* metal that has been calibrated to melt and restrict the amount of overload and short-circuit currents for copper or aluminum circuit conductors. The metal used in a fuse can be zinc, copper, silver, or an alloy, depending on the fuse type and use. Industrial fuses have copper or silver links to which a small amount of tin or tin alloy has been added to reduce their melting temperature. This principle is known as the "M" effect. Tin or tin alloy is deposited on the link, close to a restrictive segment. It thus reduces the overload melting temperature at that section of the link. The normal copper or silver used for the links has a melting temperature of approximately 1000°C which will open under short-circuit conditions, but not at the lower, safer temperature range required for protection from overloads.

When a tin alloy is added to the notched segment of a link, a molecular exchange takes place when temperatures nearing the melting point of the tin (230°C) are reached. The copper or silver link and the tin alloy combine to produce a common melting point close to that of the tin. In this way, the fuse can provide protection from *both* short circuits and overloads. Figure 17.1 illustrates a common type of link.

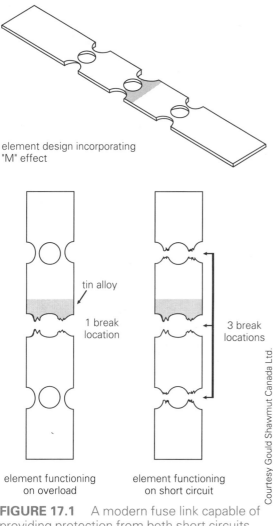

element design incorporating "M" effect

tin alloy

1 break location

3 break locations

element functioning on overload

element functioning on short circuit

Courtesy Gould Shawmut Canada Ltd.

FIGURE 17.1 A modern fuse link capable of providing protection from both short circuits and overloads

Since the link has a higher electrical resistance than the copper or aluminum conductors in the circuit, it will heat up before they do. The fuse will then automatically *open* at the restrictive segment of the link (where the tin has been added) when an overload current is passed through the conductors. Under short-circuit conditions, the fuse link can open (melt) at several notched segments in just a fraction of a second. (See Fig. 17.1)

The fuse is designed to be a "weak link in the chain." Its job is to break the circuit before any damage is done to the circuit conductors. For this reason, it is often called the *weakest thermal link* in the circuit. Although a fuse is *current* sensitive, both ambient temperature and the heat generated in the fuse determine when the fuse link will melt. Some fuses are designed to be more heat sensitive than others. Type *P* (non-time-delay) and type *D* (time-delay) fuses are used frequently in residential circuits. (See Fig. 17.2) Unlike the older fuses with zinc links (see Fig. 17.3), these fuses protect fuse panels and panel boards. In the past, panels and boards tended to be prone to overheating and possibly fire. The older fuses were not designed to react when excessive heat conditions developed in a panel board. Older zinc-link fuses should be replaced by the more protective, thermal-sensitive type P and type D fuses. See Figure 17.4 for examples of type D fuses.

Note: Remember that conductors of electricity are often conductors of heat, and so can transfer their heat on to a fuse.

Time-Delay Fuses. Circuits containing electric motors undergo a surge of current during the motor's starting

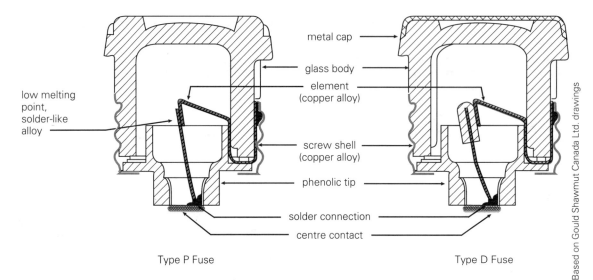

FIGURE 17.2 Type P and type D plug fuses. They provide protection from both short circuits and overloads through their thermal sensitivity.

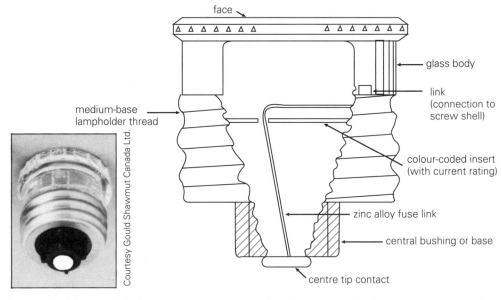

FIGURE 17.3 An older model, zinc type, screw-base plug fuse, available in 3 A, 6 A, and 10 A sizes

FIGURE 17.4 Metal-capped type D plug fuses

period, and the surge continues until the motor reaches operating speed. This sudden increase of current frequently exceeds the current rating of the circuit and could cause the fuse to blow. Air conditioners, fridges, freezers, clothes driers, and other residential, motor-driven equipment can be subject to this condition.

Little danger to circuit conductors is present when this overload lasts only briefly (up to ten seconds). The development of a fuse that would not blow under these temporary overloads but would allow the motor to start up and reach normal running amperes made the need for fuse oversizing in these circuits unnecessary. This fuse is the *time-delay* fuse which provides a much better level of protection for the circuit. Figures 17.2 and 17.4 illustrate this fuse type.

Fuse Ratings. Fuses are rated by the manufacturer in three ways, with their ratings expressed in Root Mean Square (RMS) values. RMS values are those that would be read on a standard voltmeter or ammeter if placed into an alternating current circuit.

The *continuous current rating* is the amount of circuit current the fuse will carry without blowing or interrupting the circuit. For most panel-supply cir-

cuits, this rating should be matched to the current rating of the circuit conductors. It must also comply with Canadian Electrical Code maximums for specific loads, such as motors and transformers, where currents are likely to exceed circuit ampacity.

Maximum AC rated voltage indicates to the installer the type of circuit and voltage conditions under which the fuse can safely operate. In most cases, the higher the voltage, the larger the size (length) of the fuse or distance between the fuse's contact points. Figure 17.8 illustrates differences in physical size between fuses of various voltage and current ratings.

Due to the serious, sustained arcing associated with DC circuits, special designs have been developed for this circuit type. If the DC rating is not marked on the fuse label, the installer should contact the manufacturer about suitability of the fuse on DC circuits.

Interrupting capacity is the amount of current that the fuse can safely interrupt in the circuit under short-circuit conditions. The body of the fuse must remain intact, allowing for replacement. The current flow can be very high when a short circuit occurs, and so the interrupting capacity of the fuse used must match or exceed the short-circuit current from the circuit's source.

Depending on their size, transformers supplying residential areas are capable of delivering up to ten thousand amperes under short-circuit conditions. In commercial and industrial applications, they can deliver up to *hundreds* of thousands of amperes. The Canadian Electrical Code requires that fuses protecting these circuits must be able to safely open the circuits without fuse rupture or damage to their panels or the equipment that contains them.

Screw-Base, or Plug, Fuses

The most common type of fuse used in residential buildings is the *screw-base*, usually called the *plug* fuse. (See Figs. 17.2 and 17.3 on page 252.) Plug fuses are used for all lighting and receptacle circuits operating at a maximum voltage of 125 V. Standard plug fuses are made in current ratings of 3 A, 6 A, and 10 A for the zinc-link types and 15 A, 20 A, 25 A, and 30 A for the P and D types.

Time-delay plug fuses of less than 15 A are available in fractional ampere ratings. Figure 17.2 shows a time-delay fuse in the plug-type configuration. (The section on dual-element fuses in this chapter discusses the cartridge type.)

Plug fuses are all about the same physical size. Older zinc-link types have colour-coded inserts. The inserts are visible through the transparent tops of the fuses to aid in easy recognition of current ratings, even from a distance. In the past, the current rating was often stamped on the contact point at the fuse base. Some manufacturers found that stamping the base could distort the contact point and lead to overheating in the panel. In an effort to prevent these problems, they now take more care when marking these contact points. Many modern type P fuses still use colour-coded inserts under their transparent faces, though; modern type D fuses have colour-coded metal caps over their faces and tops. Blue represents 15 A; orange, 20 A; red, 25 A; and green, 30 A.

For a period of time, a colour-coded base was installed on the fuse to match up with a fuse rejection system that could be installed in a panel's fuse opening. These systems were designed to prevent the placing of a fuse having too large a current rating into the circuit. (See Fig. 17.5) When this situation occurred, the older zinc-link fuse would allow heat to build up far beyond safe levels and not open. Modern type P and type D fuses, which have overload-sensing features built-in, will open the circuit when there is an overload heating condition. However, despite this additional protection, their ampacity should not exceed circuit conductor capacity.

Note: Some manufacturers of panels built rejection features into their panels and did not require the coloured inserts to be installed at a later time.

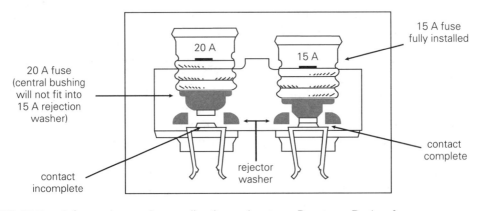

FIGURE 17.5 A fuse rejector ring application using type P or type D plug fuses

Applications of Electrical Construction

FIGURE 17.6   Plug fuse rejection rings

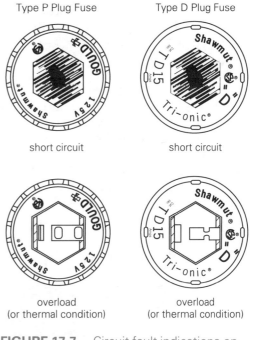

Type P Plug Fuse   Type D Plug Fuse

short circuit   short circuit

overload   overload
(or thermal condition)   (or thermal condition)

FIGURE 17.7   Circuit fault indications on screw-base plug fuses

Fuse rejections rings, or washers, are available for 15 A and 20 A plug fuses. They are sometimes colour coded and are designed to be put into panel fuse sockets. (See Fig. 17.6) The colour code helps a person to choose the right fuse for the circuit, because the fuse colour code matches that of the rejection ring. The ring itself makes it impossible to put in a fuse of the wrong current rating, because the higher the current rating of the fuse, the larger the diameter of the base. For example, using this system makes it impossible to insert a 25 A fuse into a 15 A circuit. In the past, serious electrical problems were caused whenever a person installed a fuse with too large a current rating for the protection of the circuit conductors. The fuse rejection ring helps to prevent this from happening. Modern type P and type D fuses have phenolic rejection tips or bases to co-ordinate with fuse rejection rings. (See Figs. 17.2 and 17.4)

Circuit Fault Indications

The see-through glass body of the plug fuse is a great help for finding what caused a fuse to blow. If an overloaded circuit or a motor starting up has caused the fuse to open, the glass face will still be clear. Only a small part of the link will have melted away and have caused a gap to appear. (See Fig. 17.7) If a short circuit has caused the fuse to open, the inside of the fuse's glass face may blacken. Seeing the remains of the fuse link will be difficult, because most of the link will have melted from the sudden heat. The cause of the problem should be located and corrected before the fuse is replaced.

When a fuse operates close to its rated current value for any length of time, it starts to warm up. The circuits in a distribution panel can be checked simply by running a finger over the faces of the fuses. Any fuse that feels warm is carrying current close to its rated value. On the older zinc-link fuses, the colour-coded insert indicating the current value would look brown or burned. A check to see whether the correct size and type of fuse have been installed and that the circuit has not been overloaded for the size of the conductors would be in order. Modern type P and type D fuses will not brown. Instead, they will open the circuit: they are heat sensitive and respond with this type of protection when the current and heat in the circuit go beyond safe limits. Thus, the conductors and the panel itself are protected. Standard zinc-link fuses would not open the circuit unless the current exceeded the fuse's amp rating. Considerable heat damage could result from the heat buildup being transferred to the panel and the conductors themselves.

Ferrule-Contact Cartridge Fuses

Residential equipment such as stoves, clothes driers, water heaters, and electric heaters operate at 240 V and often need a different type of fuse, usually the 250 V ferrule-contact cartridge fuse. A 600 V unit is also available for industrial use. (See Fig. 17.9)

Within the two voltage ranges there are six physical sizes as determined by ampere rating group. The groups are as follows: 1 A-30 A, 35 A-60 A, 70 A-100 A,

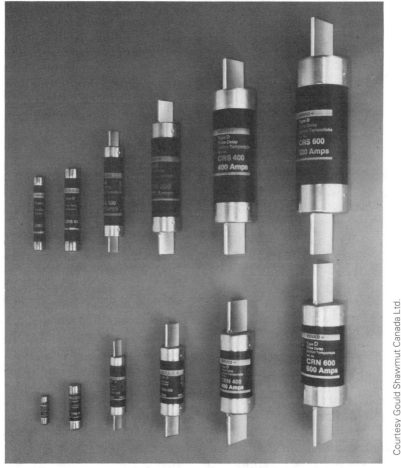

Courtesy Gould Shawmut Canada Ltd.

FIGURE 17.8 Relative fuse sizes for standard code and HRCI-R fuses

Applications of Electrical Construction

FIGURE 17.9 Ferrule-contact cartridge fuses

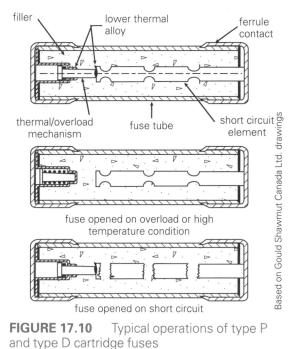

filler

lower thermal alloy

ferrule contact

thermal/overload mechanism

fuse tube

short circuit element

fuse opened on overload or high temperature condition

fuse opened on short circuit

FIGURE 17.10 Typical operations of type P and type D cartridge fuses

110 A-200 A, 225 A-400 A, 450 A-600 A. The larger the ampere rating of the group, the larger the physical diameter of the fuse. (See Fig. 17.8)

The 600 V ferrule-contact code fuse is much larger in physical size than the 250 V unit. When originally developed,

standard code fuses at the 600 V level required greater length and diameter to cope with the high arcing experienced during link openings. Standard code fuses can interrupt a current of 10 kA without rupturing. Modern fuse developments have resulted in more durable (fibreglass or porcelain) bodies and improved fillers within fuses to contain and extinguish any arcs that form during fuse openings. Figure 17.10 illustrates the internal condition of a link after a fuse has opened under short-circuit and overload conditions.

Knife-Blade Cartridge Fuses

When currents in excess of 60 A are flowing in a circuit, the larger knife-blade cartridge fuse is used to protect the conductors. (See Figs. 17.11 and 17.12) The 250 V fuse is used for residential and commercial purposes, and the 600 V knife-blade fuse is used for both industrial and commercial applications.

As with the ferrule-contact cartridge fuse, the 600 V knife-blade fuse is much larger in physical size than the 250 V knife-blade fuse. These fuses are made in four current rating groups: 70A-100 A, 110 A-200 A, 225 A-400 A, and 450 A-600 A. The length and diameter of the fuses

FIGURE 17.11 Cartridge and knife-blade fuses

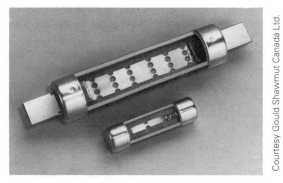

Courtesy Gould Shawmut Canada Ltd.

FIGURE 17.12 Construction of ferrule-contact and knife-blade fuses

in each group increase as the current rating increases. For example, the fuses in the 70 A-100 A group are smaller in length and diameter than those in the 110 A-200 A group.

Arc-Quenching Material

Many of the large cartridge fuses are filled with an arc-quenching material. This arc-extinguishing gypsum powder or silica sand is designed to quickly extinguish an arc and to reduce the damaging short-circuit current to zero. Arc-quenching material is therefore important. Figure 17.12 shows both the fuse and the arc-quenching material contained within. The fuses in Figure 17.19 on page 262 also show arc-quenching material.

One-Time Fuses

The common one-time fuse has for many years been made with a zinc alloy link, with an interrupting capacity of 10 kA (standard code fuse). The zinc links in some fuse applications suffered from metal fatigue at the restrictive segments (cutouts) due to the constant expansion and contraction of the metal in the link. Loads that were constantly cycling on and off, such as electric heaters,

freezers, fridges, etc., would eventually bring the link to a point where it would open the circuit for no apparent reason. The circuit had no fault or problem other than the metal fatigue in the fuse link. One reason why copper is now being used for the links in many modern fuses is that it is far less likely to suffer from metal fatigue.

If a one-time fuse is used and a short circuit or overload condition causes the fuse to blow, the entire fuse unit has to be replaced. Special one-time fuses are available with an interrupting capacity of 50 kA. This fuse has a copper alloy link and provides much added protection for a small increase in cost. These fuses are available in the 1 A-600 A and 250 V/600 V, standard code sizes. P types are available at 250 V, from 15 A to 60 A.

Renewable-Link Cartridge Fuses

In specific industrial or training installations, fuses often need to be replaced. *Renewable link cartridge* fuses accept replacement links and can be installed quickly. (See Figs. 17.13 and 17.14) These easily disassembled units are known for the ease and low cost with which their links can be replaced. The fuses themselves do not need to be replaced.

Safety Note: Be sure that the link is not oversized and do not double up on links (a practice known as *spiking*). In other words, allow the fuse to provide its intended level of protection. Since renewable-link cartridge fuses have neither an interrupting capacity higher than 10 000 A nor arc-quenching material, they should not be considered for use on circuits where higher currents than that may be encountered.

Applications of Electrical Construction

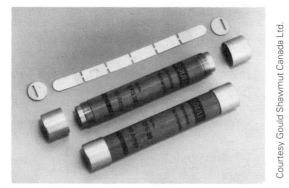

Courtesy Gould Shawmut Canada Ltd.

FIGURE 17.13 A renewable-link cartridge fuse

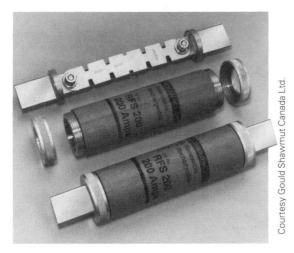

Courtesy Gould Shawmut Canada Ltd.

FIGURE 17.14 A renewable-link knife-blade fuse

High-Rupture Capacity Fuses

Section 14 of the Canadian Electrical Code states that standard code cartridge fuses must not be used in circuits that have a current in excess of 600 A and a voltage in excess of 600 V.

Another consideration is that fuses must have ratings appropriate for the handling of anticipated short-circuit currents. *High-rupture capacity (HRC)*

fuses provide good protection from both overloads and short circuits; as noted earlier in this chapter, modern distribution systems often deliver short-circuit currents far above the ability of standard code fuses to interrupt safely.

HRC fuses can be used with motors when a time-delay feature is required, as well as with other high current-consuming equipment, for example, lighting loads, transformers, etc. They provide rapid protection from short circuits and can also provide needed time delays during the moderate (temporary) overloads experienced when motors or transformers are first turned on. HRC fuses also do not deteriorate. Most have moisture-proof fibreglass or ceramic barrels, filled with high-quality silica sand for arc-quenching. Under heavy current, short-circuit conditions, the silica sand quickly turns into a glass-like material, blocking any chance of an arc forming between fuse ends. The copper links and silver- or tin-plated, copper contact blades add to the HRC fuse's overall quality. (See Fig. 17.15)

HRC fuses are designed to interrupt large short-circuit currents without rupturing and to protect cables and equipment well. They are available in two basic types. The Form I and Form II fuses are known as HRC-I and HRC-II, respectively. Electrical and physical differences exist between the two types.

HRCI (Form I).   These fuses provide protection from both overloads and short circuits. They have a high interrupting capacity (200 kA) and are made in a variety of types and classes for use in protecting cables and equipment.

HRCII (Form II).   This kind of fuse provides protection from short circuits

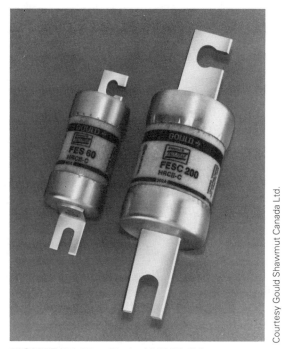

FIGURE 17.15   Typical HRCII-C fuses

Courtesy Gould Shawmut Canada Ltd.

only and must be used in conjunction with some other type of overload device (most often an overload relay). Although it has a 200 kA interrupting capacity, it must never be used to replace an HRCI fuse. An HRCI fuse can, however, be used to replace an HRCII fuse, since it provides both types of protection.

The common types of HRC fuses are as follows.

HRCII-C This HRCII fuse is of British origin, with a bolt-in design. It is available with electrical ratings of 600 V and 2 A-Ĝ00 A. It is generally used as a backup, short circuit device for motor circuits, and often used along with combination motor starters or other overload devices. (See Fig. 17.15)

Class H These fuses, of the HRCI type, had the same basic dimensions as standard code cartridge fuses of the same current and voltage ratings. The 1 A-60 A and 70 A-600 A types had rejection tabs or notches in their contact areas to prevent the installation of 10 kA standard code fuses in panels or equipment. They were produced in both the 250 V and 600 V versions with current ratings from 1 A-600 A. Fast-acting and time-delay units were also developed. However, Class H fuses are now considered obsolete and should be replaced by HRCI-R fuses.

HRCI-J These fuses are designed for use in modern equipment and have superior current limiting ability. Their specific compact dimensions prevent them from being interchanged with any other type of fuse. These fuses are made in the 1 A-60 A ferrule-contact type and the 70 A-600 A bolt-in/blade type, all with voltage ratings of 600 V. The fast-acting type is used for the protection of feeder circuits and for providing a needed short-circuit protection for circuit breakers. (See Fig. 17.16) HRCI-J time-delay types are available for both the NEMA and the smaller IEC types of motor/control contactors. They can also be used to protect transformers.

HRCI-R The HRCI-R fuse has overall dimensions similar to those of the standard fuses covered earlier in this chapter. A replacement for the Class H fuse, it has a special rejection feature —a groove on the ferrule (1 A-60 A), or a "U" shaped notch on the knife-blade contact of the larger (70 A-600 A) type. The rejection feature is found on one end of the fuse only.

The fuses are produced in 250 V and 600 V sizes with current ratings from 1 A-600 A, including fractional

Applications of Electrical Construction

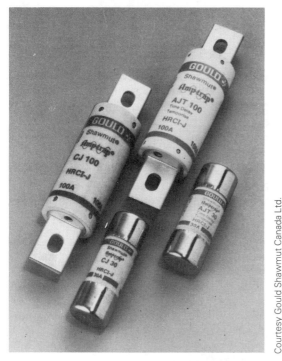

FIGURE 17.16 Typical HRCI-J fuses

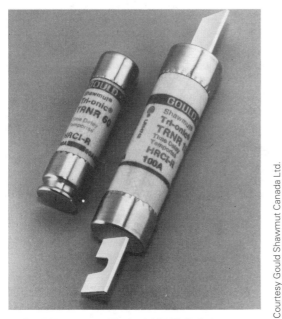

FIGURE 17.17 Typical HRCI-R fuses

Courtesy Gould Shawmut Canada Ltd.

Courtesy Gould Shawmut Canada Ltd.

ratings for time-delay fuses. The fast-acting type serves as a replacement for the protection of feeder circuits and as backup protection for circuit breakers; the time-delay type is used for motor circuit protection and for circuits where a high momentary inrush of current can be expected. (See Fig. 17.17)

HRC-L These larger, specific dimension fuses are rated at 601 A-6000 A at 600 V. They are specially designed to protect the main power supplies to large industrial complexes and apartment buildings, where services larger than 600 A are required. Commerce Court in Toronto, Ontario, is an example. (See Fig. 17.18) HRC-L fuses, including time-delay ones, are used to protect circuits feeding large motors such as chillers or air-conditioning units in high-rise apartment or office buildings. Since any loose contact with large, high ampere fuses will soon cause tremendous heat to develop at the point of contact, HRC-L fuses are of the bolt-in type: firm, positive connections with fuse panels are thus ensured and damage to the fuses and panels avoided.

Dual-Element Fuses

When referring to a fuse, "dual-element" is often confused with "time-delay." *Dual-element* is a manufacturer's term for describing the construction of the fuse link or element within the fuse body. Dual-element fuses can be made in either time-delay or non-time-delay types. All dual-element fuses do, however, provide protection from both short circuits and overloads by the use of two individual components on the same element (link). A copper element is normally used for the link part of the fuse, with restrictive notches or segments

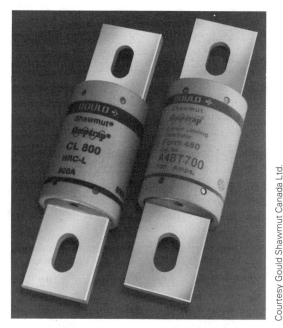

FIGURE 17.18 Typical HRC-L fuses

providing protection from short circuits. (See Fig. 17.19) One end of the element is attached to the fuse contact by a low-thermal alloy and a *plunger* mechanism. (See Fig. 17.20) The mechanism provides protection from overload by sensing an above normal temperature on the fuse element, melting the low-thermal alloy, and allowing a spring within the mechanism to open the circuit. The time required for this reaction serves as the time-delay feature on type D fuses. The thermal sensitivity of such fuses covers the requirements of a residential code cartridge fuse. Dual-element fuses are produced in both ferrule-contact and knife-blade contact types, in current ranges from 1 A - 600 A and in both 250 V and 600 V configurations.

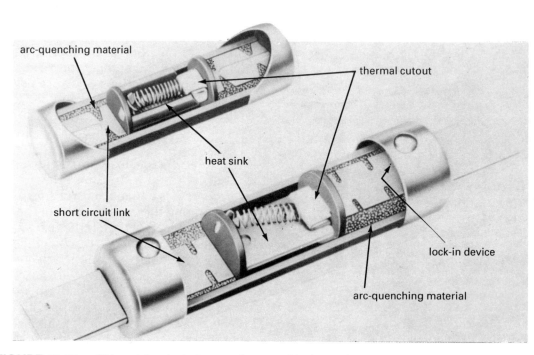

arc-quenching material

thermal cutout

heat sink

short circuit link

lock-in device

arc-quenching material

FIGURE 17.19 Older style, dual-element fuses, with thermal cutouts in centre sections

Applications of Electrical Construction

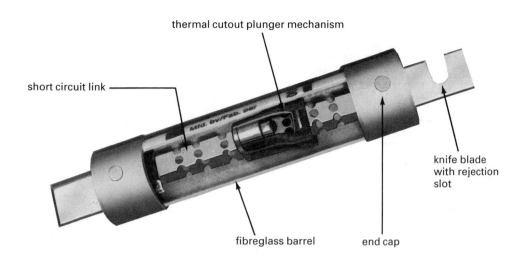

thermal cutout plunger mechanism

short circuit link

knife blade with rejection slot

fibreglass barrel

end cap

Courtesy Gould Shawmut Canada Ltd.

FIGURE 17.20 New style, dual-element fuse, with a plunger mechanism as a thermal cutout

Time-Delay

The term *time-delay* is recognized by the Canadian Standards Association (CSA) to refer to a specific *time-current overload blowing characteristic*. It generally means that a fuse, such as an HRC or standard code fuse, will carry up to 500% of its rated ampere capacity for a minimum of ten seconds. These fuses must then be marked by the manufacturer as *time-delay* or *type D*. Plug fuses designed as time-delay fuses must also be marked type D and be capable of carrying 200% of rated current for a period of twelve seconds.

Such fuses are ideal for motor circuits. Unlike standard non-time-delay fuses which often blow while motors are reaching their operating rpm, time-delay fuses can accommodate high motor-starting currents. These currents are three to five times stronger than a motor's normal running current but last only a few seconds. The time-delay characteristic allows motor-circuit fuses to be sized lower, providing better circuit protection, demanding less space within a panel or fuse box, and permitting lower equipment costs. The maximum ampere rating for these fuses, as permitted by the CSA, is 175% of motor-running current (full load); a non-time-delay fuse may be rated at 300% of the motor's full-load running current.

Fuse-rating terminology found on HRC fuses can be seen in Figure 17.21. Figure 17.22 illustrates the normal starting and running current for a 20 A motor: after approximately ten seconds the

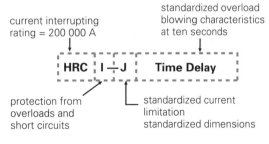

current interrupting rating = 200 000 A

standardized overload blowing characteristics at ten seconds

HRC I–J Time Delay

protection from overloads and short circuits

standardized current limitation
standardized dimensions

FIGURE 17.21 Fuse-rating terminology found on an HRC (high rupture capacity) fuse

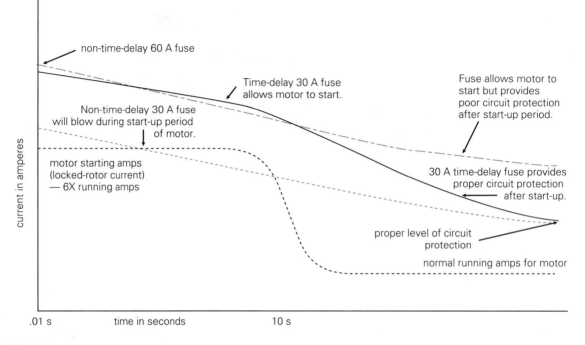

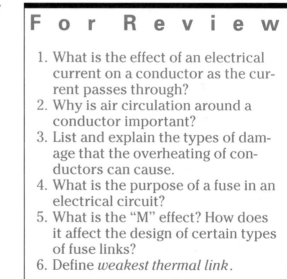

FIGURE 17.22 Time-current characteristics for a 20 A motor using time-delay and/or non-time-delay types of fuse protection

starting current drops to the much lower running current. The graph also shows that the 30 A, time-delay fuse provides the necessary delay in opening the circuit. The fuse allows the motor to start up while protecting the conductors under normal operating conditions. The 60 A non-time-delay fuse has the same starting characteristics but, unlike the 30 A time-delay fuse which would require a 30 A switch, it would need a larger, more costly 60 A switch.

For Review

1. What is the effect of an electrical current on a conductor as the current passes through?
2. Why is air circulation around a conductor important?
3. List and explain the types of damage that the overheating of conductors can cause.
4. What is the purpose of a fuse in an electrical circuit?
5. What is the "M" effect? How does it affect the design of certain types of fuse links?
6. Define *weakest thermal link*.

7. List three different types (configurations) of fuses and give their current and voltage limitations.

8. What is a fuse rejection ring? What is its purpose and how is it used?

9. List the circuit faults that can cause a fuse link to open. How does each fault show on the element?

10. Describe a simple method for deciding whether a fuse is operating close to its rated current value.

11. Why are some cartridge fuses filled with a powder?

12. What is the main danger of using a renewable-link cartridge fuse?

13. What is *spiking*, and why should it be avoided when replacing a renewable-link fuse element?

14. Describe the HRC fuse. What is its main advantage over the standard cartridge fuse?

15. Explain what time-delay fuses are and what kind of protection they provide.

16. How many types or classes of HRC fuses are there? Name them.

17. What is the purpose of rejection tabs or notches in the contact ends of HRC fuses?

18. What protection beyond that provided by HRC fuses do type P and type D fuses offer?

19. Name the type of HRC fuse that provides protection from short circuits only.

20. Name the types of HRC fuse that provide protection from both overloads and short circuits.

21. Why are large (high) ampere fuses bolted into the fuse panel rather than held in place by spring clips?

18

Residential Electric Heating

As energy costs rise, persons who design and size electrical heating systems are using more complex methods to determine the proper size of heat source for a given building. To produce accurate heat loss and gain figures for a variety of residential buildings requires many tables, charts, and formulae not suited to the purpose of this text. Specialized heat loss/gain courses are available through local electrical leagues and supply authorities, and successful completion of such courses can lead to certification as a heat loss designer/consultant. This chapter will introduce some of the principles involved in heat loss calculation by presenting a simplified calculation method. It is not intended to replace or meet the more complex standards required by heating industry professionals.

Electrical heating systems for residential buildings are discussed here. They have several advantages over systems using fossil fuels such as oil or natural gas.

Advantages of Electrical Heating

Electrical heating systems are produced in a variety of types to match building structure requirements. Central heating systems, capable of replacing existing gas or oil furnaces, are available in forced air and hydronic (hot water) configurations. They are especially useful when a building is already equipped with duct work or radiators. Central air conditioning is an option for an electric furnace when a ducted, forced air system is selected for the building. (For the unitary heating system described next, central air conditioning is *not* possible unless special duct work is installed.)

A different and major type of electric heating is the *unitary* or baseboard heating system which provides independent temperature control for each room or area to be heated. A *thermostat*, which is placed in the room, permits this separate control. In the event of equipment failure, heat from adjoining rooms will flow in to the one with the defective heater unit. Heating systems that use furnaces as a heat source, on the other hand, let the entire building cool off when equipment fails.

A second major advantage of the unitary system is the saving of space in the basement area. The area previously required for the furnace and associated duct work and plumbing can be put to good use by the home-owner, and the "finishing off" of a basement area can be simplified. When converting from an oil heating system, additional space saving is achieved by the removal of the oil storage tank. Furthermore, the house does not need a chimney for the removal of fumes and hot gases because none are created. All heat produced by the system is directed into the building itself, providing 100% efficiency.

Electrical heating systems have other advantages: they are noise-free, odour-free, do not give off combustible fumes, and are clean to operate. Improvements in building standards and construction materials should mean that recently designed houses are well built, making the installation of electrical heating systems even more cost-effective.

Costs of Electrical Heating

Heat is often calculated in terms of joules. One watt is one joule per second (1 W = 1 J/s).

Local utilities that supply electrical energy usually agree that the electrically produced heat unit is one of the more expensive ones. When comparing the cost of heating systems, however, other factors must be taken into account. A house designed with an electrical heating system must be of high-quality construction. Windows, doors, and other places where air can enter or leave must be properly fitted, sealed with caulking compound, and equipped with storm or double glass window units. In some communities there are heating inspectors who check all such installations. Their inspections are in addition to those made by the electrical inspection authority.

The fact that the owner of an electrically heated house does not have to pay for a furnace and duct work and their maintenance costs must also be considered. Capital is freed to be spent on the individual heating units required for each area of the home. For houses that already have duct work, some heating manufacturers make electric furnaces that take advantage of it, which allows the customer to convert to electrical heating with a minimum of inconvenience and expense. (In some cases, though, a customer may need to increase the size of the main service equipment to handle the increase in electrical load.)

The average consumer who is using electrical energy to heat a house uses much more electricity than would otherwise be used. As a result, the *end rate* (lower cost to consumer) is reached much sooner, and both heating and lighting energy is obtained at the cheaper end rate. The end rate is also applied to cooking and water-heating units.

Electrical energy is in good supply, which is not always the case with other forms of energy. The advantage of availability should be considered.

When a cost analysis is prepared and all factors are taken into account, a properly installed electrical heating system is usually found to be competitive in terms of price with other forms of heating.

Insulation

The main function of insulation is to trap *dead air* (air that is stationary) between fibres or cells. Dead air retards heat from escaping. Heat travels *from* hot *to* cold (that is, from inside the house to outside). Storm doors and windows simply trap dead air in the space between the

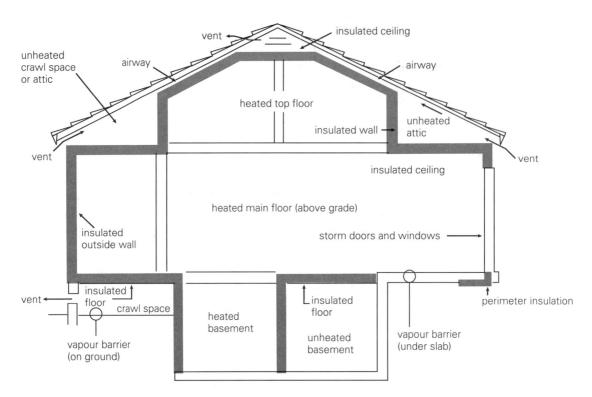

FIGURE 18.1 Installing residential thermal insulation. (*Note*: All walls, ceilings, roofs, and floors separating heated from unheated spaces should be insulated.)

layers of glass. It is this *dead air layer* that is responsible for saving heat.

Insulation for the electrically heated house must be properly located (see Fig. 18.1), be of high quality and be properly installed. It must also conform to a minimum thermal resistance as shown in Tables 18.1, 18.2 and 18.3. Table 18.1 lists the minimum insulation requirements for various parts of a building.

The *RSI value** of insulation is assigned to a product by its manufacturer. It refers to the product's heat-retaining ability. The higher the RSI

*The insulation industry is adopting metric RSI values; however, imperial R values are still evident. To change RSI values to imperial R values, multiply by 5.7. For example, RSI-4.9 equals R-28.

TABLE 18.1 The Canadian Building Code's Minimum Insulation Requirements for Buildings	
Building Element Exposed to the Exterior or to Unheated Space	**RSI Value Required**
Ceiling below attic or roof space	5.40
Roof assembly without attic or roof space	3.52
Wall other than foundation wall	2.11
Masonry or concrete foundation wall	1.41
Frame foundation wall	2.11
Floor, other than slab-on-ground	4.40
Slab-on-ground containing pipes or heating ducts	1.76
Slab-on-ground not containing pipes or heating ducts	1.41

Applications of Electrical Construction

TABLE 18.2 Thermal RSI Values for Various Insulating Products		
Description of Insulation	**Per mm**	**Per in.**
Mineral wool and glass fibre	0.0208	2.99
Cellulose fibre	0.0253	3.65
Vermiculite	0.0144	2.08
Wood fibre	0.0231	3.33
Wood shavings	0.0169	2.44
Sprayed asbestos (health hazard)	0.0201	2.90
Expanded polystyrene complying with CGSB 51-GP-20M (1978)		
1	0.0257	3.71
2 bead board	0.0277	3.99
3	0.0298	4.30
4 extruded	0.0347	5.00
Semi-rigid glass fibre sheathing	0.0305	4.40
Rigid glass fibre roof insulation	0.0277	3.99
Natural cork	0.0257	3.71
Rigid urethane or isocyanurate board	0.0420	6.06
Mineral aggregate board	0.0182	2.62
Compressed straw board	0.0139	2.00
Fibreboard	0.0194	2.80
Phenolic thermal insulation	0.0304	4.38

value, the better the heat-retaining ability of the product. Table 18.2 lists a variety of insulating products and their RSI values. Once a wall, ceiling, or other part of a building has been completed, all of the products used in the building's construction add to that section's heat retention. The Canadian Building Code has set minimum standards for the various parts of a building, and these can be seen in Table 18.3.

There are several forms of insulation: batts, rolls, loose insulation which is poured or blown, and rigid sheets (slabs).

Batts. Blanket insulation in this form, packaged for shipping in large plastic enclosures, can be seen in Figure 18.2. It comes in a variety of lengths, widths, and RSI values and is designed to form a pressure fit between studs and joists or similar framing members. The RSI value of a batt depends on its thickness—the thicker the batt, the higher the value. A form of vapour barrier, for example, polyethylene film, must be used on the warm side of the batt to prevent moisture infiltration into the fibreglass and the reduction of the batt's RSI value over a period of time. (See Fig. 18.5 on page 273.)

Rolls. This form of blanket insulation can be seen in Figure 18.3. Produced in long lengths, it is available in a variety of thicknesses, widths and RSI values. Like the batt, it relies on a friction fit. This product is used wherever unobstructed runs of insulation are possible: floors and basement walls are two examples. (See Fig. 18.6 on page 273.)

Poured or Blown Insulation. This type of insulation is made of loose, nodulated wood, vermiculite, or cellulose. Its RSI value depends on its thickness, as

TABLE 18.3 Minimum RSI Values for Various Assemblies in a Building as Listed by the Canadian Building Code (W/m²·°C)

Building Assembly	Maximum Number of Celsius Degree Days	
	Up to 5000	Above 5000
Exposed walls	3.0	3.4
Exposed roof or ceiling		
—frame	5.6	6.4
—solid	3.0	3.4
Foundation walls		
—frame	3.0	3.4
—solid	1.5	1.5
Exposed floors		
—frame	4.7	4.7
—solid	3.0	3.4
Slab-on-ground at grade		
—unheated	1.3	1.7
—heated	1.7	2.1

Notes on Table 18.3:
(1) "Exposed" means exposed to outdoor temperature or unheated area.
(2) "Solid" means brick concrete blocks or concrete.
(3) "Frame" means a wood or steel stud frame to which interior and exterior cladding is applied.
(4) The RSI value shown for slab-on-ground at grade is for rigid insulation.
(5) Slab-on-ground at grade: "heated" means a concrete floor containing heating ducts or pipes; "unheated" means a concrete floor not containing heating ducts or pipes.

(a) Friction Fit Batts

Based on a Fiberglas Canada Inc. illustration

(b)

Value		Nominal Thickness		Standard Widths		Standard Lengths	
RSI	R	mm	in.	mm	in.	mm	in.
1.4	8	65	2.5	381, 584	15, 23	1219	48
1.7	10	89	3.5	381, 584	15, 23	1219	48
2.1	12	89	3.5	381, 584	15, 23	1219	48
2.4	14	89	3.5	381, 584	15, 23	1206	47.5
3.5	20	152	6	381, 584	15, 23	1219	48
4.9	28	202	8	406, 610	16, 24	1219	48
5.4	31	222	8.75	406, 610	16, 24	1219	48
6.1	35	251	9.87	406, 610	16, 24	1219	48
7.0	40	265	10.37	406, 610	16, 24	1219	48

FIGURE 18.2 Batt insulation is available in a wide variety of RSI values, thicknesses and widths. (*Note*: Some dimensions appear in millimeters rather than in any larger metric unit because the insulation industry uses millimetres.)

(a) Friction Fit Rolls

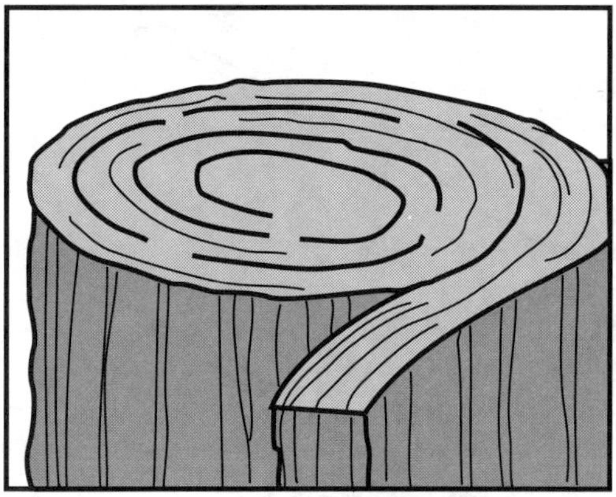

Based on a Fiberglas Canada Inc. illustration

(b)

Value		Nominal Thickness		Standard Widths		Standard Lengths	
RSI	R	mm	in.	mm	in.	m	ft.
1.4	8	70	2.75	381, 584	15, 23	23	75.5
1.7	10	89	3.5	381, 584	15, 23	17	55.8
2.1	12	89	3.5	381, 584	15, 23	17	55.8

FIGURE 18.3 Roll insulation is well suited to areas where long open runs exist between joists or studs.

indicated in Figure 18.4, and it is sold by the bag. Poured or blown insulation is usually installed by an insulation contractor using a fibre-blowing machine. It is most suited for horizontal spaces such as attics or roof areas.

Home-owners often rely on this form of insulation to boost the RSI value of their existing insulation. To achieve this end, they need to be sure to choose a material that will maintain its RSI value over the years. Cellulose insulation is a form of paper product (similar to newsprint) that is chemically treated (usually with dry chemicals) to provide some moisture- and fire-proofing. This type of insulation, when dry, is a good insulating material. Like many other paper products, however, the cellulose insulation has a tendency to absorb moisture after it has been installed, thus reducing the RSI value drastically. Due to its compara-

tively low initial cost, many home-insulating companies select this material for residential use. But if the moisture-proofing chemicals are not properly applied and adequate ventilation in the soffit and peak areas allowed for, the installation of this material is virtually a waste of time and money. Figure 18.7, on page 273, illustrates poured or blown insulation installed in an attic area.

Rigid. This type of insulation is produced in slab or sheet form and is made from *polystyrene* or *polyurethane* foam. Both foams are excellent insulating products and vapour barriers: moisture is unable to penetrate them. Figures 18.8 and 18.9, both on page 274, illustrate rigid glass fibre boards, with available thicknesses, widths, and RSI values listed.

(a)

Based on a Fiberglas Canada Inc. illustration

(b)

Value		Min. Thickness		Max. Net Coverage		Min. Bags per Net	
RSI	R	mm	in.	m²/bag	ft.²/bag	100 m²	1000 ft.²
3.5	20	191	7.5	5.9	63	17	16
4.9	28	267	10.5	4.2	45	24	22
5.3	30	286	11.25	3.9	42	25.5	24
5.6	32	305	12	3.7	39	27.5	25.5
6.0	34	324	12.75	3.4	37	29	27
6.3	36	343	13.5	3.3	35	31	28.5
6.7	38	362	14.25	3.1	33	32.5	30
7.0	40	381	15	2.9	31	34	32
8.8	50	470	18.5	2.4	26	42	39

FIGURE 18.4 A nodulated form of glass fibre insulation is frequently blown in attic areas to upgrade the RSI value in older buildings.

These slabs are attached to masonry walls by an adhesive. The adhesive is necessary in commercial buildings where no wood framing is present. Installation of a protective surface of 13 mm gypsum board or similar fire-resistant product over the polystyrene insulation is also necessary. Polystyrene is combustible and gives off a dangerous gas when ignited.

Vapour Retarders

Insulation traps dead air only when it is dry and free from moisture. If the insulation is damp, it will conduct heat out of the building quickly. The greatest threat is from excess moisture inside the building trying to escape outside through the walls. Moisture from cooking, bathing (showers), and people must be removed

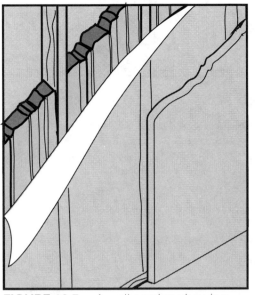

Based on a Fiberglas Canada Inc. illustration

FIGURE 18.5 A wall section showing friction-fit batt or roll insulation installed beneath a vapour barrier and plaster board

Based on a Fiberglas Canada Inc. illustration

FIGURE 18.6 Installation of friction fit rolls in open run areas. (*Note*: Be sure to wear a protective mask and protective clothing.)

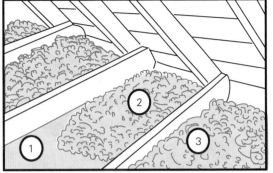

Courtesy Ontario Electrical League

Step 1. Place air-vapour barrier at bottom of space between joists. Lap to ensure that there is a good barrier against the flow of air and vapour.

Step 2. For required thermal resistance value, pour or blow insulation between joists to depth recommended by manufacturer. Take care insulation does not cover soffit or under-eave ventilators.

Step 3. Rake or smooth insulation to ensure equal thickness over entire area. To obtain proper settled density and thickness, refer to CMHC acceptance listing for manufacturer's product.

FIGURE 18.7 Procedure for installing loose insulation

safely. Otherwise, it will condense on the walls, form drops, and run down, which often damages wall surfaces.

To keep insulation reasonably dry, the vapour retarding barrier must be installed on the warm side of the insulation, that is, between the plaster or panelling of the wall and the insulation itself. A continuous, 0.2 mm polyethylene film is fastened over the insulation prior to covering the wall with plaster board or a similar finish. Figure 18.10 illustrates the proper location of a vapour retarding barrier.

When installing the film, do not puncture it in any way and be sure to secure it properly with staples. In this way you will prevent air flow through tiny openings or cracks. Air tends to find any weak spot, such as the hole in a balloon, regardless of the spot's location.

(a)

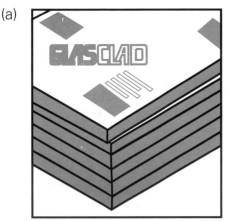

Glass Fibre, Rigid Board on an Above-Ground Wall

Exterior Insulating Sheathing

(b)

Value		Nominal Thickness		Standard Widths		Standard Lengths	
RSI	R	mm	in.	mm	in.	mm	in.
0.77	4.4	25	1	1219	48	2438, 2743	96, 108
1.18	6.7	38	1.5	1219	48	2438, 2743	96, 108

FIGURE 18.8   Glass fibre, rigid board used in conjunction with batt insulation to increase the RSI value of walls

(a)

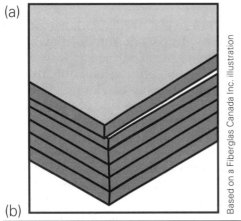

Air flow will be surprisingly high there and the air may be full of moisture.

Once the house is properly sealed with vapour-retarding film, artificial ventilation means, such as vents and fans, are needed to exhaust the moist air that accumulates in the home daily. These measures are so important that some localities will call for an insulation inspection to ensure that the insulation and vapour-retarding film have been

(b)

Value		Nominal Thickness		Standard Widths		Standard Lengths	
RSI	R	mm	in.	mm	in.	mm	in.
1.5	8.5	50	2	1220	48	2440	96
2.3	13	75	3	1220	48	2440	96

FIGURE 18.9   Rigid fibreglass board for insulation on exterior of basement walls, down to the weeping tile level

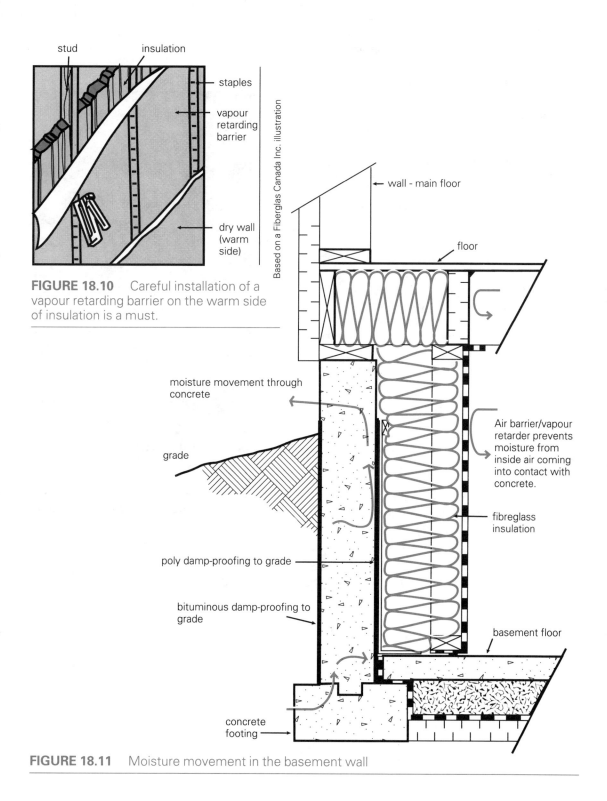

stud   insulation

staples

vapour
retarding
barrier

Based on a Fiberglas Canada Inc. illustration

dry wall
(warm
side)

FIGURE 18.10   Careful installation of a
vapour retarding barrier on the warm side
of insulation is a must.

wall - main floor

floor

moisture movement through
concrete

grade

Air barrier/vapour
retarder prevents
moisture from
inside air coming
into contact with
concrete.

fibreglass
insulation

poly damp-proofing to grade

bituminous damp-proofing to
grade

basement floor

concrete
footing

FIGURE 18.11   Moisture movement in the basement wall

properly installed and not damaged by other building trade areas.

Remember basement walls because if they are insulated there will be a higher degree of comfort in the home. These walls must be protected from moisture produced both inside and outside the building, however. Figure 18.11 illustrates a basement wall section and the methods used to prevent dampness from entering and spoiling the wall's insulation.

Ventilating Devices

Ventilating fans are placed over stoves and in bathrooms to remove excess moisture. They also, however, remove some of the heat from the house, and so

method A

method B

Methods of providing adequate air-flow ventilation into attic areas

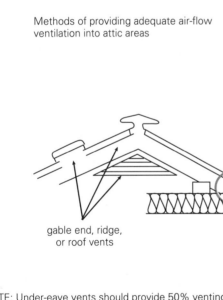

gable end, ridge, or roof vents

under-eave vents

Courtesy Ontario Electrical League

Based on Fiberglas Canada Inc. Illustrations

NOTE: Under-eave vents should provide 50% venting area.
Gable end, ridge, or roof vents should provide other 50% venting area.

FIGURE 18.12   Attic ventilation. Provide a minimum of 1 m² free (unobstructed) ventilating area for each 300 m² insulated ceiling area. Cathedral-type and low-pitched roofs require 1 m² ventilation for each 150 m² insulated attic area. Distribution of vents must provide cross-ventilation in attic from end to end and from top to bottom.

should not be used continuously. These fans are an important part of the heating installation and are looked for by electrical heating inspectors.

Attic insulation must also be kept moisture-free. Having a system of vents placed in various parts of the attic will achieve this. One square metre of vent is required for each 150/300 m² of attic area. (See Fig. 18.12) The circulation of air tends to keep the insulation dry and moisture-free. Remember that a cold attic during the winter indicates that the insulation is indeed working and keeping heat in the house.

Baseboard Heater

One of the most common types of electrical heating unit is the baseboard heater. It consists of an enclosed heating element, fitted with heat-radiating *metal fins* and supported in a metal frame. It operates on the principle that air heated by the element will rise and flow out of the top of the frame, while cooler air will be drawn from the floor area into the bottom of the heater. This process is called *convection of air*. (See Fig. 18.13)

Baseboard heaters are rated in both wattage and voltage. The most common unit for residential use is the *240 V* heater, simply because a 240 V heater of a given wattage requires half as much current as a 120 V heater of the same wattage. The need for extra large conductors in the heater circuits is reduced.

Baseboard heaters are made in *low* and *standard wattage densities*. A 1000 W heater in a *low-watt* density will be approximately 1.8 m long and low enough in surface temperature that it may be installed safely under drapes and other *heat-sensitive* materials. It is suitable for use under heat-sensitive synthetic fabrics, which shrink, lose

their shape, or discolour when exposed to heat for a prolonged period of time.

A 1000 W heater in a *standard watt* density will be approximately 1.2 m long and reach a higher surface temperature. The heat is more concentrated, and so this type of heater should *not* be used under or near delicate fabrics. It is excellent, however, for use in areas where space is limited and higher surface temperatures are not a problem.

Draperies should be hung with the nearest fold at least 5 cm away from the heater and 4 cm off the floor. Proper air circulation through the heater is then possible.

Baseboard heaters are produced in a variety of *wattage ranges*, starting as low as 250 W and going as high as 3000 W. Manufacturers provide specifications for their own products.

Installation of Baseboard Heaters.

These heaters should be mounted at the finished floor level to allow the cool air to enter the heater easily. Also, they must be mounted on a flat surface, without bending or distorting the heater frame. If the frame is distorted in any way, the heater will produce some noise as it expands and contracts during its heat cycles.

The heater is secured to the wall with several *wood screws*. It needs nearly no maintenance, except for an occasional vacuuming to remove lint collected on the heater fins. Lint slows the convection of air through the heater and lowers its efficiency. (See Figs. 18.14 and 18.15)

Radiant Heating for Ceilings

Another type of electrical heating unit is the *radiant-heating ceiling cable*. This

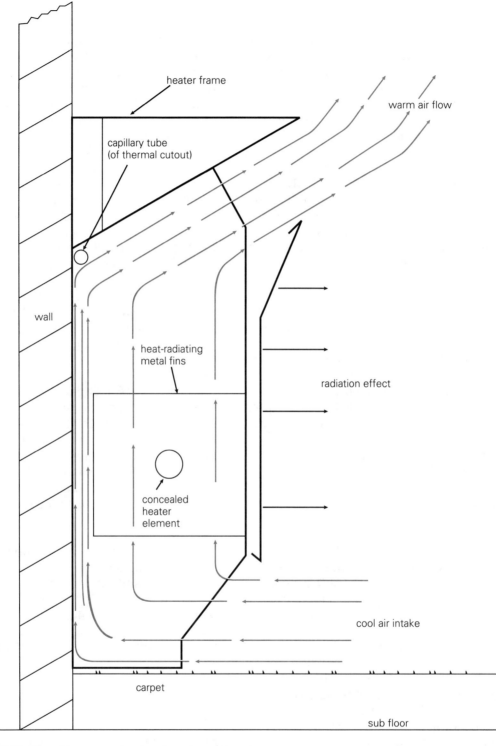

FIGURE 18.13 Air convection in a baseboard heater

FIGURE 18.14 A baseboard heater with a built-in thermostat

FIGURE 18.15 A baseboard heater with an air conditioner switch and receptacle attachment

slender, insulated wire must be carefully installed on the ceiling before plaster is applied. It is sold in calibrated lengths, with wattage and voltage ratings stamped on each cable reel or container. (See Fig. 18.16) A tag or label on the ends of the cable further reminds the installer of the wattage and voltage rating for that cable length.

The cable must not be cut or shortened in any way, but used as it comes from the supplier on the voltage indicated on the cable reel. Any installation-site change in the length of the heating cable will alter the wattage output and current flow through the cable. Any attempt to splice the cable where it has been shortened might establish a failure point and cause considerable expense and inconvenience after the cable has been covered over with plaster.

Recent developments in the heating industry have led to the introduction of

FIGURE 18.16 Reel of radiant-heating ceiling cable

a new product, *radiant-heating foil*. This consists of thin metal heating elements which are electrically insulated and waterproof embedded in strong plastic laminates. The product is available in a variety of widths, lengths, and wattages. It is easier to install than the ceiling cable and requires no special tools or equipment.

Any ceiling heating system radiates its heat energy into a room, warming any heat-absorbing object present. The entire ceiling, therefore, becomes a heat source. Objects in the room start to give away their heat to the surrounding air and room temperature becomes even and comfortable.

Installation of a Ceiling Cable. If a layer of drywall plaster has been installed first, the ceiling cable can be *stapled* directly to the drywall with special staples. Take care not to staple through the cable or damage it in any way. The cable should be kept a minimum of 20 cm away from ceiling light fixtures and about 15 cm from walls. Taking these safety measures will prevent driving a hook or fastener into the cable when hanging drapes or new light fixtures.

If a building has poured concrete walls and ceilings or a similar masonry construction, the ceiling cable will need *hanger strips*. These plastic strips are run across the ceiling at each end of the room, secured to the ceiling with adhesive, and then fitted with the heating cable. The cable is looped from end to end. The plastic supporting strips provide the proper spacing between the cable runs. Cables should be kept at least 5 cm apart, depending on cable spacing calculations. Manufacturers will provide the proper spacing requirements for their cables. Often, adhesive tape is needed at intervals in a long room to prevent sagging of the cables before plastering.

Each cable set comes with a length of heavily insulated conductor at each end of the cable. Such a length is called a *cold lead*. Cold leads are run from the

ceiling down to an electrical box on the wall where the thermostat and/or power supply is located. The length of these leads should *not* be changed.

To ensure that the cable is not nicked or damaged with a trowel, sand-based *gypsum plaster* must be installed in the ceiling carefully. The plaster must be allowed to dry fully before the cable is energized. Take care to check the cable for continuity both before and after plastering. Using the cable to dry the plaster will only result in shrinking and/or cracking of the plaster around the heating cable.

Once the plaster is dry (allow at least a week), the cable can be energized and allowed to heat the room. Increase the temperature setting of the thermostat slowly, and do not pass the midpoint setting of the thermostat for at least two weeks. Taking this safety precaution is advisable because the plaster may not be thoroughly dry in all ceiling areas.

Installation of Radiant-Heating Foil. The foil is rolled out to the proper length as supplied by the manufacturer, without cuts or splices made to it. The unit is then stapled to a joist, using the securing strips built into its edge. Staples should be applied at 30 cm or 40 cm (12 in. or 16 in.) intervals. Electrical connections are made in a clear-cover connection box for easy inspection prior to covering with the ceiling material. (See Fig. 18.17)

Specific Heat Loss Areas

Windows are major heat loss areas. However, proper caulking around window frames will stop infiltration of cold air and drafts and save heat over a period of time.

Applications of Electrical Construction

FIGURE 18.17 Installation of radiant-heating foil

Courtesy Chromalox Inc.

For many years, it was common practice to install wooden or aluminum storm windows over existing window openings in older homes. Aluminum-framed windows were neat and easily removed for cleaning and maintenance. They also did not require painting and did not swell or jam from moisture intake. The aluminum frame, however, was a good conductor of heat and passed this heat from the building's interior to the outside. Wood-framed windows required more maintenance, were heavier to remove and install, and were often difficult to open and close after a few years. But the wooden frame did retain heat much better than an aluminum frame.

Now, with window developments keeping pace with other design improvements in the building industry, *double* and *triple pane* window units are available. The hermetically sealed panes of glass create a dry air space between the panes. They make heat retention possible and the inconvenient removal, cleaning and installation of storm window units unnecessary.

A further design improvement is specially treated glass which can be purchased as part of double and triple pane window units. The glass limits outward heat flow in the winter and inward heat flow during the summer even more than the standard double or triple pane units.

Some window units have aluminum frames designed to support the glass without being in actual contact with it. These frames are low in maintenance and can be obtained with sections that open for cleaning or fresh air entry.

Modern window units may also have wooden frames. In many cases, these frames are covered with a plastic material to eliminate moisture intake and the need for painting the wood.

Fireplaces are attractive and give off much heat when operating. When not operating, however, they allow a great deal of heat to be drawn out of the room, up the flue, and out the chimney. A brisk wind across the chimney will speed heat removal from the room. To prevent this, the fireplace *damper* should always be well-fitted and kept closed when the fireplace is not in use. A set of well-fitting doors to cover the fireplace's opening will further aid heat retention when no fire is burning in the unit. (See Fig. 18.18) Many types of fuel-efficient fireplace inserts are available for installation in new or existing fireplaces. These modern products enhance a room's appearance and help heat the home when they are in use. They also provide superior protection from heat loss when not in use.

Another possible cause of heat loss is carelessly installed insulation around *pipes and electrical boxes* within walls and ceiling areas. Cold air will infiltrate these weak spots; therefore, extra care should be taken to ensure a proper fit around boxes and plumbing pipes. As

FIGURE 18.18 A fireplace should have a tight damper and a glass screen for effective heat retention.

FIGURE 18.19 When installing plumbing and wiring, particular attention should be given to the installation of insulation. Insulation should completely fill any cavities.

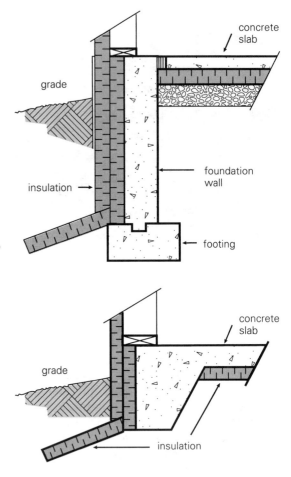

FIGURE 18.20 Insulation on the outside face of a slab on grade

mentioned in Chapter 7, special plastic forms are available to prevent air leakage around wiring boxes. Figure 18.19 illustrates the insulation required around such a box.

A house built on a *concrete slab* or *pad* is vulnerable to heat loss, and requires placement of a rigid exterior type of insulation around various sections of the slab and its footings. (See Fig. 18.20) The insulation will prevent heat from radiating out of the slab and into the cool earth.

Applications of Electrical Construction

Basement Heating

Residential basements are given special treatment. For the purposes of heat loss calculation, the basement is divided into two parts. (See Fig. 18.21)

In a basement with a 2.4 m (8 ft.) ceiling, the area from the ceiling down to grade level is called the *above grade* portion. The rest of the basement is called the *below grade* portion.

Provincial building codes often

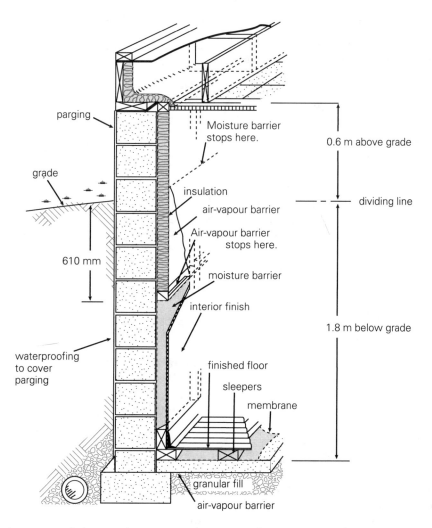

parging

Moisture barrier stops here.

0.6 m above grade

grade

insulation

air-vapour barrier

dividing line

Air-vapour barrier stops here.

610 mm

moisture barrier

interior finish

1.8 m below grade

waterproofing to cover parging

finished floor

sleepers

membrane

granular fill

air-vapour barrier

NOTE: The basement wall above grade must meet the same requirements as any other exterior wall exposed to weather. The wall below grade should be moisture-proofed on the inside. Thermal insulation should be applied on exposed wall to a distance of 610 mm below grade (with an air-vapour barrier). Vapour barriers and damp-proofing must be in accordance with requirements detailed in Residental Standards Canada, sections 12E(3), 20E and 20F.

FIGURE 18.21 Basement wall treatment

require basement walls that are 50% above grade to have a minimum of RSI-2.1 value insulation installed over the entire wall. Basement walls that expose less than 50% of their surface to above-grade temperatures must be insulated (RSI-2.1) to a depth of 61 cm below grade level.

Energy-conscious home-owners, however, may wish to insulate the complete basement wall, thus reducing heat loss and lowering energy costs for this area of the home. If there is any chance of water leaking into the basement area, the insulation should be kept approximately 30 cm from the floor to prevent the insulation from soaking (wicking) up the water.

A check of local building codes is advisable to determine the exact amount of below-grade insulation required in a given area of the country.

Often, rigid insulation is fastened to the above-grade walls with an adhesive. Wooden strapping and RSI-1.4 value batts can also be installed for heat retention. Sometimes, small amounts of heating cable are placed in the floor to help the main heaters warm up the area.

Basic Heat Loss Calculations

The First Step: *Heat-Retaining Walls and Ceilings.* Walls and ceilings exposed to outdoor temperatures must be examined to determine their ability to resist the transmission of heat from inside the building to the outside. The ability of various building products and their installation techniques to resist the flow of heat will have a considerable effect on the heat loss for a given wall or ceiling. Figure 18.22 shows a wall cross-section and the RSI numbers

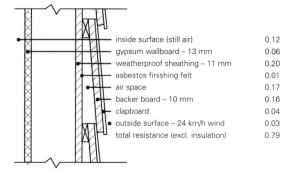

inside surface (still air)	0.12
gypsum wallboard – 13 mm	0.06
weatherproof sheathing – 11 mm	0.20
asbestos finishing felt	0.01
air space	0.17
backer board – 10 mm	0.16
clapboard	0.04
outside surface – 24 km/h wind	0.03
total resistance (excl. insulation)	0.79

FIGURE 18.22 Wall section RSI values

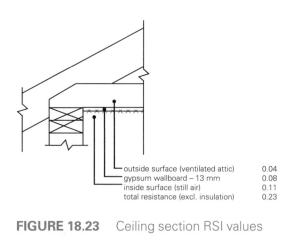

outside surface (ventilated attic)	0.04
gypsum wallboard – 13 mm	0.08
inside surface (still air)	0.11
total resistance (excl. insulation)	0.23

FIGURE 18.23 Ceiling section RSI values

(resistance to heat flow) for each of the materials used. Figure 18.23 shows a ceiling installation and its RSI values. The total RSI value in each figure represents the heat-retaining ability of the materials used without the aid of insulation.

Table 18.3, on page 270, lists the minimum thermal resistance required for the various building assemblies in a house, regardless of the type of heating system used. Table 18.1, on page 268, provides the minimum thermal resistance of insulation that must be installed in each building assembly. Table 18.2 lists a variety of insulation products that can be used to establish the minimum

Building Material	Thickness (mm)	Resistance Factor	Heat Loss Factor (W/m²·C)
TABLE 18.4 Resistance Values for Various Building Materials			
Insulating fibreboard sheathing	25	0.419	2.38
	15	0.262	3.82
	13	0.209	4.78
	13	0.056	17.85
Gypsum board	25	0.220	4.54
Plywood	13	0.110	9.09
Hardwoods (Maple, oak, etc.)	25	0.160	6.25
Softwoods (Pine, fir, spruce, etc.)	25	0.220	4.54
Western cedar (18% moisture content)	25	0.274	3.64
Loose fill insulation Macerated paper *(Cellulose fibre)	25	0.628 to 0.704	1.59 to 1.42
Mineral wools (32 to 80 kg/m³ density)	25	0.586	1.71
Vermiculite (Expanded mica) 112 kg/m³ density	25	0.366	2.73
Air spaces (in walls)	20 to 100	0.171	5.85
Foamed styrene (Density 26 kg/m³) (See manufacturer's specifications)	25	0.608 to 0.704	1.64 to 1.42
Polyurethane	25	1.04 to 1.06	0.96 to 0.94
Flooring (Hardwood)	20	0.120	8.33
Brick (Clay or shale)	100	0.060	16.66
Concrete block (Sand and gravel, 3 oval cores)	100	0.125	8.00
	200	0.195	5.13
	300	0.225	4.44
Cinder block	100	0.195	5.13
	200	0.303	3.30
	300	0.333	3.00

Values shown above are taken from the Acceptable Building Material Systems & Equipment—Central Mortgage and Housing Corporation.

*Illinois Institute of Technology

thermal resistance as listed in Table 18.1. Additional information on the quality and RSI value of the insulation can be obtained from the manufacturers on request.

For comparison purposes, the RSI value of various building products can be located in Table 18.4. More complete tables can usually be obtained from the local hydro utility, electrical league, or building product supplier.

The Second Step: *Heat Loss Through Walls and Ceilings.* The second step in determining the amount of heat needed for a house is to calculate the heat lost through the walls and ceilings. To do this, the RSI total of the wall or ceiling must be mathematically converted to a *U factor*. The U factor (overall co-efficient of heat transfer) is the amount of heat transmitted through a heat barrier in one hour per square metre of surface for each 1°C of temperature difference between the inside and outside of the barrier.

Table 18.5 lists outside temperatures experienced in selected Canadian locations, along with the number of degree days below 18°C. These figures are based on 2.5% of the January outdoor temperatures recorded in those areas of the country.

For example, an exposed wall listed in Table 18.3 must have an RSI value of at least 3.0 when its insulation and all assembly products are considered. The U factor equals one third, or 0.333 W/m² for each 1°C of *design temperature difference*.

If the house with this wall were in the Toronto region of Ontario, the outdoor design temperature would be −18°C, and the degree days would be 4082 (under 5000 as required by Table 18.3). If the occupants want an indoor temperature of 23°C, the design temperature differ-

ence would be 23°C − (−18°C) = 41°C.

If the U factor for a 1°C temperature difference is 0.333 W/m², then the heat loss for a 41°C temperature difference will be 41 × 0.333 W/m², which equals 13.65 W/m² or approximately 14 W/m².

The heat lost through the ceilings is calculated in much the same way as heat lost through the walls.

To find the total heat loss through walls of a room, the total area of wall exposed to cold air must be calculated. To do so, multiply the height of the room by the length of the exposed wall. (Remember that a corner room has two exposed walls.) Figure 18.24, on page 290, shows a wall with a gross area 2.4 m high times 4.6 m long, which is about 11.04 m². Since the window occupies 0.72 m² of space, the net exposed wall is about 10.3 m². The amount of heat loss through this wall in an hour is equivalent to 14 W/m² × 10.3 m², which is 144.2 W.

The Third Step: *Heat Loss Through Doors and Windows.* Table 18.6, on page 291, lists the heat loss per square metre through various types of door and window installations for each degree of design temperature difference. If the room shown in Figure 18.24 has a wood-framed window, that is, single-glazed with a storm, the heat loss will be 2.90 W/m² of window for each degree of design temperature difference. The total heat loss through the glass will be as follows. The area of the window equals 0.6 m × 1.2 m, which is 0.72 m². Therefore, the heat loss per degree Celsius will be 0.72 m² × 2.90 W/m², which is 2.09 W/m². With a temperature difference of 41°C, the heat loss will be 41 × 2.09 W/m², which is 85.7 W. This heat loss is added to the 144.2 W lost through the exposed wall.

Applications of Electrical Construction

TABLE 18.5 Design Data for Selected Locations in Canada

Province and Location	Design Temperature 2.5% °C	Degree Days Below 18°C
British Columbia		
Abbotsford	−10	3150
Agassiz	−13	2960
Alberni	− 5	3180
Ashcroft	−25	4060
Beatton River	−37	7010
Burns Lake	−30	5720
Cache Creek	−25	4080
Campbell River	− 7	3200
Carmi	−24	5210
Castlegar	−19	3747
Chetwynd	−35	5890
Chilliwack	−12	2970
Cloverdale	− 8	3030
Comox	− 7	3203
Courtenay	− 7	3250
Cranbrook	−27	4762
Crescent Valley	−20	4320
Crofton	− 6	3140
Dawson Creek	−36	5890
Dog Creek	−28	5110
Duncan	− 6	3200
Elko	−28	4900
Fernie	−29	4980
Fort Nelson	−40	7063
Fort St. John	−36	6119
Glacier	−27	5730
Golden	−28	4950
Grand Forks	−20	4050
Greenwood	−20	4520
Haney	− 9	3280
Hope	−16	3150
Kamloops	−25	3756
Kaslo	−23	4110
Kelowna	−17	3680
Kimberley	−26	4890
Kitimat Plant	−16	4110
Kitimat Townsite	−16	4130
Langley	− 8	2980
Lillooet	−23	4130
Lytton	−19	3220
Mackenzie	−35	5950
McBride	−34	5720
McLeod Lake	−35	5720
Masset	− 7	3720
Merritt	−26	4190
Misson City	− 9	2980
Montrose	−17	4080
Nakusp	−24	4130
Nanaimo	− 7	3010
Nelson	−20	3920
New Westminster	− 8	2930
North Vancouver	− 7	3090
Ocean Falls	−12	3520
100 Mile House	−28	4900
Osoyoos	−16	3530
Penticton	−16	3514
Port Alberni	− 5	3180
Port Hardy	− 5	3661
Port McNeill	− 5	3480
Powell River	− 9	2900
Prince George	−33	5388
Prince Rupert	−14	4117
Princeton	−27	4560
Qualicum Beach	− 7	3250
Quesnel	−33	4940
Revelstoke	−26	4073
Richmond	− 7	2920
Salmon Arm	−23	4090
Sandspit	− 6	3650
Sidney	− 6	3090
Smithers	−29	5290
Smith River	−46	7610
Squamish	−11	3140
Stewart	−23	4710
Taylor	−36	5890
Terrace	−20	4430
Tofino	− 2	3250
Trail	−17	3650
Ucluelet	− 2	3250
Vancouver	− 7	3007
Vernon	−20	4040
Victoria	− 5	3076
Williams Lake	−31	5105
Youbou	− 5	3360
Alberta		
Athabasca	−35	6280
Banff	−30	5719
Barrhead	−34	6000
Beaverlodge	−35	5820
Brooks	−32	5290
Calgary	−31	5345
Campsie	−34	6010
Camrose	−33	5720
Cardston	−30	4830
Claresholm	−31	5120
Cold Lake	−36	6450
Coleman	−31	5120
Coronation	−31	5906
Cowley	−31	5150
Drumheller	−31	5570
Edmonton	−32	5589
Edson	−34	5910
Embarras Portage	−41	7490
Fairview	−38	6170
Fort Saskatchewan	−32	5890
Fort Vermilion	−41	7170
Grande Prairie	−36	6145
Habay	−41	7050
Hardisty	−33	5950
High River	−31	5320
Jasper	−32	5532
Keg River	−40	6820
Lac La Biche	−35	6140
Lacombe	−33	5740
Lethbridge	−30	4718
McMurray	−39	6778
Manning	−39	6600
Medicine Hat	−31	4874
Peace River	−37	6424
Penhold	−32	5845
Pincher Creek	−32	5010
Ranfurly	−34	5980
Red Deer	−32	5700
Rocky Mountain House	−31	5550
Slave Lake	−36	6220
Stettler	−32	5590
Stony Plain	−32	5780
Suffield	−32	5360
Taber	−31	4750
Turner Valley	−31	5700
Valleyview	−37	6110
Vegreville	−34	6000
Vermilion	−35	6140
Wagner	−36	6180
Wainwright	−33	6000
Wetaskiwin	−33	5670
Whitecourt	−35	6130
Wimborne	−31	5620
Saskatchewan		
Assiniboia	−32	5340
Battrum	−32	5400
Biggar	−34	5890
Broadview	−34	6080
Dafoe	−36	6360
Dundurn	−35	5840
Estevan	−32	5542
Hudson Bay	−37	6470
Humbolt	−36	6280
Island Falls	−39	7100
Kamsack	−35	6290
Kindersley	−33	5710
Loydminster	−35	6280
Maple Creek	−31	5180
Meadow Lake	−36	6550
Melfort	−37	6390
Melville	−34	6170
Moose Jaw	−32	5400
Nipawin	−38	6550
North Battleford	−34	6050
Prince Albert	−37	6562
Qu'Appelle	−34	6060
Regina	−34	5920
Rosetown	−33	5860
Saskatoon	−35	6077
Scott	−34	6260
Strasbourg	−34	5890
Swift Current	−32	5482
Uranium City	−44	8210
Weyburn	−33	5720
Yorkton	−34	6239
Manitoba		
Beausejour	−33	5830
Boissevain	−32	5610
Brandon	−33	6037
Churchill	−39	9213
Dauphin	−33	6150
Flin Flon	−38	6780
Gimli	−34	6030
Island Lake	−36	7210
Lac du Bonnet	−34	5950
Lynn Lake	−40	7820
Morden	−31	5490
Neepawa	−32	5950
Pine Falls	−34	6000
Portage la Prairie	−31	5890
Rivers	−34	5940
St. Boniface	−33	5830
St. Vital	−33	5830
Sandilands	−32	5890
Selkirk	−33	5890
Split Lake	−38	7880
Steinbach	−33	5830
Swan River	−36	6280
The Pas	−36	6852
Thompson	−42	7930

Place	Temp	Value
Transcona	−33	5830
Virden	−33	5890
Whiteshell	−34	5950
Winnipeg	−33	5889

Ontario

Place	Temp	Value
Ailsa Craig	−17	3980
Ajax	−20	4080
Alexandria	−24	4580
Alliston	−23	4520
Almonte	−26	4740
Ansonville	−33	6220
Armstrong	−39	6892
Arnprior	−27	4800
Atikokan	−34	6040
Aurora	−21	4300
Bancroft	−27	4960
Barrie	−24	4470
Barriefield	−22	4240
Beaverton	−24	4580
Belleville	−22	4190
Belmont	−17	3980
Bowmanville	−20	4130
Bracebridge	−26	4800
Bradford	−23	4410
Brampton	−19	4200
Brantford	−17	3920
Brighton	−21	4240
Brockville	−23	4300
Brooklin	−20	4240
Burks Falls	−26	5070
Burlington	−17	3700
Caledonia	−17	3920
Cambridge	−18	4130
Campbellford	−23	4410
Camp Borden	−23	4470
Cannington	−24	4580
Carleton Place	−25	4690
Cavan	−22	4470
Centralia	−17	3940
Chapleau	−35	5950
Chatham	−16	3530
Chelmsford	−28	5290
Chesley	−19	4240
Clinton	−17	4130
Coboconk	−25	4740
Cobourg	−21	4190
Cochrane	−34	6230
Colborne	−21	4190
Collingwood	−22	4580
Cornwall	−23	4470
Corunna	−16	3810
Deep River	−29	5180
Deseronto	−22	4080
Dorchester	−18	4030
Dorion	−33	5890
Dresden	−16	3700
Dryden	−34	6080
Dunbarton	−19	4030
Dunnville	−15	3810
Durham	−20	4620
Dutton	−16	3750
Earlton	−33	5866
Edison	−34	6000
Elmvale	−24	4580
Embro	−18	4130
Englehart	−33	5950
Espanola	−25	5070
Exeter	−17	4080
Fenelon Falls	−25	4690
Fergus	−20	4610
Fonthill	−15	3700
Forest	−16	3830
Fort Erie	−15	3590

Place	Temp	Value
Fort Frances	−33	5830
Gananoque	−22	4240
Georgetown	−19	4250
Geraldton	−35	6550
Glencoe	−16	3810
Goderich	−16	4190
Gore Bay	−23	4910
Graham	−37	6470
Gravenhurst	−26	4740
Grimsby	−16	3580
Guelph	−19	4220
Guthrie	−24	4520
Hagersville	−16	3920
Haileybury	−32	5830
Haliburton	−27	4920
Hamilton	−17	3710
Hanover	−19	4350
Hastings	−23	4470
Hawkesbury	−25	4800
Hearst	−34	6500
Honey Harbour	−24	4580
Hornepayne	−37	6580
Huntsville	−26	4760
Ingersoll	−18	4030
Jarvis	−16	3860
Jellicoe	−36	6450
Kapuskasing	−33	6366
Kempville	−25	4540
Kenora	−33	5932
Killaloe	−28	4940
Kincardine	−17	4240
Kingston	−22	4266
Kinmount	−26	4800
Kirkland Lake	−33	6150
Kitchener	−19	4110
Lakefield	−24	4630
Lansdowne House	−39	7110
Leamington	−15	3560
Lindsay	−24	4580
Lions Head	−19	4350
Listowel	−19	4630
London	−18	4068
Lucan	−17	4030
Mailand	−23	4300
Markdale	−20	4690
Martin	−36	6330
Matheson	−33	6220
Mattawa	−29	5340
Midland	−23	4580
Milton	−18	4080
Milverton	−19	4520
Minden	−26	4850
Mississauga	−18	3810
Mitchell	−18	4400
Moosonee	−36	6931
Morrisburg	−23	4410
Mount Forest	−21	4755
Muskoka Airport	−26	4837
Nakina	−35	6540
Napanee	−22	4130
Newcastle	−20	4130
New Liskeard	−32	5830
Newmarket	−22	4350
Niagara Falls	−16	3740
North Bay	−28	5318
Norwood	−24	4520
Oakville	−18	3640
Orangeville	−21	4650
Orillia	−25	4610
Oshawa	−19	4130
Ottawa	−25	4673
Owen Sound	−19	4220

Place	Temp	Value
Pagwa River	−34	6330
Paris	−17	4030
Parkhill	−16	3980
Parry Sound	−24	4620
Pembroke	−28	4960
Penetanguishene	−23	4580
Perth	−25	4520
Petawawa	−29	5010
Peterborough	−23	4520
Petrolia	−16	3750
Picton	−21	4080
Plattsville	−18	4130
Point Alexander	−29	5180
Porcupine	−34	6220
Port Burwell	−15	3810
Port Colborne	−15	3640
Port Credit	−18	3700
Port Dover	−15	3830
Port Elgin	−17	4240
Port Hope	−21	4190
Port Perry	−22	4410
Port Stanley	−15	3810
Prescott	−23	4350
Princeton	−17	4030
Raith	−35	6060
Red Lake	−34	6220
Renfrew	−27	4790
Ridgeway	−15	3590
Rockland	−26	4800
St. Catharines	−16	3550
St. Marys	−18	4130
St. Thomas	−16	3850
Sarnia	−16	3840
Sault Ste. Marie	−25	5180
Schreiber	−35	6070
Seaforth	−17	4240
Simcoe	−17	3962
Sioux Lookout	−34	6180
Smiths Falls	−25	4520
Smithville	−16	3920
Smooth Rock Falls	−34	6280
Southampton	−17	4250
South Porcupine	−34	6220
South River	−27	5180
Stirling	−23	4340
Stratford	−18	4300
Strathroy	−17	3920
Streetsville	−18	4080
Sturgeon Falls	−27	5180
Sudbury	−28	5447
Sundridge	−27	5120
Tavistock	−18	4190
Thamesford	−18	4030
Thedford	−16	3860
Thunder Bay	−31	5746
Tillsonburg	−17	3920
Timagami	−30	5570
Timmins	−34	6189
Toronto	−18	4082
Trenton	−21	4116
Trout Creek	−27	5230
Trout Lake	−38	7680
Uxbridge	−22	4450
Vanier	−25	4690
Vittoria	−15	3860
Walkerton	−18	4160
Wallaceburg	−16	3620
Waterloo	−19	4110
Watford	−16	3810
Wawa	−35	5640
Welland	−15	3640
West Lorne	−16	3750
Whitby	−20	4080

Location			Location			Location		
White River	−39	6380	Matane	−24	5400	Campbellton	−26	5100
Wiarton	−18	4412	Mégantic	−27	5280	Chatham	−24	4884
						Edmundston	−27	5340
Windsor	−16	3590	Mont Joli	−24	5353			
Wingham	−18	4240	Mont Laurier	−29	5340	Fredericton	−24	4699
Woodstock	−18	4100	Montmagny	−25	5340	Gagetown	−23	4490
Wyoming	−16	3810	Montreal	−23	4471	Grand Falls	−27	5250
			Montreal Nord	−23	4470	Moncton	−22	4709
Quebec						Oromocto	−23	4740
Acton Vale	−24	4690	Mount Royal	−23	4470			
Alma	−30	5830	Nitchequon	−38	7880	Sackville	−21	4590
Amos	−34	6300	Noranda	−33	6220	Saint John	−22	4771
Ancienne Lorette	−25	5110	Outremont	−23	4470	St. Stephen	−22	4580
Arvida	−29	5740	Percé	−22	5290	Shippigan	−22	5180
						Woodstock	−26	4770
Asbestos	−29	4800	Pierrefonds	−23	4470			
Aylmer	−25	4740	Pincourt	−23	4520	**Nova Scotia**		
Bagotville	−31	5776	Plessisville	−26	5120	Amherst	−21	4580
Baie Comeau	−27	5981	Pointe Claire	−23	4470	Antigonish	−20	4580
Beaconsfield	−23	4470	Pointe Gatineau	−25	4740	Bridgewater	−15	4190
			Port Alfred	−29	5720	Canso	−17	4410
Beauport	−25	4850	Port Cartier	−29	6000	Dartmouth	−16	4200
Bedford	−23	4470	Port Harrison	−38	9070			
Beloeil	−24	4580	Preville	−24	4520	Debert	−22	4580
Brossard	−24	4520	Québec	−25	5080	Digby	−15	3850
Buckingham	−26	4900				Greenwood	−17	4130
			Richmond	−25	4740	Halifax	−16	4123
Cacouna	−25	5400	Rimouski	−25	5400	Kentville	−18	4240
Campbell's Bay	−28	4850	Rivière du Loup	−25	5533			
Camp Valcartier	−25	5120	Roberval	−30	5740	Liverpool	−14	4010
Chicoutimi	−30	5510	Rock Island	−24	4900	Lockeport	−14	3980
Coaticook	−24	5010				Louisburg	−15	4410
			Rosemere	−24	4580	Lunenburg	−15	4190
Contrecoeur	−24	4800	Rouyn	−33	6220	New Glasgow	−21	4580
Cowansville	−24	4580	Ste. Agathe des Monts	−27	5380			
Dolbeau	−31	5950	Ste. Anne de Bellevue	−23	4520	North Sydney	−16	4410
Dorval	−23	4470	St. Canut	−25	4900	Pictou	−21	4580
Drummondville	−25	4740				Port Hawkesbury	−19	4470
			St. Félicien	−31	6000	Springhill	−20	4580
Farnham	−24	4590	Ste. Foy	−25	4900	Stewiacke	−21	4520
Fort Chimo	−39	8460	St. Hubert	−24	4540			
Fort Coulonge	−28	4850	St. Hubert de Témiscouata	−26	5780	Sydney	−16	4459
Gagnon	−33	7490	St. Hyacinthe	−24	4650	Tatamagouche	−21	4580
Gaspé	−23	5340				Truro	−21	4704
			St. Jean	−24	4630	Wolfville	−19	4300
Gatineau	−25	4740	St. Jérôme	−25	5060	Yarmouth	−13	4024
Gentilly	−25	4850	St. Jovite	−27	5290			
Gracefield	−28	5070	St. Lambert	−23	4470	**Prince Edward Island**		
Granby	−25	4580	St. Laurent	−23	4470	Charlottetown	−20	4623
Great Whale River	−36	8133				Souris	−19	4580
			St. Nicholas	−25	4850	Summerside	−20	4600
Harrington Harbour	−25	6110	Schefferville	−38	8229	Tignish	−20	4850
Havre St. Pierre	−27	6110	Senneterre	−34	6220			
Hemmingford	−23	4580	Seven Islands	−30	6135	**Newfoundland**		
Hull	−25	4740	Shawinigan	−26	5110	Argentia	−13	4600
Iberville	−24	4630	Shawville	−27	4850	Bonavista	−17	5010
			Sherbrooke	−28	5242	Buchans	−21	5530
Joliette	−25	4880	Sillery	−25	4900	Cape Harrison	−29	6880
Jonquière	−29	5720	Sorel	−24	4840	Cape Race	−14	5010
Kenogami	−29	5730	Sutton	−24	4690			
Knob Lake	−38	8229				Cornerbrook	−19	4900
Knowlton	−24	4630	Tadoussac	−26	5380			
			Temiskaming	−30	5220	Gander	−18	5039
Kovik Bay	−38	9550	Thetford Mines	−26	5350	Goose Bay	−31	6522
Lachine	−23	4470	Three Rivers	−25	5070	Grand Bank	−14	4560
Lachute	−25	4850	Thurso	−26	4850	Grand Falls	−21	5100
Lafleche	−24	4520						
La Malbaie	−26	5340	Val d'Or	−33	6146	Labrador City	−35	7770
			Valleyfield	−23	4520	Port aux Basques	−15	4800
La Salle	−23	4470	Varennes	−24	4630	St. Anthony	−24	5940
La Tuque	−29	5350	Verchères	−24	4740	St. John's	−14	4804
Laval	−24	4580	Verdun	−23	4470	Stephenville	−17	4783
Lennoxville	−28	4850						
Léry	−23	4520	Victoriaville	−26	5040	Twin Falls	−35	7820
			Ville D'Anjou	−23	4470	Wabana	−15	4850
Les Saules	−25	5010	Ville Marie	−31	5760	Wabush Lake	−35	7770
Lévis	−25	4900	Waterloo	−24	4580			
Loretteville	−25	5120	Westmount	−23	4470	**Yukon Territory**		
Louiseville	−25	5010				Dawson	−50	8274
Magog	−26	4730	Windsor Mills	−25	4630	Whitehorse	−41	6879
Malartic	−33	6110	**New Brunswick**			**Northwest Territories**		
Maniwaki	−29	5319	Alma	−21	4580	Frobisher Bay	−40	9845
Masson	−26	4850	Bathurst	−23	5160	Yellowknife	−43	8593

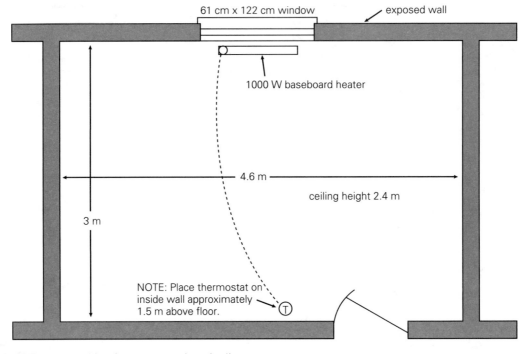

FIGURE 18.24   Heating a room electrically

The Fourth Step: *Infiltration Loss.* Cold air enters rooms through cracks around doors and windows. Heat loss through such infiltration is calculated by the *air change* method. Air in a room or a building changes constantly, entering or exiting through doors, windows, cracks, etc. The amount of the air change is influenced by the size of room, height of ceiling, and ventilating devices, such as fans. Table 18.7 lists various room conditions and the expected air change over a period of one hour.

If, for example, the room shown in Figure 18.24 is in a new house and has no doorway that is exposed to outside temperatures, then, according to Table 18.7, 0.75 air changes per hour can be expected.

Table 18.8 on page 293 lists the heat loss factors (watts per square metre) for various air changes and temperature

differences. Under the 0.75 Air Change heading, with a 2.4 m ceiling height and a design temperature difference of 41°C, the loss is 25 W/m² of floor space. The floor area of the room is 4.6 m × 3 m, which is 13.8 m². The infiltration loss is 13.8 m² × 25 W/m², which is 345 W.

This heat loss is added to the wall and window losses.

The Fifth Step: *Heat Loss Through the Ceiling.* According to Table 18.3 on page 270, an exposed roof or ceiling of frame construction must have a thermal resistance of 5.6 RSI if it is in an area of the country under 5000 degree days. The U factor for the ceiling is 1/5.6 or approximately 0.178 W/m² per degree Celsius of temperature difference. If the temperature difference is 41°C, the heat loss is 41 × 0.178 W/m², which is 7.30 W/m² for each hour of operation.

TABLE 18.6	Heat Loss Factors for Windows and Doors		
Heatload Factor per Degree of Temperature Difference			
Windows and Doors		**Rt**	**W/m²·°C**
Single glass—metal frame		0.16	6.25
Single glass—metal frame with storm		0.32	3.18
Double glass—metal frame (6 mm air space)		0.23	4.43
Double glass—metal frame (13 mm air space)		0.25	3.97
Triple glass—metal frame (13 mm air space) + storm		0.41	2.44
Single glass—wood frame		0.17	5.79
Single glass—wood frame with storm		0.35	2.90
Double glass—wood frame (6 mm air space)		0.28	3.52
Double glass—wood frame (13 mm air space)		0.30	3.29
Triple glass—wood frame (13 mm air space) + storm		0.52	1.93
Skylight—metal frame—single glass		0.14	6.93
Skylight—metal frame—double glass (6 mm air space)		0.21	4.77
Skylight—wood frame—single glass		0.16	6.25
Skylight—wood frame—double glass (6 mm air space)		0.27	3.75
Doors			
Patio Doors—single glass—metal frame		0.16	6.42
Patio Doors—double glass—wood frame		0.20	4.94
Solid wood—40 mm		0.36	2.78
Solid wood—40 mm (wood storm)		0.65	1.53
Solid wood—40 mm (metal storm)		0.53	1.87
Metal (Flat)		0.16	6.42
Metal (Corrugated)		0.13	7.95

Rt means resistance total in RSI values.

The ceiling area of the room shown in Figure 18.24 is 13.8 m². This is the area through which heat can pass. The total heat loss for the room over a period of one hour is 13.8 m² × 7.30 W/m², which is approximately 100.7 W.

The Sixth Step: *Total Heat Loss for the Room.* To calculate the total heat loss for the room over a period of one hour, the loss through each of the four areas must be added.

For example:

Heat Loss	
Through the wall	144.2 W
Through the window	86.0 W
Infiltration loss	345.0 W
Through the ceiling	100.7 W
Total	675.9 W

Determining Heater Size

Winter weather can be unpredictable and have little in common with carefully recorded norms. Care should be taken to match the actual heater size (wattage) with the calculated heat loss for the room or area.

The sample room shown in Figure 18.24 has a calculated heat loss of 675.9 W. Table 18.9 lists low- and standard-watt density baseboard heater units. The closest heater unit in the standard-watt density column to our calculated heat loss is the 750 W one. Our chosen heater unit has approximately 75 W of extra heating ability: it can therefore provide a margin of comfort if there is a winter season that is colder than normal. Care must be taken not to oversize

TABLE 18.7 Air Changes per Hour for Various Rooms and Conditions

Infiltration Factors	Air Change Method
New Houses (Weatherstripped, storm doors and windows)	
Rooms without exposed doors	0.75 air changes/hour
Rooms with exposed door/s (Entrance area)	1.0 air changes/hour
Basements	
Walls less than 50% above grade	0.5 air changes/hour
Walls more than 50% above grade or fully exposed	0.75 air changes/hour
Existing Houses (Not weatherstripped, but with storm doors and windows)	
Rooms without exposed doors	1.0 air changes/hour
Rooms with exposed door/s (Entrance area)	1.5 air changes/hour
Basements	
Walls less than 50% above grade	0.75 air changes/hour
Walls more than 50% above grade or fully exposed	1.0 air changes/hour

Coefficients for Infiltration per Square Metre per Degree Celsius Temperature Difference (2.4 m ceiling height)

	No Exposed Doors ($W/m^2 \cdot {}^\circ C$)	With Exposed Doors Entrance Area ($W/m^2 \cdot {}^\circ C$)
New Houses		
Areas other than basements	0.61	0.82
Basements	0.41	0.61
Existing/Conversion Houses		
Areas other than basements	0.79	1.23
Basements	0.62	0.79

the heater units too much, though: extra load will be placed on the service equipment if and when a sudden drop in temperature causes many of the heaters in the building to come on at the same time. As a general rule, the selected heater unit should be within 10% of the calculated heat loss for the room or area.

Ceiling cable or radiant-heating foil of the same wattage rating could be installed in the room as an alternate heating source.

Heater Location

Heater units should always be located as close as possible to the major heat loss area of a room. Window areas are usually responsible for the greatest heat loss, and so baseboard heaters are normally located under them. Warm air rising out of a heater tends to offset the cooling effect of cold air entering the room around the window. (See Figs. 18.25 and 18.26)

Thermostats and Location

Thermostats are heat-sensitive switches designed to regulate the on/off cycles of heater units. They do not control the amount of heat (wattage) put out by the heater unit. They merely turn the heater *on* or *off*. The length of time that the heater remains on determines the temperature in the room. The thermostat cycles the heater unit to keep the room

Applications of Electrical Construction

TABLE 18.8

TABLE 18.8 Heat Loss Factors per Square Metre Floor Area and Air Change Values (Infiltration)

	Ceiling Height (m)	Heat Loss Factors Per Square Metre (W/m^2)							
Values for Various Temperature Differences									
		1°C	**38°C**	**41°C**	**44°C**	**47°C**	**50°C**	**53°C**	**56°C**
0.5 Air Change									
	2.4	0.41	15	17	18	19	20	22	23
	2.7	0.47	18	19	20	22	23	25	26
	3.0	0.50	19	21	22	24	25	27	28
0.75 Air Change									
	2.4	0.62	24	25	27	29	31	33	35
	2.7	0.70	27	29	31	33	35	37	39
	3.0	0.78	29	32	34	36	39	41	43
1.0 Air Change									
	2.4	0.81	31	33	36	38	41	43	46
	2.7	0.93	35	38	41	44	47	49	52
	3.0	1.0	39	42	45	48	51	54	58
1.5 Air Change									
	2.4	1.2	46	50	54	57	61	65	68
	2.7	1.4	52	56	61	65	69	73	77
	3.0	1.5	58	63	67	72	77	81	86

TABLE 18.9 Low- and Standard-Watt Density Heater Specifications

Volts (specify)	Watts	Approximate Length (cm)	Approximate Shipping Mass (kg)
Specifications: Single Phase: Low-Watt Density			
120 208 240	500	96	6
120 208 240	750	127	7
120 208 240	1000	188	10
120 208 240	1250	250	13
208 240	1500	250	13
Specifications: Single Phase: Standard-Watt Density			
120 208 240	500	66	4
120 208 240	750	96	6
120 208 240	1000	127	7
120 208 240	1250	188	10
120 208 240	1500	188	10
208 240	1750	250	13
208 240	2000	250	13
208 240	2250	250	13

FIGURE 18.25   A residential baseboard installation

Courtesy Ontario Hydro

FIGURE 18.26   A baseboard heater corridor installation

at the temperature selected on the thermostat.

For many years, conventional thermostats had basic mercury bulbs for their switch contacts. The advent of central air conditioners meant that thermostats had to become more sophisticated—they had to be able to control both heating and cooling cycles. Now there are thermostats that can sense temperatures accurately through electronic circuitry.

An *electronic* thermostat smoothes out the cycling process of the heater

unit, providing a more consistent and even temperature throughout the room. As a result, the home-owner can often choose a lower thermostat setting, resulting in lower heating costs over a period of time.

Thermostats, as mentioned earlier, are highly temperature sensitive and should not be mounted on cold, outside walls. They should be located on *inside walls*, away from any draft or heat source that might cause them to open and close the heater circuit in an abnormal manner. They can be obtained in both line and low voltage configurations to match the type of heating system used. Figure 18.27 shows several types of thermostat controls.

Temperature regulation can be handled differently in foyers and entrance halls, which are often subject to cold blasts of air from door openings. If a *forced air unit* is installed in the wall near the entrance, the effect of cold blasts can be counteracted. Cold air entering the area will activate the thermostat on the heater. The fan then forces enough warm air through the heater to return the area to inside design temperature.

Such a compact unit can also sometimes be used to advantage in bathrooms with limited wall space at floor level.

A forced air unit consists of a metal enclosure or tub with a detachable fan and heater unit. A grille placed over the unit directs the air flow and protects the heater coils. (See Figs 18.28 and 18.29)

Heating Cost

Once the heat loss (heating load) has been calculated for each area, the *total heat load* figure can be worked out. The total heat load is the sum of the heat losses calculated for each area. An

Applications of Electrical Construction

Low
voltage
thermostat

Line voltage thermostat

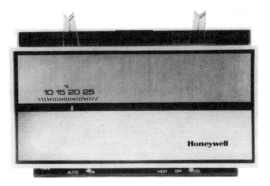
Heating and cooling thermostat

Courtesy Honeywell Limited

FIGURE 18.27 Commonly used thermostats

enclosure
(or tub)

fan and heater unit
(fits into tub)

grillwork (directs heated air
and protects heater element)

Courtesy Chromalox Inc.

FIGURE 18.28 Forced air heater unit

example for calculating heating costs follows. It is based on estimated heat losses (in round figures) for a five-room bungalow in Toronto or a similar area.

Heated Area	Heat Loss
Bedroom #1:	1000 W
Bedroom #2:	1000 W
Kitchen:	1000 W
Living and dining room:	2500 W
Bathroom:	500 W
Basement:	3000 W
	9000 W, or 9 kW

FIGURE 18.29 Forced air heater unit applications (near cold air entry and where wall space is limited)

Courtesy Chromalox Inc.

Total heat load:

Annual kilowatt hour consumption =

$$\frac{HL \times DD \times C}{DTD}$$

Where:

HL = heat loss in kilowatts
DD = annual degree days for the area (See Table 18.5)
C = a constant (15.3 for the Toronto area; consult local utility)
DTD = design temperature difference between indoors and outdoors

Therefore, kilowatt hour consumption is

$$\frac{9 \times 4082 \times 15.3}{40} = 14\ 052\ \text{kW·h}$$

Typical utility charges are as follows:

First 500 kW·per month @ 7.9¢/kW·h
Remaining kilowatt hours
@ 5.5¢/kW·h (end rate)

Electrical heating customers approach the end rate with lighting and cooking energy. The annual heating cost, therefore, is calculated by multiplying the kilowatt hour consumption by the end rate.

Total annual cost:

14 052 kW·h × 5.5¢/kW·h = $772.86

Overheating Protection

Baseboard heaters are equipped with a slender, liquid-filled or vapour-filled copper tube that travels the length of the heater. If for some reason the heater reaches an abnormally high temperature, the expanding liquid or vapour will cause a *relay* at one end of the tube to open the circuit. This prevents heat damage to the walls of the home. (See Fig. 18.30)

Snow Melting Heaters

One type of radiant-heating cable is designed for use in driveways to keep the wheel track areas clear of snow. It is installed under the concrete pavement and comes in pre-assembled lengths from the manufacturer. It is also used to melt snow on ramps and stairways. (See Figs. 18.31 and 18.32)

Snow that has melted, run into eaves, and refrozen also causes problems. Eavestrough heaters are used to keep this part of the roof clear. (See Fig. 18.33)

Pipes can be kept from freezing by wrapping them with *pipe-heating cable*. A thermostat that fits along the side of the pipe controls the heating process. (See Fig. 18.34)

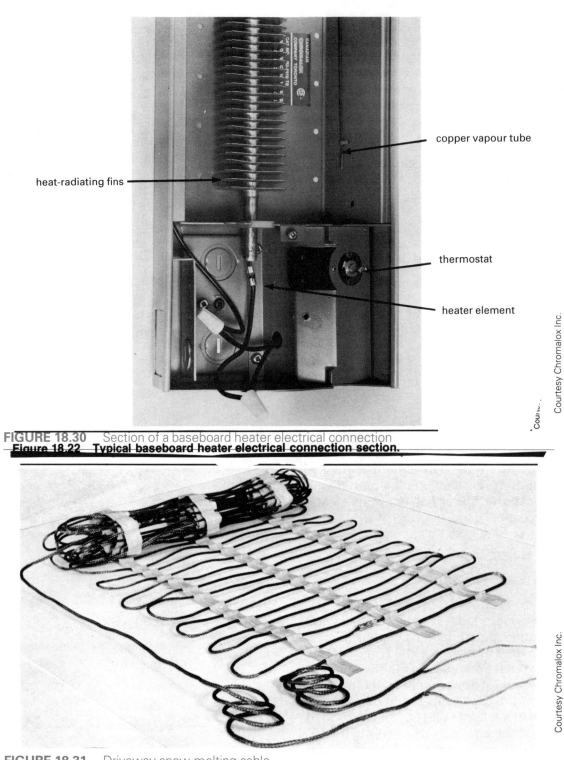

heat-radiating fins

copper vapour tube

thermostat

heater element

FIGURE 18.30   Section of a baseboard heater electrical connection

Figure 18.22   Typical baseboard heater electrical connection section.

FIGURE 18.31   Driveway snow-melting cable

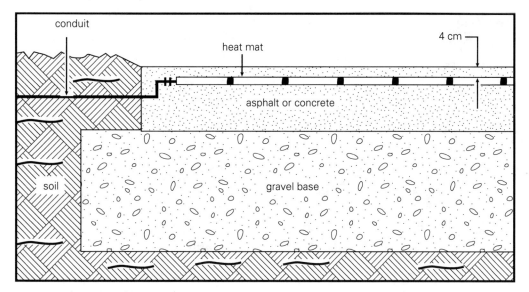

NOTE: For asphalt installations, place bituminous binder on base course both under mat and over mat before placing final course.

FIGURE 18.32 A driveway installation of snow-melting cable

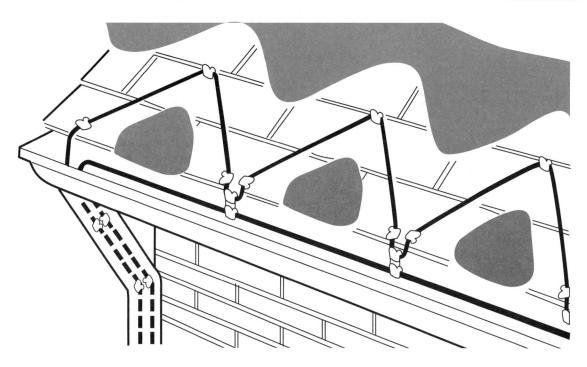

FIGURE 18.33 An eavestrough heating cable installation

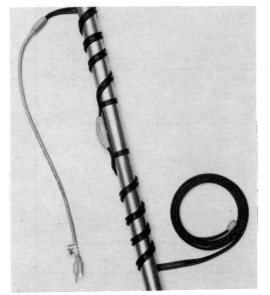

FIGURE 18.34 Pipe-heating cable with thermostat

Courtesy Chromalox Inc.

1. List six advantages that electric heating has over other heating systems.
2. What is the main purpose of insulation?
3. What are the minimum insulation requirements for an electrically heated house?
4. Explain what is meant by an insulation's RSI designation.
5. List the four types of insulation that are suitable for electrically heated houses.
6. What is a vapour retarding barrier? Explain its purpose and identify where it is installed.
7. Why is it important to ventilate an attic?
8. Describe the general construction of a baseboard heater.
9. Why are baseboard heaters made in two wattage densities?
10. List three factors that must be taken into account when installing baseboard heaters.
11. Describe briefly two types of radiant heating systems.
12. List three precautions that must be taken when installing a radiant heating system in the ceiling.
13. List four major residential heat loss areas.
14. What two types of window units should be used for an electrically heated home? Which type would be the best?
15. Why are heaters usually placed under windows?
16. What is the advantage of placing a forced air heater unit in an entrance hall or foyer?
17. Why are baseboard heaters equipped with cutout relays?
18. What precaution should be taken when insulating basement walls of a home that is likely to have water entering in the basement?
19. Name one major advantage of using a forced air electric furnace to heat a house instead of baseboard heaters.
20. Why must outside design temperature be taken into consideration when calculating a heat loss?

Discharge Light Sources

Fluorescent lighting was first developed during the 1930s. Its principle is simple enough, but it took years of research before it was developed into the highly developed tube found in modern lighting fixtures.

Advantages of Fluorescent Lighting

Fluorescent lights have many advantages over standard incandescent lamps.

Fluorescent lamps last approximately *twenty times* as long as standard incandescent lamps. The higher initial cost of the fluorescent tube is more than offset by its long life.

As a result, *maintenance costs* are much lower, because fluorescent lamps do not need to be replaced as often as standard incandescent lamps. Modern 40 W fluorescent tubes have an average life expectancy of 20 000 h. The average incandescent bulb has a life expectancy of 1000 h.

Fluorescent tubes produce more than *five times* as much light per watt of electricity consumed than the incandescent bulb. A 120 cm, 40 W fluorescent tube produces nearly as much light as a 150 W incandescent bulb. The fluorescent tube also keeps its *brightness* for a longer period of time.

Fluorescent tubes produce *less heat* than incandescent bulbs of the same size (wattage output). Large incandescent bulbs will burn anyone trying to remove them while they are still in operation. In contrast, most fluorescent tubes can be handled safely, regardless of how long they have been in operation. Also, when a building is air conditioned, fluorescent tubes provide adequate light without placing as large a heat load on the air conditioner as do incandescent bulbs of the same light output.

Note: Tube life is greatly influenced by the number of times the fixture is turned on and off. The current that surges through the lamp circuit when the fixture is turned on tends to shorten the life of the lamp. A lamp that is turned on and off frequently will need to be replaced far sooner than a lamp that is allowed to operate for a number of hours between starts. As a result, lights in many industrial plants are left burning during lunch hours, coffee breaks, and shift changes. As the cost of electrical energy continues to increase, a more realistic compromise has been reached between tube life and energy costs. It is now considered more economical to turn the lamps off when the off period is expected to be more than 20 min in duration.

Disadvantages of Fluorescent Lighting

In some places, such as clothing, fabric, and meat stores, *colour* is critical and the design of the lighting system important. Fluorescent lights give off an abundance of blue and green tones, but are low in reds and yellows. Their use in, for example, a clothing or meat store, would prevent a customer from seeing the true colour of the product. Walking outside to natural light is not always practical, and so in these cases both incandescent and fluorescent lights are used to help bring out true colours. However, fluorescent tubes for "colour critical" areas have now been developed.

Using discharge lamps over rotating machinery can sometimes be dangerous. The 60 Hz AC supply system of fluorescent fixtures makes the light flicker at a very high speed. This flickering is almost invisible to the naked eye, except when the light falls directly on rotating machinery. If the machine is rotating at a speed close to the speed at which the light is flickering, the machine will appear to be standing still. This *stroboscopic effect* deceives the human eye. There have been cases where an operator has absent-mindedly reached in to touch a rotating part, thinking that the machine has come to a standstill.

Fortunately, this problem can be remedied by installing a small incandescent lamp over the moving part or assembly. This filament type of lamp will cancel out any strobe effect produced by the fluorescent tube fixture.

Fluorescent Tube Parts

The heart of the fluorescent fixture is the tube itself. The tube has several components. (See Fig. 19.1)

Glass Tube. The tube provides an airtight enclosure in which the mercury, gas, and phosphor can function.

Base. As the end of the tube, the base connects the lamp to the electrical circuit. Several pin configurations are available.

Cathode. These small, oxide-coated filaments heat up and emit electrons in to the tube.

Mercury. Droplets of mercury are placed in the tube. They vapourize during the operation of the lamp and emit (give off) ultraviolet energy.

Filling Gas. A small amount of highly purified argon gas is also placed in the tube. This gas ionizes (producing electrical conductivity in gases) when sufficient voltage is applied. Current can then flow readily through the tube.

Phosphor Coating. All of the light energy produced by the mercury is ultraviolet and invisible to the naked eye. The phosphor coating reacts to the ultraviolet rays and turns this energy into visible light.

The Starter

Some fluorescent fixtures (preheat type) need a small *starting* mechanism to establish an electron emission from the *cathode* (filament). (See Fig. 19.2)

Current enters the starter through one of the *contact pins*. The neon gas in

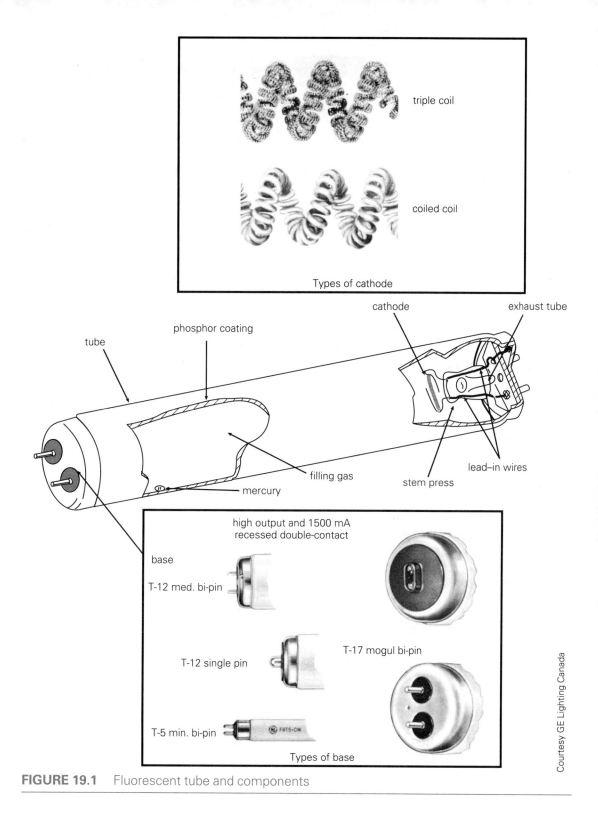

triple coil

coiled coil

Types of cathode

cathode

exhaust tube

phosphor coating

tube

lead–in wires

filling gas

stem press

mercury

base

high output and 1500 mA
recessed double-contact

T-12 med. bi-pin

T-12 single pin

T-17 mogul bi-pin

T-5 min. bi-pin

F8T5-CW

Types of base

Courtesy GE Lighting Canada

FIGURE 19.1   Fluorescent tube and components

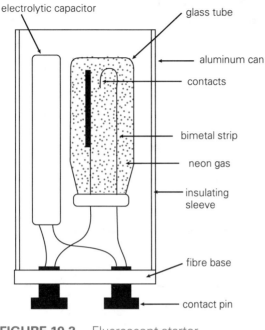

electrolytic capacitor

glass tube

aluminum can

contacts

bimetal strip

neon gas

insulating sleeve

fibre base

contact pin

FIGURE 19.2 Fluorescent starter

the glass bottle provides a high-resistance path from one contact to the other for a small amount of current. This causes the gas to *glow*. *Heat* from the glowing gas warms the *bimetal strip*, which then bends and closes the *contacts*. This action creates a low-resistance path through the *starter*. Once the contacts have closed, the gas ceases to glow, the bimetal strip cools, and the contacts open. All this takes about two or three seconds.

The *capacitor* in the starter controls the amount of *arcing* between the contacts when they open and close. It thus prolongs the life of the starter.

The Ballast

Mercury droplets in the tube vapourize during the operation of the lamp. Current then flows through the tube more and more easily. This *ionization* process

could allow the current flow to increase to the point were the tube would destroy itself. The *ballast* controls and regulates this current flow in much the same way that ballast in a ship controls the stability of the vessel. It is a coil of wire wound on a laminated steel core.

Operation of the Fixture

When a fluorescent lamp is switched on, current flows through the ballast, filaments, and starter. (See Fig. 19.3) The high-resistance *neon* gas in the starter glows (heats up) and bends the *bimetal contacts* until they touch. This closing of the contacts provides a low-resistance path, and the *filaments* heat up quickly as a result of the extra current flow. Heating of the filaments causes *electrons* to be emitted in to the tube. The neon gas in the starter stops glowing, and the bimetal contacts cool and open the low-resistance path.

The increase in current flow through

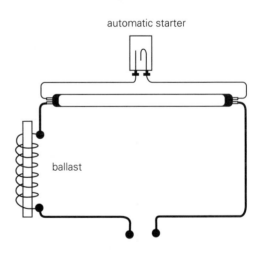

automatic starter

ballast

FIGURE 19.3 A circuit for the operation of 1 lamp

Applications of Electrical Construction

the filaments also causes a strong *magnetic field* around the ballast. When the bimetal contacts open, the high-resistance path is re-established. This causes the strong magnetic field around the ballast coils to collapse. As the magnetic field collapses, it induces a high-voltage *kick* in to the ballast coils. This voltage is high enough to strike an arc across the tube, using the emitted electrons, argon gas, and mercury vapour as a current path.

Once an arc is established across the tube, most of the *circuit current* flows through the tube. Not enough current flows to the starter to re-establish a starting cycle in the neon gas.

As mentioned before, current flow in the tube tends to increase. When the current flow increases, a stronger magnetic field is established around the ballast. This magnetic field produces (induces) a voltage in the ballast, which flows in the *opposite* direction to the applied voltage from the source. The more the current tries to increase in the tube, the more the ballast tries to hold back the applied voltage and current. The current flow in the fixture quickly stabilizes itself.

When the *ultraviolet light energy* from the mercury vapour strikes the *phosphor coating* on the inside of the tube, a cool, comfortable, visible light is produced. As the tube nears the end of its life expectancy, an *oxide coating* from the filaments gradually appears on the ends of the tube. Such a darkened area indicates that the tube is about ready to be replaced.

A tube that is flickering but not starting should be replaced before the starter and/or ballast are damaged by the continuous starting currents.

Rapid-Start Fixtures

This fluorescent fixture does *not* need a separate starting device. Approximately 4 V is continuously supplied to the filaments of the tube by special *heater windings* in the ballast while the tube is in operation. The constantly heated filaments emit a steady stream of electrons, thereby allowing the ballast voltage to easily strike an arc across the tube.

Rapid-start fixtures, which were developed after the pre-heat and instant-start types, take advantage of the voltage that exists between the cathodes and the metal frame of the fixture. For this reason, *all* rapid-start fixtures *must be grounded*. Otherwise, in cool weather they will often fail to start.

As a result of the continuous heating of the cathodes (filaments), a lower open-circuit voltage can be used and the physical size and weight of the ballast reduced.

Figure 19.4 shows a single-tube, rapid-start circuit. These popular fixtures are available in two-lamp units. (See Fig. 19.5) The two-lamp fixture starts one tube slightly ahead of the other, with the help of a capacitor in the ballast.

Many industrial plants with a large number of electrical motors have a poor *power factor*. They receive large quantities of electrical energy and require large conductors and control equipment, but waste much of the power. (This situation is like paying for only two flavours of ice cream from a 3-flavour brick, because you want only two of the flavours. The cost of the wasted third flavour would then have to be covered by the supplier.) The capacitor in fluorescent fixture ballasts helps to overcome this problem. That is why industries with a poor power factor often switch to fluorescent fixtures. They can then reduce their electrical bills.

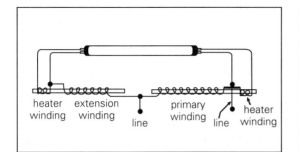

FIGURE 19.4 A single-tube, rapid-start circuit. (The rapid-start ballast supplies a small amount of voltage continuously to the cathodes.)

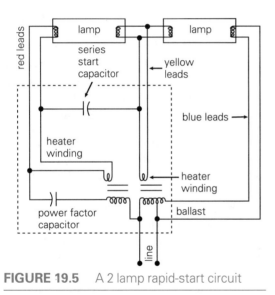

FIGURE 19.5 A 2 lamp rapid-start circuit

Ballast Overheating

In the process of limiting tube currents and providing starting voltages, the ballast tends to warm up. As a ballast ages, the *laminated steel core* of the transformer often loosens, vibrates (producing a humming sound), and penetrates the insulation on the windings. At times, this will short circuit some of the wind-

ings, causing an excess current flow in the remaining windings. As a result of this breakdown in insulation, a damaging amount of heat is produced in the ballast. Today, manufacturers are equipping ballasts with a *thermal protector (cutout)* to open the ballast circuit automatically when the heat reaches a dangerous level.

Ambient temperature (the temperature of the surrounding air) also plays an important role in the life span of the ballast. Figure 19.6 shows the life expectancy of a ballast under various temperature conditions. Figure 19.7 shows a rapid-start ballast and its construction.

The Canadian Electrical code requires that all fixtures mounted on a combustible surface be equipped with a *thermally protected ballast*. Ballasts are rated by the manufacturer to operate at temperatures up to 90°C.

Instant-Start Fixtures

The tubes for this fixture are generally of the *single-pin* type and cannot be

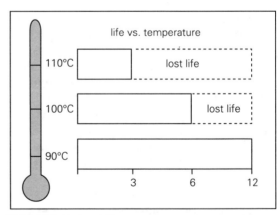

FIGURE 19.6 Life versus temperature. The generally accepted "lost life rule" is that "for every 10°C increase in temperature, insulation life is cut in half."

Applications of Electrical Construction

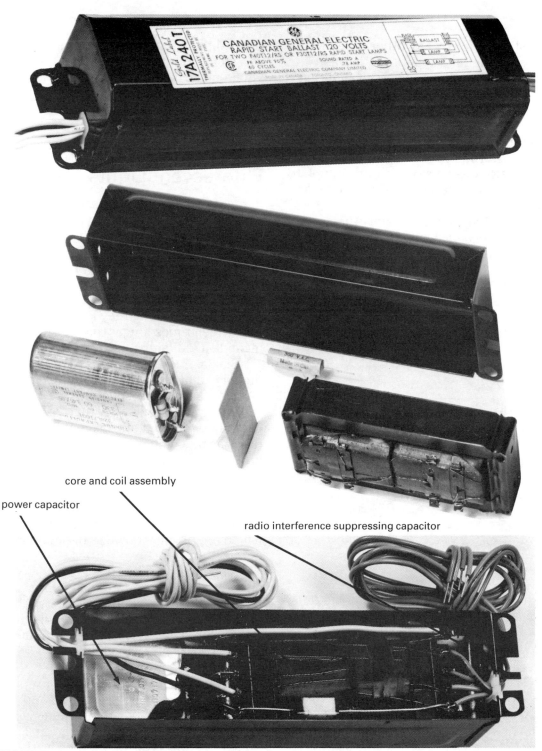

CANADIAN GENERAL ELECTRIC
RAPID START BALLAST 120 VOLTS
FOR TWO F40T12/RS OR F30T12/RS RAPID START LAMPS

17A240T

power capacitor

core and coil assembly

radio interference suppressing capacitor

FIGURE 19.7　Rapid-start ballast

substituted for by any other type of tube. At one time, a bi-pin tube was made and is still found in some lighting installations.

This fixture does not require preheating. It is started by creating a higher ballast output voltage with a step-up transformer in the larger ballast. The lamp cathodes must be strong enough to withstand this sudden and forceful starting technique.

Instant-start fixtures start their lamps *in sequence*, much like that of the rapid-start unit. (See Fig. 19.8) Because of this higher starting voltage, many of the tube sockets are built so that they open the ballast circuit when a tube is removed from the fixture.

Instant-start lamps are made in lengths ranging from 60 cm (2 ft.) to 240 cm (8 ft.).

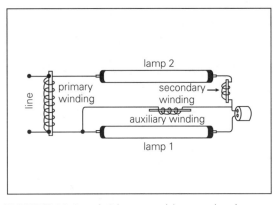

FIGURE 19.8   A 2 lamp rapid-start circuit. Series operation reduces size, weight, wattage loss, and cost.

Metric Lamps

A metric lamp in a 166 cm length operates satisfactorily with the F40 (40 W lamp) ballasts available. Light output and lamp life are the same as for a 40 W lamp. This lamp can fit into a previ-

ously dimensioned 48 in. fixture because of a small adapter at one end of the lamp which makes up the difference in length. Fixtures specifically produced for use with such lamps are referred to as *1200 mm* units.

Low Wattage Biaxial Fluorescent Lights

In recent years, a newly designed, miniature, twin-tube-and-globe type of fluorescent lamp has been increasingly in demand. These new lamps have incandescent-like colour and may be used in areas previously illuminated by incandescent lamps. With an estimated useful life span of 10 000 h, they last from four to thirteen times longer than normal incandescent lamps.

Biaxial fluorescent lamps are produced in a number of sizes, for example, 7 W, 9 W, and 13 W. While the rated lamp wattage is low, remember that fluorescent lamps put out approximately four times as much light per watt as a normal incandescent lamp of the same size. Not only do these lamps combine high efficiency, long life, and warm, incandescent-like colour, they also save energy. Figure 19.9 shows these new lamps, which are produced with a bi-pin plug-in connection. (See Fig. 19.10)

A specially designed screw-in adapter allows the 7 W and 9 W biaxial lamps to be installed in almost any incandescent socket having a medium-sized lamp base. The adapter contains a small ballast and can be fitted with a retaining collar and screws. It can thus be used in areas subjected to vibration and should be able to prevent a lamp from falling out of its mount when installed in a base up configuration. Figure 19.11 shows this mounting kit. To accommodate the various shapes of

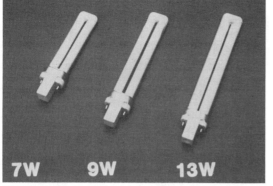

Courtesy GE Lighting Canada

FIGURE 19.9 Biaxial fluorescent lamps in 7 W, 9 W and 13 W sizes

lighting fixtures, a more compact version of the 13 W lamp is now being produced. Figure 19.12 shows the standard size and compact 13 W biaxial lamp units.

An equally new and clever lamp design further simplifies the replacement of incandescent lamps with low wattage fluorescent units. These energy-efficient, long-life (9000 h), one-piece lamp-and-ballast combination lamps screw in to most incandescent sockets without further adjustment to the socket or the circuit wiring. (See Fig. 19.13) They use 15 W, and if installed in place of a 60 W lamp of the same light output, a savings of $24.30 will be made over the lamp's life when calculating energy costs at 6¢/kW·h. The warm colour of these

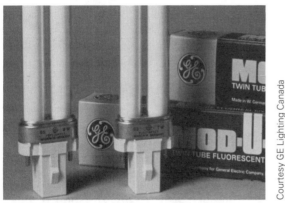

Courtesy GE Lighting Canada

FIGURE 19.10 A bi-pin mounting cap for a low wattage biaxial fluorescent lamp

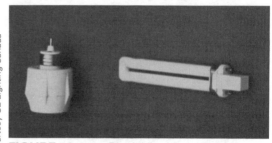

Courtesy GE Lighting Canada

FIGURE 19.11 Biaxial lamp and screwbase adapter with built-in ballast for use in medium-base lampholders

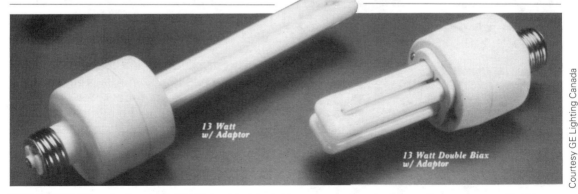

Courtesy GE Lighting Canada

13 Watt w/ Adaptor

13 Watt Double Biax w/ Adaptor

FIGURE 19.12 A standard 13 W double biaxial lamp compared with an ultra compact 13 W lamp of the same type. A screw-in adapter base allows either lamp to be used in an incandescent socket.

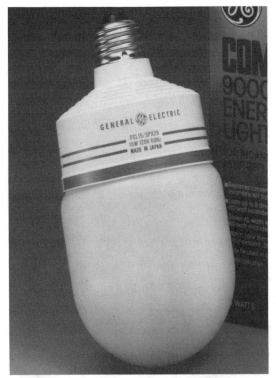

FIGURE 19.13 A combination lamp and ballast, globe type, for use in a medium-base lampholder

Courtesy GE Lighting Canada

FIGURE 19.14 Typical application of a globe type fluorescent lamp

Courtesy GE Lighting Canada

lamps will blend well with other incandescent lamps that may be installed nearby. Figure 19.14 shows typical use of these units.

The lamps do have some limitations, though. Some manufacturers recommend that they not be used with dimmer switches, be allowed to come in contact with moisture, or be used outdoors or in enclosed fixtures where temperature changes may be too extreme for safe and proper lamp use.

Power Groove Lamps

During the late 1950s, a fluorescent tube that used indentations, or *grooves*, on one side of the glass tube to increase light output was developed. Later research showed that grooving both sides of the tube produced an even higher light output. (See Figs. 19.15 and 19.16)

The grooves in the tube bring the phosphor-coated glass *closer* to the arc stream. This squeezing of the arc results in more visible light energy. In the power groove tube, the arc stream must follow a *wavy* path as it travels the length of the tube. (If the arc is straightened out, there is approximately 2.7 m of arc contained in the 2.4 m tube. This increase and concentration of the arc length allows the tube to produce more light than a conventional fluorescent tube.

The name *power groove* was given to this tube by one manufacturer. Power groove lamps are useful light sources for industrial applications. (See Fig. 19.17)

High-Intensity Discharge Lamps

The term *high-intensity discharge (HID)* describes a wide variety of lamps. But all

Applications of Electrical Construction

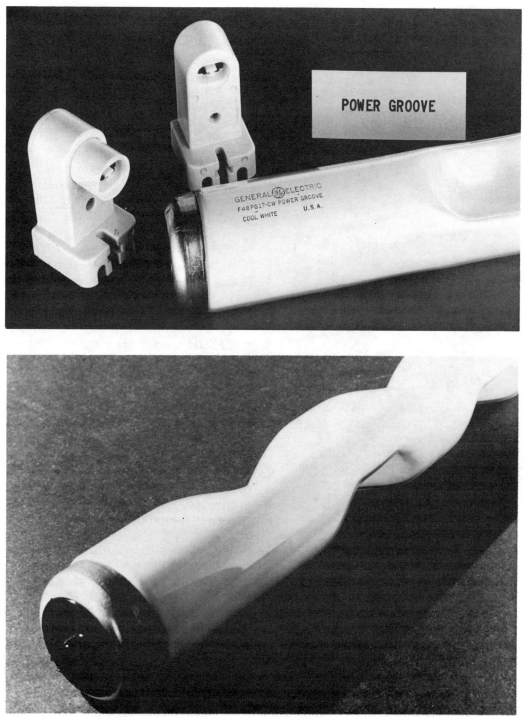

FIGURE 19.15   Power groove fluorescent tubes and sockets

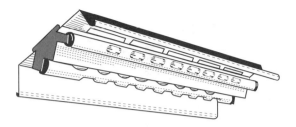

FIGURE 19.16 A power groove fluorescent fixture for industrial applications

Courtesy GE Lighting Canada

FIGURE 19.17 Power groove lighting installation

HID lamps are alike in one way: they produce light from a gaseous discharge arc tube and operate at a pressure above that found in regular fluorescent lamps.

HID lamps were first introduced in 1934. They can be divided in to three major categories: the mercury vapour lamp, the metal halide lamp, and the high-pressure sodium lamp.

Mercury Vapour Lamps

In a standard, low-pressure fluorescent tube, most of the light energy is in the ultraviolet range. Phosphor-coated tubes must be used to produce visible light. The higher-pressure mercury vapour tube produces visible light *directly*. (See Figs. 19.18 and 19.19)

Some mercury vapour lamps have a phosphor coating on the *inside* of the *outer* bulb. This coating reacts to the ultraviolet energy produced by the lamp and modifies the colour of the light output.

The mercury vapour lamp is produced in sizes ranging from 50 W to 1000 W output. Average life expectancy is more than 24 000 h, which makes this an ideal light source where re-lamping is both costly and time-consuming. The mercury vapour lamp is used for street and road lighting, area lighting (for example, parking lots), and industrial lighting in factories and warehouses.

It has several components.

Base. The mogul screw-base found on most mercury vapour lamps connects the lamp to the ballast and external circuits. Letters (matching the months of the year) and numbers are imprinted in the base. They help in keeping a record of lamp life.

Starting Resistor. This tiny, heat-withstanding resistor limits current in the starting circuit to a safe value.

Starting Electrodes. The arc is established between the main and starting electrodes, ionizing the argon gas and helping to strike the main arc.

Main Electrodes. Made of a double layer of tungsten wire and coated with rare earth oxides, they act as terminal points for the main arc.

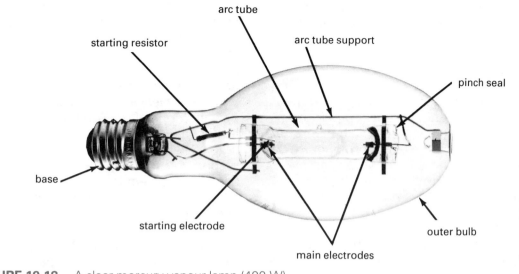

FIGURE 19.18   A clear mercury vapour lamp (400 W)

FIGURE 19.19   A phosphor-coated mercury vapour lamp (1000 W)

Arc Tube Support.   This polished metal frame supports the arc tube and conducts current to the upper main electrode.

Arc Tube.   This pure quartz tube is called the *heart* of the mercury lamp. It contains a precise quantity of mercury and a small amount of argon gas. Some manufacturers coat the ends of this tube (around the electrodes) with platinum. Doing so ensures that the tube will start in cold weather.

Pinch Seal.   It seals the ends of the arc tube and prevents both the escape of argon and the entry of nitrogen. (Nitrogen is used between the arc tube and the outer tube.)

Outer Bulb. Heat- and weather-resistant glass is used to protect the internal parts and maintain a nearly constant arc-tube temperature. Maintaining a high arc-tube temperature is important if the lamp is to operate efficiently. These bulbs are sometimes phosphor-coated, and they filter ultraviolet light energy.

How a Mercury Vapour Lamp Produces Light. The operation of the mercury lamp begins with the *arc tube*. There is a *starting electrode* beside the *main electrode* at one end of the tube. Increased starting voltage from the *ballast* (approximately double the line voltage) strikes an *arc* between these two electrodes, using the argon gas as a current path. The *argon* ionizes (breaks up) and spreads rapidly throughout the tube. As the ionized argon reaches the main electrode at the opposite end of the tube, the light-producing *arc* is formed.

The small *starting resistor* (approximately 40 k Ω) limits the arc current during starting. As a direct result of its high resistance, current flow quickly shifts to the lower resistance, *main arc* stream. Once the main arc has been established, current flow along its path is approximately 1000 times greater than the flow between the starting electrodes.

Heat from the main arc continues to vapourize the *mercury* in the arc tube for several minutes after starting. As more and more mercury is vapourized, current flow increases. Only the stabilizing, current-limiting feature of the ballast (similar to a fluorescent ballast) prevents the lamp from destroying itself.

Light energy is produced by the ionized *argon* gas particles colliding with the *mercury* atoms. As electrons in the mercury atom are jarred out of orbit and replaced by electrons from a nearby atom, *radiation* is given off. The *colour* of light *(wave length)* produced depends on which ring of the orbiting electrons has been hit by the colliding particles of argon. High pressures in the arc tube are responsible for deeper penetration than in the fluorescent tube. This results in more visible and less ultraviolet light energy.

Horizontal Operation of Mercury Vapour Lamps. These lamps are slightly more efficient when operated while in a *vertical* position. Horizontal operation reduces the lamp's efficiency slightly because the arc will float upwards in the arc tube.

Restarting Mercury Vapour Lamps. If the power supply to a mercury vapour lamp is interrupted, the arc will extinguish and not restart for several minutes. This is because sufficient pressure will have built up in the tube during operation to prevent the arc from re-establishing itself immediately. Once the tube has cooled and the pressure lowered, the arc will restart automatically as usual. There is not enough ballast voltage to restart the lamp until it cools and the pressure in the arc tube decreases.

Lamp life is shortened by continuous starting. If a mercury vapour lamp is allowed to operate for long periods of time, its life span will be much longer.

Mercury Vapour Ballast. Although bulbs able to operate without ballasting have been designed, most mercury vapour lamps need a ballast. The ballast does for HID lamps what it does for fluorescent lamps. Line voltage is boosted slightly, lamp current is steadied, and the power factor is corrected.

Figures 19.20, 19.21, and 19.22 show a ballast circuit, a cutaway view of an outdoor weatherproof unit, and an indoor model with the same circuitry as the outdoor unit.

Mercury Vapour Applications. For many years mercury vapour lighting was used to replace incandescent lighting systems over roadways. Although more recent lamp developments are now replacing mercury vapour in these areas, shopping malls, commercial buildings, and safety and security systems—places where dusk to dawn lighting is required—still make use of mercury vapour lighting.

Figure 19.23 shows the difference in street light output.

Figures 19.24, 19.25, and 19.26 show mercury vapour lighting used for area

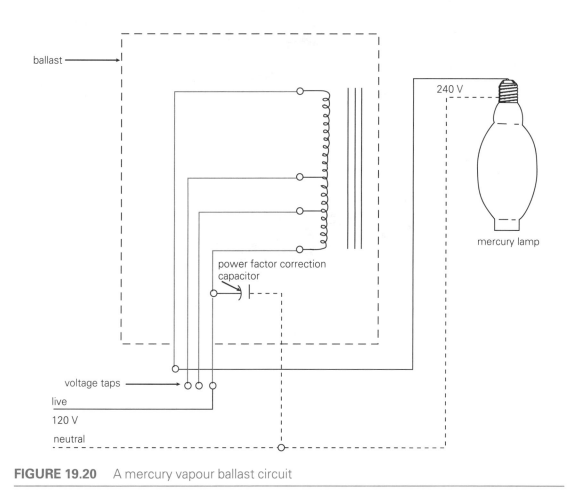

FIGURE 19.20   A mercury vapour ballast circuit

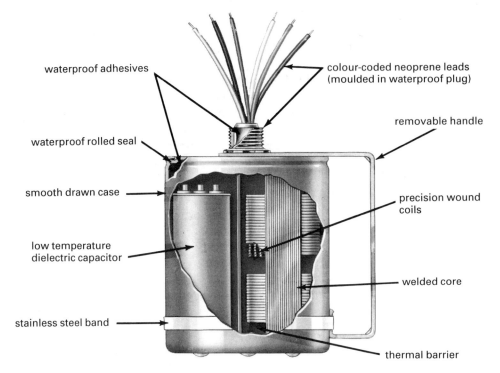

waterproof adhesives

colour-coded neoprene leads (moulded in waterproof plug)

removable handle

waterproof rolled seal

smooth drawn case

precision wound coils

low temperature dielectric capacitor

welded core

stainless steel band

thermal barrier

Courtesy GE Lighting Canada

FIGURE 19.21   A weatherproof mercury vapour ballast

Courtesy GE Lighting Canada

FIGURE 19.22   An indoor mercury vapour ballast

NOTE: 500 W incandescent luminaires.
Mounting height 7.5 m; poles 24 m staggered.

NOTE: 400 W Lucalox luminaires.
Mounting height 11 m; poles 15 m staggered.

FIGURE 19.23 A difference in light output on a 12.8 m roadway width

FIGURE 19.24 A lighting fixture suitable for use with mercury vapour and other high intensity discharge light sources

FIGURE 19.25 Mercury vapour lighting for outdoor decorative use

Courtesy GE Lighting Canada

FIGURE 19.26 Mercury vapour and incandescent lighting used in combination for outdoor area lighting

Courtesy GE Lighting Canada

lighting (for example, in parking lots) and decorative lighting.

Figures 19.27 and 19.28 show a 250 W lamp used for area lighting around a swimming pool.

Figures 19.29 and 19.30 show 1000 W lamps and fixtures used for a large parking lot.

Figure 19.31 shows this lamp used as a decorative, residential, street lighting unit to brighten the area around a driveway.

HID Lamp Sizes and Shapes

HID lamps are made in many shapes and sizes. (See Fig. 19.32)

Metal Halide Lamps

During the early 1960s, experiments with other metals produced a lamp with better light radiation characteristics than the mercury vapour lamp. The problem

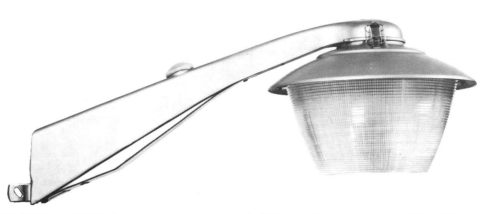

FIGURE 19.27 A 250 W mercury vapour, metal halide, and high-pressure sodium light fixture (luminaire)

Courtesy GE Lighting Canada

318

Applications of Electrical Construction

FIGURE 19.28 An outdoor area lit by a 250 W lamp unit

FIGURE 19.29 Typical 400 W or 1000 W high intensity discharge fixture for use in roadway lighting. This ``Cobra head'' unit is equipped with a photo-electric control unit.

FIGURE 19.30 An outdoor area lit by 1000 W mercury vapour lamps

FIGURE 19.31 A decorative street lighting unit

arbitrary (A)

parabolic aluminized reflector (PAR)

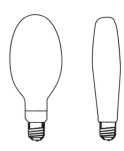

elliptical (E)

reflector (R)

bulged-tubular (BT)

reflector (R)

FIGURE 19.32 HID lamp shapes

was to find metals that could be easily vapourized but would remain chemically stable. It was finally solved when metals in the form of their *halide salts* (usually iodides) were added to the basic mercury arc tube.

Today, lamps use iodides of *sodium, thallium*, and *indium* along with the mercury in the arc tube. The result is a metal halide lamp that can produce 50%

more light than a mercury lamp, with greatly improved colour. (See Figs. 19.33, 19.34, and 19.35)

How a Metal Halide Lamp Produces Light.
The operation of this lamp is similar to that of the mercury lamp, and most mercury lamp fixtures will readily accept the halide lamps. A *metal halide ballast* must be used, however, because

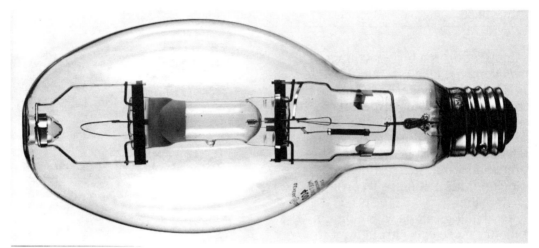

FIGURE 19.33   A 400 W metal halide lamp

Courtesy GE Lighting Canada

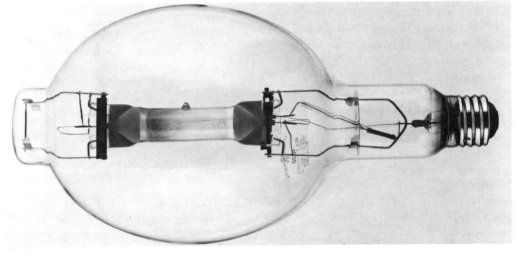

FIGURE 19.34   A 1000 W metal halide lamp

Courtesy GE Lighting Canada

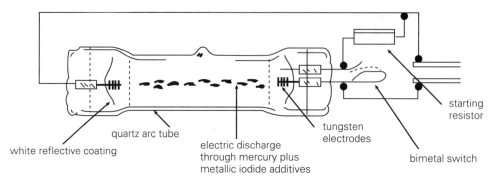

quartz arc tube

white reflective coating

electric discharge
through mercury plus
metallic iodide additives

tungsten
electrodes

starting
resistor

bimetal switch

FIGURE 19.35   Internal components of a metal halide lamp

the lamp requires a higher voltage. The *fused quartz arc tube* of a metal halide lamp is slightly smaller than that of the mercury lamp and has a special coating on the ends to maintain proper electrode temperature during operation. A small *bimetal switch* is built in to the lamp to short out the starting contacts during operation. It helps to prolong electrode life.

Metal halide lamps will maintain about 40% more light throughout their useful life. This averages out to 15 000 h, rated at 10 h operation per start.

Metal Halide Lamp Applications.   The clear, heat-resistant glass bulbs need no phosphor coating, provide a better colour than mercury vapour lamps, and are used extensively to light sports stadiums where their colour and light output is compatible with the requirements of colour TV cameras. Made in 175 W to 1500 W sizes, these lamps are also used for stores and other commercial applications. Due to their higher efficiency, they are gradually replacing mercury vapour lamps as a light source.

Figure 19.36 shows a typical fixture. It is used for ball park lighting and other area lighting systems with many of the

Courtesy GE Lighting Canada

FIGURE 19.36   A P-1000 floodlight with a trunnion-type mount. For use with mercury vapour and metal halide lamps.

HID lamps. There must be a ballast in the circuit supplying this fixture.

Figure 19.37 shows a modern industrial fixture design. The ballast is built directly over the reflector on this unit.

Applications of Electrical Construction

FIGURE 19.37 A metal halide lamp indoor ballast and reflector unit

Courtesy GE Lighting Canada

Figure 19.38 shows a double-lamp unit. It is useful for industrial lighting applications.

High-Pressure Sodium Lamps

The high-pressure sodium lamp is quite different from other HID lamps. It is much simpler in design, because of the tremendous research and development that went in to the production of materials used in its construction. (See Fig. 19.39) It is regarded as the most efficient source of white light artificially produced. The 35 W, 50 W, 70 W, 100 W, 150 W, 200 W, 250 W, 400 W, and 1000 W lamps put out approximately 50% more light than either mercury vapour or metal halide lamps of the same wattage ratings. (See Figs. 19.40, 19.41, and 19.42)

One reason for the success of the high-pressure sodium lamp is the

FIGURE 19.38 A metal halide lamp indoor ballast and double reflector unit

Courtesy GE Lighting Canada

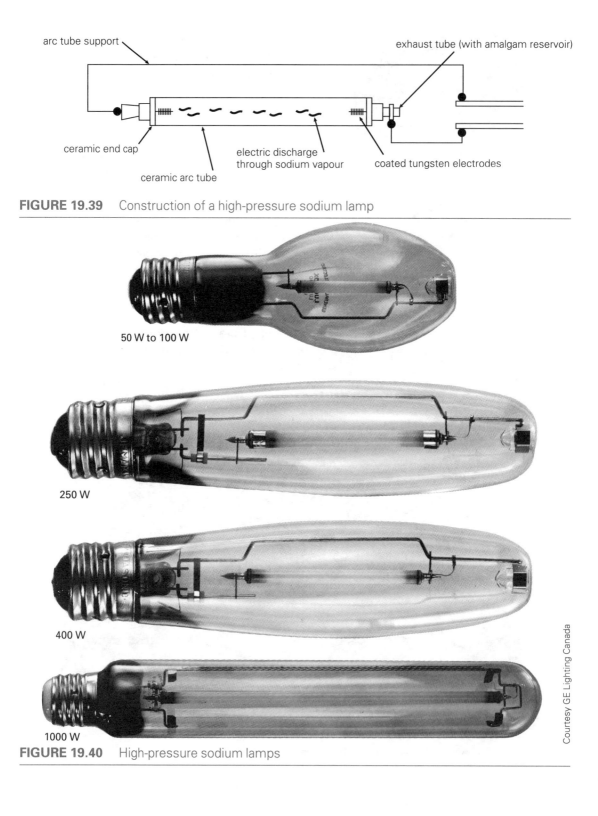

arc tube support

exhaust tube (with amalgam reservoir)

ceramic end cap

electric discharge
through sodium vapour

coated tungsten electrodes

ceramic arc tube

FIGURE 19.39   Construction of a high-pressure sodium lamp

50 W to 100 W

250 W

400 W

1000 W

FIGURE 19.40   High-pressure sodium lamps

Courtesy GE Lighting Canada

Applications of Electrical Construction

FIGURE 19.41   Deluxe Lucalox high-pressure sodium lamps

FIGURE 19.42   An outdoor floodlight application of a high-pressure sodium lamp

ceramic material used for the *arc tube*. A special method for *sealing* off the tube, which allows it to contain the high-pressure sodium discharges, was also developed.

The ceramic is made of *translucent aluminum oxide* and was developed specially for this lamp. One trade name is Lucalox.

Like many ceramics, Lucalox can withstand operating temperatures as high as 1300°C. Unlike many other ceramics, it is virtually free of tiny *pores*, so a high percentage (92%) of visible light can pass through. Lucalox contains few if any *impurities*, which makes it highly resistant to the corrosive effects of hot sodium. Quartz, on the other hand, deteriorates rapidly when exposed to sodium.

The ceramic *end caps* are joined to the arc tube in such a way that they will maintain the seal during the expansion and contraction (heating and cooling) cycles of the tube. The *oxide-embedded electrodes* and the *end caps* are also designed to withstand the corrosive effects of hot sodium.

The tube is filled with a mixture of *xenon* and *mercury* as well as *sodium*. A special circuit consisting of a small electronic board produces a high *voltage pulse* (2500 V) for about 1 μ s during each half of the alternating current cycle. This voltage pulse is strong enough to ionize the xenon across the main electrode gap, making starting electrodes unnecessary. Once the arc has been established, the high voltage pulsing is discontinued and normal ballast voltage maintains the arc.

These lamps reach full brilliancy in a shorter period of time than mercury or metal halide lamps. Also, they may be restarted without waiting for the tube pressure to drop.

Colour changes in the lamp can be seen as the sodium becomes fully vapourized. A dim, *bluish white* light is given off as the xenon ionizes. This changes quickly to a brighter *mercury-blue* glow as temperature rises in the tube. A *yellow* shade takes over from the blue, showing that the sodium has reached low-pressure temperature. As temperature and pressure in the tube reach normal operating levels, the colour changes to the *white* light seen during operation. The arc temperature is over 2000°C when the lamp is at normal operating temperature.

Lucalox lamps were made at one time in two different models for *base up* or *base down* operation. Each could operate in a near horizontal position. The reason for this was that the excess

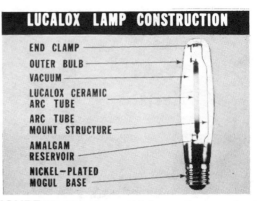

FIGURE 19.43 A new high-pressure sodium lamp, suitable for use in any burning position

sodium mixture collected at the coolest point in the arc tube. A special reservoir was fitted to the arc tube at the coolest end of the lamp to collect it. In base up lamps, the reservoir was placed at the end farthest from the base. In base down units, the reservoir was placed near the base of the lamp.

Modern Lucalox lamps are universal burning, that is, they burn in any position, and they have an external *amalgam reservoir* located at one end of the arc tube. This special reservoir contains the excess sodium amalgam, keeping it away from the arc stream, and thereby extending the life span of the lamp. These lamps are available in both clear and diffused coated versions to accommodate the various light distribution and luminaire requirements. The clear lamps provide the best optical control of the light energy, while the difuse-coated lamps provide a smoother, but lower brightness light in low-mounted, decorative applications. Figure 19.43 illustrates this new lamp.

Deluxe Colour High-Pressure Sodium Lamps. The most recent lamps provide a major improvement in the appearance of people, foliage, and

Applications of Electrical Construction

furnishings. The broader, more balanced colour spectrum provides richer reds, blues, and greens than the standard high-pressure sodium lamps. These deluxe colour high-pressure sodium lamps are produced in 70 W medium-base and 250 W mogul-base configurations. (See Fig. 19.41) They can be operated in temperatures as low as –29°C. They can also be operated satisfactorily with standard high-pressure sodium ballasts, allowing for easy conversion from existing high-pressure sodium lamp installations.

High-Pressure Sodium Lamp Applications.

These highly efficient lamps produce more light per watt of electrical energy consumed than most other forms of lighting. They are being used to floodlight outdoor areas such as that shown in Figure 19.44 and the outside of buildings as in Figure 19.42. They are also being used to illuminate the inside of manufacturing and repair buildings where there is a high ceiling (high bay) as shown in Figures 19.45, 19.46, and 19.47. Warehouses (see Fig. 19.48) make excellent use of these light

FIGURE 19.45 A high bay lighting installation using metal halide and high-pressure sodium lamps

FIGURE 19.46 High bay lighting at a boiler assembly plant

FIGURE 19.44 An outdoor floodlight application of a high-pressure sodium lamp

FIGURE 19.47 High bay lighting in a CPR repair shop

FIGURE 19.48 A warehouse storage area lighting system

sources as well. The tremendous light output of this form of lighting makes it suitable for roadway lighting installations as seen in Figure 19.49.

Multi-Vapour Lamps

New high efficiency, multi-vapour lamps provide users with 50% more light over the life of the lamp than the older mercury vapour lamps. These lamps will operate with 80% of the mercury vapour ballasts now in use, indoors or outdoors, and function in all burning positions.

The lamps, rated at 400 W, are produced in clear and phosphored glass units. The clear lamp provides good colour and is useful where good light control and light "cut off" are important considerations in lighting. The phosphor coated lamp provides even better colour when a softer, diffused light source is needed.

These lamps, which operate on a mixture of metal halides rather than just one, such as the mercury or sodium lamps, are suitable for use in all weather conditions where the temperature is −18°C or above. The lamps require between two to four minutes to reach full brilliancy and will restart after shut off in 10 to 15 min. This type of lamp is recommended for use by energy conscious users of lighting systems where funds are not available for complete system change-over to newer light forms. The lamps have an average rated life span of 15 000 h based on 10 h operation per start.

As well as being used to modernize the less efficient mercury vapour lamp systems, multi-vapour lamps are suitable for new system installations. Universal (any position) burning, greatly increased light output, and lower initial cost are just a few of the reasons for selecting this type of lamp. It is highly

FIGURE 19.49 Highway 400 lighting system with Lucalox lamps

acceptable for use in floodlighting buildings, merchandise displays, and sports-playing fields.

Standard multi-vapour lamps are used for new equipment installations, while I-line (Interchangeable) lamps are used as replacements for mercury vapour lamp units.

The most recent development in the multi-vapour lamp family is that of a *metal halide* lamp containing a unique blend of phosphors. The lamp provides a warm, rich colour which adds to the appeal of the products and areas it illuminates. It was designed to complement other light sources, such as fluorescent, incandescent and halogen lamps, while providing uniform colour on the areas it illuminates.

Metal halide lamps can be installed in existing sockets, operate in any position, and provide the advantages associated with long life. They are produced in three sizes—175 W, 250 W, and 400 W. (See Fig. 19.50)

Courtesy GE Lighting Canada

FIGURE 19.50 Multi-vapour metal halide lamps

Tungsten-Halogen Lamps

The tungsten-halogen lamp is not a member of the metallic vapour/arc family, but it is often confused with the vapour/arc lamps.

Like other incandescent lamps, the tungsten-halogen lamp has a *tungsten filament*. It produces light energy by passing current through this filament, causing it to glow brightly. *Argon* and a small amount of *halogen* gas are combined under relatively high pressure within the quartz filament tube to produce a brighter, whiter light.

A unique *cleaning cycle* is responsible for the long life and continued high output of this lamp. As the tungsten filament reaches operating temperature, small particles are boiled off (emitted) into the tube. The halogen gas picks up these particles and returns them to the filament. This *circulating process* keeps the tube wall clear, prevents deterioration of the filament, and maintains high output and colour rendition during the life span of the lamp.

Tungsten-halogen lamps are available in standard, screw-base sizes and in a linear arrangement of tube and filament. (See Figs. 19.51 and 19.53) The screw-base lamps can replace standard incandescent lamps without changing either the fixture or the wiring. Take care when handling lamps with an exposed filament tube. Perspiration from the hands will erode the quartz tube. A pair of cloth gloves will prevent damage to the tube during installation.

Most tungsten-halogen lamps are made for specific uses, such as for stage, studio, and photographic lighting where control of colour and direction are required. They are also used for decorative lighting around buildings. Figure 19.52 shows a floodlight fixture.

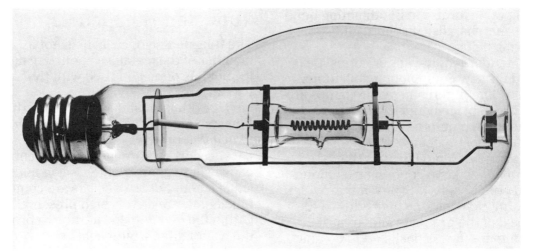

Mogul base lamp (for area lighting)

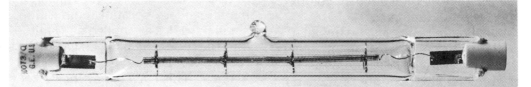

Linear quartz filament tube (for floodlights)

Medium-base
flood lamp

FIGURE 19.51 Tungsten-halogen lamps

Courtesy GE Lighting Canada

FIGURE 19.52 A quartz-flood floodlight
fixture

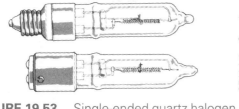

FIGURE 19.53 Single-ended quartz halogen lamps

Courtesy GE Lighting Canada

In recent years, the use of halogen light sources for merchandise display and residential area lighting and highlighting has increased. This is due to some extent to the development of smaller, more compact light sources, such as the PAR 20 and PAR 30 lamps.

Lamp Sizing. Parabolic-shaped lamps as shown in Figures 19.32 and 19.54 are sized according to the diameter of the lamp's lens face, in eighths of an inch. The PAR 20 size lamp unit would then be $\frac{20}{8}$ or 2.5 in. in diameter. The PAR 30 lamp would then be $\frac{30}{8}$ or 3.75 in. in diameter.

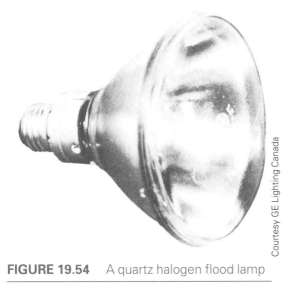

Courtesy GE Lighting Canada

FIGURE 19.54 A quartz halogen flood lamp

Energy Saving

Conservation conscious manufacturers are now producing a range of fluorescent tubes that provide nearly as much light output as conventional tubes but reduce power consumption between 14 and 20%. Use of these lamps can add up to considerable savings over the period of a year. These new lamps are able to fit into existing fixtures, providing complete interchangeability with existing tube sizes and wattage ratings. There is no loss in average lamp life when using these tubes.

Many commercial users of lighting, where large numbers of units are operating for extended periods of time, are updating their systems to the newer, more efficient forms of lighting such as the High Intensity Discharge lamps. Some companies have claimed complete coverage of their change-over costs within several years. The more efficient the light source, the more energy (and money) can be saved. Figure 19.55

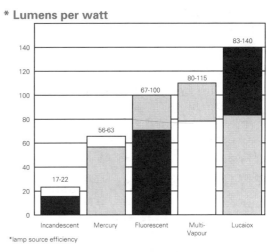

* **Lumens per watt**

*lamp source efficiency

FIGURE 19.55 Newer, more efficient light sources provide more light at a lower rate of power consumption and costs.

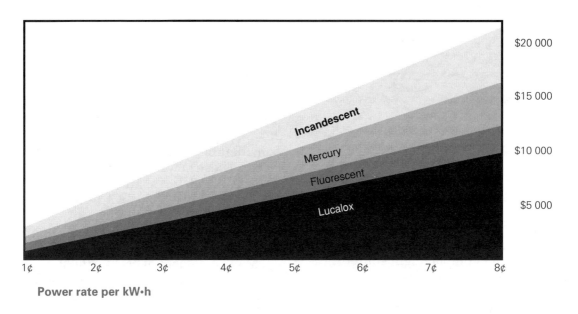

$20 000

$15 000

$10 000

$5 000

Incandescent

Mercury

Fluorescent

Lucalox

1¢   2¢   3¢   4¢   5¢   6¢   7¢   8¢

Power rate per kW·h

FIGURE 19.56 A comparison of the cost to illuminate 2000 m² to 540 lx for a year by lighting type. (The symbol ''lx'' is the shortform for ''lux,'' the metric unit for measuring illumination of a source of light per unit area on a surface.)

illustrates the differences in lamp efficiency for the various types of light sources available. The cost of lighting can be compared for the various forms of lighting in Figure 19.56.

Lamp Maintenance

Lamps are rated by the manufacturers to give the user an idea of the expected average life from the lamp installed. Fluorescent lamps, for instance, are rated at 20 000 h average life. If the lamps were to burn 20 h a day for 6 d a week, it would take over three years for the lamps to burn out. In large industrial or commercial lighting layouts, the cost of labour for lamp replacement often outweighs the cost of the actual lamp. It is recommended by manufacturers that the

lamps be replaced when they have been burning for 75% of their rated life span.

On the average, only 15% of the lamps will have actually burned out at the 0.75 life span level, and so replacement costs will not be too high. Once the lamps reach 75% of their life span, the remaining lamps (85% of them) can be expected to burn out in fairly rapid succession. The labour cost from there on in can be quite heavy.

If all the lamps are replaced at the rated 75% life span level and the fixtures cleaned and washed to increase reflection, the amount of light will be increased considerably and maintenance costs kept within reasonable limits.

1. List three main advantages that fluorescent lighting has over incandescent lighting.
2. List the disadvantages of fluorescent lighting.
3. List and explain the use of the parts of the fluorescent tube.
4. Explain how a starter operates in a preheat lamp.
5. What are the two functions of the ballast? How does it work?
6. Why must rapid-start fixtures be grounded?
7. What effect does fluorescent lighting have on power factor in industrial plants?
8. What are manufacturers now doing to prevent damage from ballast overheating?
9. What change has been made in the design of tube sockets for instant-start fixtures? Why?
10. List and explain the use of the parts of a mercury vapour lamp.
11. Explain briefly in your own words how a mercury vapour lamp produces light.
12. What are the disadvantages of mercury vapour lighting?
13. List four areas of use for mercury vapour lighting.
14. How does a metal halide lamp differ from a mercury vapour lamp?
15. What are the advantages of metal halide lamps?
16. What special material was developed for use with the high-pressure sodium lamp?
17. What are the advantages of high-pressure sodium lamps?
18. Describe the colour sequence of a high-pressure sodium lamp during its warm-up period.
19. To what family of lamps does the tungsten-halogen lamp belong?
20. Explain briefly how a tungsten-halogen lamp works.
21. What are the advantages of the tungsten-halogen lamp?
22. What type and size of metric lamp is available?
23. What two methods can be used to conserve energy when using a lighting system?
24. At what point in a fluorescent lamp's rated life span should it be replaced? Explain why.
25. What type of light source can be used to replace mercury vapour lamps for improved lighting costs and energy conservation?

Motor Control

Electrical motors are used for a wide variety of residential, commercial, and industrial operations. Because of this, guidelines are necessary. This chapter deals with the need for motor control and some of the more common control systems.

The Need for Special Control Equipment

Electrical motors operate on *magnetism*. The amount of current needed to create the magnetism depends on the size and design of the motor.

Nearly all motors tend to draw much more current during the *starting* period (*starting* current) than when rotating at *operating* speed (*running* current). Motors are rated in horsepower or watts. The higher the rating of the motor, the higher the starting and running currents will be.

Every motor tends to produce a counter voltage and current within its windings. This *generator* action produces a voltage and current that flows *opposite* to the applied current; in fact, the generated current flow helps to control the flow of incoming current through the motor. Manufacturers design motors with this current in mind. Proper control

of incoming current extends the life span of the motor. Excessive input current will severely damage or burn the motor's windings.

One main factor in determining the amount of generated voltage and current in the motor is its *speed*. If the load placed on a motor reduces the speed, less generated current will be developed and more applied current will flow. That is, the *greater* the load on the motor, the *slower* it will rotate and the *more* applied current will flow through its windings. This is why a motor requires more current during the starting period: most electrical motors have a starting current that is three to five times' the normal running current.

As the speed picks up, the generated voltage within the motor gradually increases. However, the instant the starting switch is closed, no generated voltage exists and the applied current becomes very high. If the motor is jammed or prevented from rotating in any way, a *locked rotor* condition is created. The part of the motor that rotates is often called the *rotor*. When the rotor fails to turn, the excessive applied current is called the *locked rotor current*. This high current will cause the motor to burn out quickly.

Motor Control Switches

There are two basic methods for starting motors: *across-the-line* starting, in which full-line voltage is applied to the motor, and *reduced voltage* starting. This section will discuss only the across-the-line starting method.

Any switch or control device used to start (and stop) a motor *must* be able to withstand a higher inrush of current during the starting period. Switch contacts must close rapidly, make a sure connection, and prevent arc damage to the switch. Since another important function of the switch is to open the circuit, allowing the motor to stop, any switch or control device must also be able to open the circuit under locked rotor conditions. (See Section 28, Canadian Electrical Code.) The switch or control device must have a strong *spring action* to open the contacts quickly. Otherwise, there may be considerable arc damage to the switch.

A control switch without the appropriate design features cannot safely control the motor's starting or running current. It also cannot provide safe stopping under abnormal overcurrent conditions caused by locked rotors or short circuits. All these conditions can lead to serious arc damage within the switch, resulting in the destruction of the switch contacts and their control mechanism.

Safety Note: For this reason, experienced electricians usually do *not* stand in front of a switch when it is being operated. The *safest* way is to keep the face and body off to one side and use the left hand to operate the switch. Although severe arcing does not happen often, the danger of serious damage to the switch and injury to the operator is always there.

Motor starting current and running current vary with the motor's power. As the power increases, the starting and running currents also increase. Any switch used to control a motor, whether in a basement workshop or for industrial duty, should have a *power rating* marked on it. This rating will indicate whether the switch can start and stop the motor safely. (See Section 28, Canadian Electrical Code.)

Location of Control Devices

Section 28 of the Canadian Electrical Code recommends that a control switch be located within sight of the motor. The operator can then check that there is no danger to equipment or persons before starting the motor.

Overcurrent Protection

Like any other electrical circuit, motor circuits *must* be protected from overcurrent conditions. Otherwise, a locked rotor condition or a short circuit will damage both the wiring and control devices. The Canadian Electrical Code lists the fuse or circuit breaker devices that can be used. (See Table 20.1)

As a general rule, overcurrent protection should not exceed 300% of the motor's running current, which is listed as the *full load current* on the nameplate of the motor. High starting currents are responsible for fusing the motor at a value above its rated current. The Canadian Electrical Code gives more detailed figures for the various types of motors. (See Table 20.2)

Motor Conductor Sizes

The Canadian Electrical Code lists ampacity ratings for conductors supplying current to a single motor. (See Table 20.1) Motors with larger current ratings than those listed require conductors

TABLE 20.1

Full-Load Current Rating of Motor	Minimum Allowable Ampacity of Conductor	Overload Protection for Running Protection of Motors		Overcurrent Protection Maximum Allowable Rating of Fuses and Maximum Allowable Setting of Circuit Breakers of the Time-Limit Type for Motor Circuits					
		Maximum Rating of Type D Fuses	Maximum Setting of Overload Devices	Single Phase All Types and Squirrel Cage and Synchronous (Full Voltage, Resistor and Reactor Starting)		Squirrel Cage and Synchronous (Auto-transformer and Star-Delta Starting)		DC or Wound Rotor AC	
Amperes		Amperes	Amperes	Fuse Amperes	Circuit Breaker Amperes	Fuse Amperes	Circuit Breaker Amperes	Fuse Amperes	Circuit Breaker Amperes
1	15	1.125	1.25	15	15	15	15	15	15
2	15	2.225	2.50	15	15	15	15	15	15
3	15	3.5	3.75	15	15	15	15	15	15
4	15	4.5	5.00	15	15	15	15	15	15
5	15	5.6	6.25	15	15	15	15	15	15
6	15	7	7.50	20	15	15	15	15	15
7	15	8	8.75	25	15	15	15	15	15
8	15	9	10.00	25	20	20	15	15	15
9	15	10	11.25	30	20	25	15	15	15
10	15	12	12.50	30	20	25	20	15	15
11	15.00	12	13.75	30	30	30	20	20	15
12	15.00	15	15.00	40	30	30	20	20	15
13	16.25	15	16.25	40	30	35	30	20	20
14	17.50	17.5	17.50	45	30	35	30	25	20
15	18.75	17.5	18.75	45	30	40	30	25	20
16	20.00	17.5	20.00	50	40	40	30	25	20
17	21.25	20	21.25	60	40	45	30	30	30
18	22.50	20	22.50	60	40	45	30	30	30
19	23.75	20	23.75	60	40	50	40	30	30
20	25.00	25	25.00	60	50	50	40	30	30
22	27.5	25	27.5	60	50	60	40	35	30
24	30.0	30	30.0	80	50	60	40	40	30
26	32.5	30	32.5	80	70	70	50	40	40
28	35.0	35	35.0	90	70	70	50	45	40
30	37.5	35	37.5	90	70	70	50	45	40
32	40.0	40	40.0	100	70	70	70	50	40
34	42.5	40	42.5	110	70	70	70	60	50
36	45.0	45	45.0	110	100	80	70	60	50
38	47.5	45	47.5	125	100	80	70	60	50
40	50.0	50	50.0	125	100	80	70	60	50
42	52.5	50	52.5	125	100	90	70	70	70
44	55.0	50	55.0	125	100	90	100	70	70
46	57.5	50	57.5	150	100	100	100	70	70
48	60.0	60	60.0	150	100	100	100	80	70
50	62.5	60	62.5	150	125	100	100	80	70
52	65.0	60	65.0	175	125	110	100	80	70
54	67.5	60	67.5	175	125	110	100	90	70
56	70.0	70	70.0	175	125	125	100	90	70
58	72.5	70	72.5	175	125	125	100	90	100
60	75.0	70	75.0	200	150	125	100	90	100
Col. 1	Col. 2	Col. 3	Col. 4	Col. 5	Col. 6	Col. 7	Col. 8	Col. 9	Col. 10

For full information of conditions that may change the values in this table, see the corresponding table in the Canadian Electrical Code.

Applications of Electrical Construction

TABLE 20.2	Rating or Setting of Overcurrent Devices for the Protection of Motor Branch Circuits		
	Per Cent of Full Load Current		
Type of Motor	**Fuse Rating**	**Maximum Circuit-Breaker Setting**	
		Instan- taneous Type	**Time- Limit Type**
Alternating Current			
Single-phase all types	300	—	250
Squirrel-cage and Synchronous:			
Full-voltage Starting	300	700	250
Resistor and Reactor Starting	300	—	250
Auto-transformer Starting:			
Not more than 30 A	250	—	200
More than 30 A	200	—	200
Wound Rotor	150	—	150
Direct Current			
Not more than 40 kW	150	250	150
More than 40 kW	150	175	150

For full information of conditions that may change the values in this table, see the corresponding table in the Canadian Electrical Code. (Note that the 40 kW motors were formerly rated as 50 horsepower motors.)

with ratings equal to *125%* of the motor's full load current.

Motors that are operated for short periods can be supplied with conductors that have a somewhat lower current rating. The Canadian Electrical Code shows how to determine conductor ampacities. (See Table 20.3)

When conductors are used to supply two or more motors on the same circuit, conductor ampacity can be determined by adding the full load currents of all the motors in the circuit; then *25%* of the *largest* motor's full load current is added to the total. (See Section 28, Canadian Electrical Code.)

Thermal Overload Relay Protection

Often a motor is loaded *beyond* its designed capacity. Motors in wood- and metal-cutting machines, pumps, hoists, and fans are examples of where overloading can occur. Overloading slows down the motor, which results in less generated voltage and an increase of input current. For example, when using a saw, if the board is damp or the cut is too deep, the motor will be overloaded and slow down. The current flow in the windings will increase and heat the motor beyond its design temperature. A jammed pump or an extra-heavy load on a hoist will have the same effect on a motor. Most electrical motors have *cooling* fans or blades built in, but such a cooling system becomes less and less effective as the motor's speed is reduced. If nothing is done, there may be permanent damage to the motor's windings and expensive repairs will be needed.

It is human nature to try for "one

TABLE 20.3 Data for Determining Conductor Sizes for Motors of Different Duty Types

Classification of Service	Percentage of Nameplate Current Rating of Motor			
	5 Min Rating	15 Min Rating	30 and 60 Min Rating	Continuous Rating
Short-time Duty. Operating valves, raising or lowering rolls, etc.	110	120	150	
Intermittent Duty. Freight and passenger elevators, tool heads, pumps, drawbridges, turntables, etc.	85	85	90	140
Periodic Duty. Rolls, ore- and coal-handling machines, etc.	85	90	95	140
Varying Duty.	110	120	150	200

more cut" or "one last load." Knowing this, manufacturers have designed a heat-sensitive mechanical device. When a motor has been overloaded for a certain length of time, this device, called a *thermal overload relay*, opens the circuit. (See Fig. 20.1) It does not cause "nuisance tripping," however, by opening the circuit every time there is a brief or minor overload. The relay opens the circuit *only* when there is a prolonged overload condition.

Modern 3 phase motor control units require three overload relays to provide full protection for the motor. On some manufacturers' motor controllers, there are three heater units with a mechanical linkage. These relays operate a single set of overload contacts. (See Fig. 20.26 on page 354.) Other manufacturers produce motor controllers with bimetallic relays, each having individual contacts. These contacts are connected in a series circuit: any one set of contacts opening will shut down the motor starter if an overload condition exists on the motor.

Relay Class Designations. There are

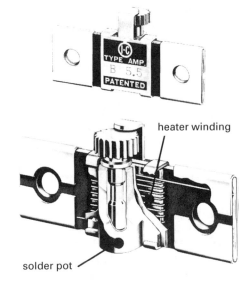

heater winding

solder pot

NOTE: The solder pot is heat sensitive. The thermal unit provides an accurate response to the overload current.

NOTE: The heater winding is heat producing. It is permanently joined to the solder pot to ensure a proper heat transfer.

Courtesy Square D Canada

FIGURE 20.1 Front and cutaway views of a thermal overload relay unit

three class designations for overload relays—Class 10, Class 20, and Class 30. These class sizes are based on the amount of time required to trip the relay or current element. The most common of these is the Class 20, which refers to a relay that trips within twenty seconds of the motor current being at 600% of its rating. Similarly, a Class 10 relay must trip within ten seconds, and a Class 30 relay must trip within thirty seconds under 600% current conditions.

Operation of the Thermal Overload Relay. The relay has several parts. Each part passes its energy on to the next.

The *heater* part is connected to the motor *in series* with the *supply conductor*. Under normal load conditions, the heater remains at a moderate temperature. As the load increases beyond the motor's designed capacity, more current flows and the temperature of the heater rises. Within a short period of time, the heater melts the *solder* in the solder pot. The small *ratchet wheel* can then rotate, triggering the *mechanical* part of the relay. These *eutectic alloy relays* are designed with precision heater elements which raise the solder alloy's temperature until the alloy liquefies at a predetermined temperature. (A eutectic alloy is a mixture of several metals that will consistently turn from a solid to a liquid upon reaching the lowest possible melting point for that particular mixture.) A variety of interchangeable heater elements can be used, thereby providing a choice of trip times for the motor controller. In this manner, the overload relay's trip characteristics can be tailored to specific motor requirements. Figures 20.2 and 20.3 illustrate the operation of this relay type. Motors requiring a longer start-up

time should have a relay unit that accommodates the starting process and provides overload protection to the motor under running conditions. The movement of the spring-loaded mechanical section cuts off current to the motor in one of two ways.

In the *manual* motor starter, the action is a simple mechanical operation. The main switching mechanism of the starter is released and allowed to spring or trip to an open position by the operation of the overload relay. The circuit delivering current to the motor then opens, preventing further operation of the motor. The manual motor starter must be *reset* by the operator once the motor and OL relay have cooled down.

The magnetic motor starter uses the OL relay action in a slightly different manner. The relay's mechanical action, set in motion by the heating of the relay's solder pot, opens a set of contacts which are connected *in series* with the starter's control circuit. When these contacts open, they cut off the current flow to the magnetic coil of the starter. As the coil loses its magnetism, it in turn releases the main contacts from their closed position. Current flow to the

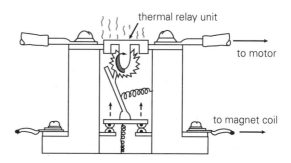

The operation of a melting alloy overload relay is shown. As heat melts the alloy, the ratchet wheel is free to turn – a spring then pushes the contacts open.

FIGURE 20.2   An overload relay mechanism

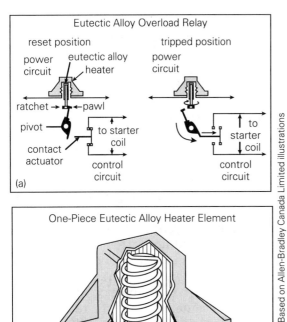

Eutectic Alloy Overload Relay

reset position — tripped position

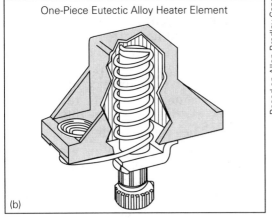

One-Piece Eutectic Alloy Heater Element

(b)

FIGURES 20.3A AND B   A eutectic alloy solder-pot relay with a ratchet-wheel trip mechanism

Based on Allen-Bradley Canada Limited illustrations

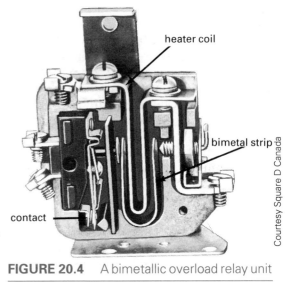

heater coil

bimetal strip

contact

Courtesy Square D Canada

FIGURE 20.4   A bimetallic overload relay unit

motor is cut off. These OL relay units must be reset by the operator once they and the motor have cooled down, but only the control circuit is re-established by the resetting action. A separate *start* button must be operated to get the motor running again. Figure 20.2 illustrates an overload (OL) relay mechanism.

Many manufacturers produce a solder-pot type of OL relay for use in their motor control equipment, but two other types of relay can be found in various motor control units.

Bimetallic thermal overload relays make use of a simple, U-shaped strip of metal (fabricated from two different types of metal) which bends or deflects when heated by the heater portion of the relay. This bending action will open a set of contacts and prevent further operation of the motor. In many cases these relays are adjustable over a range of 85% to 115% of nominal heater ratings. (See Fig. 20.4)

Unlike eutectic alloy relays, *bimetallic* units are often sensitive to shock and vibration, tripping needlessly. The two different metals that make up the bending strip move gradually to release the trip mechanism. This gradual movement means that the electrical contacts open slowly. Contacts opened slowly are prone to arc damage from the current in the circuit. They can flutter or fluctuate open and close, causing even more arc damage.

Some manufacturers produce *multiple heater* units that monitor each of the three phases in a polyphase system. A *differential* mechanism is used to provide *phase-loss sensitivity*. This, simply stated, means that the motor will not be allowed to run continuously if one of the three phases is opened or disconnected. Three phase motors will run on any two of the three phases, but at approximately half their horsepower. They are likely to heat up and burn out if not prevented from running.

The differential mechanism is necessary on bimetallic 3 phase relays because they lose some of their force on the trip bar when one of the sections has actuated. The mechanism uses the cooling action of the tripped relay section to help initiate a tripping action on one of the remaining two sections. The motor can therefore be stopped at or below the normal 3 phase tripping current level. See Figures 20.5 and 20.6.

Some manufacturers produce a bimetallic relay that makes use of a *replaceable* heater element, similar to that used with the solder-pot relay type. In the manual position, this relay behaves much like the relays discussed previously. Once set in the automatic position, the relay will restore power to the circuit as soon as it cools down to the proper level. This ability is quite convenient if the relay is installed in an inaccessible place but can also cause several problems.

If the cause of the overload has not been removed from the motor, the relay will allow the motor to restart each time it cools down, and the motor will eventually be damaged as a result of the continuous inrush currents from the many restarts.

A second and more serious problem for the operator is the fact the relay will allow the motor to restart as soon as it cools down, even if the operator is working on the machine at the time.

Safety Note: Great care must be taken, when using the automatic setting, to open the main switch for the motor.

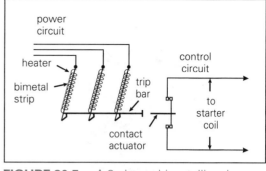

FIGURE 20.5 A 3 phase, bimetallic relay mechanism

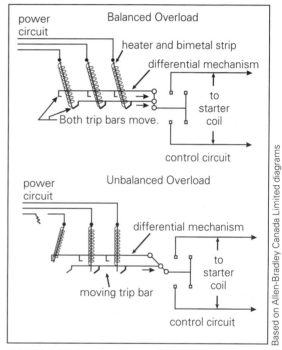

FIGURE 20.6 A differential overload relay mechanism for use on 3 phase motor control systems

To prevent injury to the operator, the motor must not be allowed to restart while investigation or repairs are in progress. Figures 20.4, 20.5, and 20.6 illustrate this type of relay.

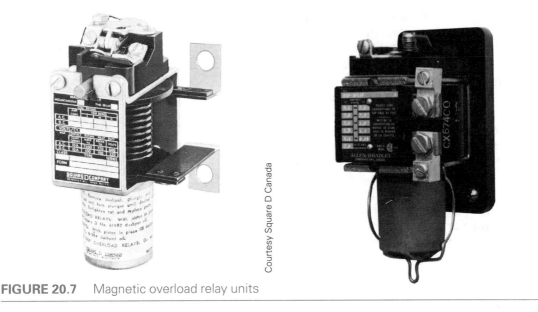

FIGURE 20.7   Magnetic overload relay units

Courtesy Square D Canada

Courtesy Allen-Bradley Canada Limited

Magnetic overload relays have a moveable, magnetic core inside a coil which is connected in series with the motor supply leads that carry current to the motor. Normal operation of the motor will allow the core to rest in such a position that the contacts of the relay are closed. When an overload condition is experienced by the motor, more current will be drawn by the motor and the coil. The magnetic attraction of the coil will become stronger than normal, and it will draw the core further into the relay.

As the core is pulled into the relay, it opens the contacts in the relay that are used to turn the motor starter off. To prevent nuisance tripping, a piston-in-oil or piston-in-air unit attached to the core slows the movement of the core. This *dash pot*, as it is called, works like an automobile shock absorber and prevents rapid core movement. The time taken for the core to open the contacts depends on the adjustment of a screw that opens or closes an oil or air by-pass. The amount of current required to draw the

core into the coil (tripping current) is adjusted by a threaded rod, which positions the core inside the magnetic coil.

This type of relay is often used to protect motors that have unusual duty cycles or require longer than normal periods of time to reach operating speed. Figure 20.7 illustrates magnetic OL relays.

Determining Overload Relay Size.   The size of thermal overload relay used is determined by the current requirements of the motor. The time it takes to open the circuit is determined by the extent of the overload condition.

Method 1.   Under normal operating conditions, motors can handle 125% of their rated full load amperes (*nameplate current*). Currents larger than this are likely to damage the motor if allowed to continue for more than a few minutes. The overload relay should be capable of opening the circuit when current *exceeds* 125% of the full load amperes.

Applications of Electrical Construction

For example, if a motor needs 8 A to operate, the overload relay should be capable of handling 10 A (125% × 8 A). This slight, extra allowance prevents nuisance tripping during minor overloads.

Method 2. Manufacturers usually place an overload relay *selection chart* inside the covers of control switches. (See Fig. 20.8) The manufacturer's nameplate on the motor will show the full load amperes.

For example, if the full load amperes is 8 A, look for the value that is mathematically closest under the Motor Amps column in Figure 20.8. In this case, the value 8.23 is closest. Now find the heater number for this manufacturer's model under the Heater Element heading. The heater number is opposite the current value, 8.23. It is W50 and is the manufacturer's catalogue number. Use it when placing an order, but remember that although it *may* be quite close to the 125% calculation, the heater number is not a current rating.

Restarting After an Overload. To restart a manual motor starter equipped with a thermal overload device, move the operating handle (*toggle*) from the central *trip* to the *reset* position. (The overload condition will have moved the toggle handle from the *on* to the central *trip* position.) The toggle handle *must* be moved to the *off* position to reset the spring-loaded mechanical section of the relay. The control switch can then be turned *on*.

The solder pot in the thermal relay will require several minutes to cool off before the solder solidifies and grips the ratchet wheel. As a result, the motor has a cooling out period before it is placed back in service.

The mechanical section of the relay

Overload Heater Element Selection			
Motor Amps	Heater Element	Motor Amps	Heater Element
0.19	W10	2.60	W38
0.21	W11	2.86	W39
0.23	W12	3.16	W40
0.25	W13	3.48	W41
0.28	W14	3.84	W42
0.30	W15	4.22	W43
0.33	W16	4.65	W44
0.36	W17	5.12	W45
0.39	W18	5.63	W46
0.43	W19	6.20	W47
0.48	W20	6.82	W48
0.52	W21	7.51	W49
0.57	W22	8.23	W50
0.62	W23	9.07	W51
0.69	W24	9.95	W52
0.76	W25	10.8	W53
0.83	W26	11.9	W54
0.91	W27	13.3	W55
1.01	W28	14.6	W56
1.12	W29	16.0	W57
1.22	W30	17.4	W58
1.34	W31	19.0	W59
1.47	W32	20.7	W60
1.62	W33	22.7	W61
1.78	W34	24.7	W62
1.96	W35	27.0	W63
2.15	W36		
2.36	W37		

FIGURE 20.8 Overload relay selection chart for fractional horsepower motors

on some older models of motor controllers would not reset *fully* until the solder had cooled completely. If the resetting mechanism was pressed while the solder was cooling and only in a semi-hardened state, the solder would not grip the ratchet wheel. The relay had to be removed from the controller, reheated with a match, allowed to cool properly before resetting, and then reinstalled in the controller.

Relays on modern motor controllers do not have to be removed and reheated. They use a solder that will only reset when the relay has cooled properly. The solder will even reset when there have been premature attempts at resetting during the cooling process.

Manual Motor Control Switches

Figure 20.9 shows a compact, *manual motor control* switch, which is an *across-the-line* type of starter. The overload relay is on the left-hand side (marked A2.57). This unit fits into a special enclosure and can control most fractional kilowatt, single-phase motors. (See Fig. 20.10) It is ideal for use in home workshops and in industry.

Three-phase systems require *3 pole switches* capable of opening all three live conductors of the circuit.

Figure 20.11 shows the more complex manual type 3 phase switch with thermal overload protection. There are two relays visible on this switch. In some provinces (for example, Ontario), however, an overload relay is required for *each* phase (that is, three relays). The unit shown in Figure 20.11 can control 600 V motors with ratings up to 7.5 kW.

Magnetic Motor Control

A second type of *across-the-line* motor starter is the *magnetic controller* which uses an *electromagnet* to activate the motor control switch.

This type of motor control has three main advantages. The first is *low-voltage-protection*. With this protection, the electromagnet will disengage and open the motor circuit when there is a significant reduction, or loss, of line voltage.

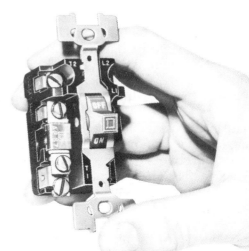

FIGURE 20.9 A manual motor control switch

FIGURE 20.10 A single-phase fractional kilowatt motor control switch with thermal overload relay and locking cover

Applications of Electrical Construction

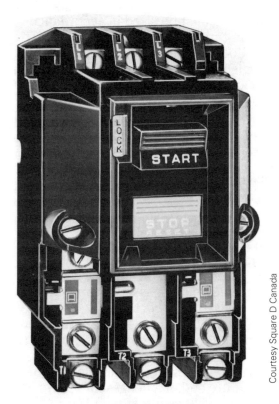

FIGURE 20.11 A 3 phase manual control switch with thermal overload protection

Courtesy Square D Canada

The second advantage of magnetic motor control is that a great variety of *activating devices* are available. Activating device styles include push buttons, limit switches, float switches, pressure switches, and foot controls.

Magnetic motor control is extremely *versatile* and can perform many operations. Also, it can be located anywhere required for convenience or safety, and it can have any number of smaller control stations in the circuit to activate the starter. In contrast, a manual motor starter must be placed within easy reach of the operator. If the starter is bulky, if damaging liquids or vapours are present, or if multiple control points are required, placing the starter near the operator may present a problem.

A third major advantage of magnetic motor control is the use of *low voltage* in the control circuits. Service and maintenance to the control circuit is less dangerous if 120 V is used to control a 600 V motor circuit. Controllers with this feature use a *step-down transformer* in the controller to arrive at the proper voltage for the control circuit. Figure 20.12 shows a circuit breaker-combination starter, step-down transformer, and magnetic starter in one cabinet.

Motors that continue to operate when there has been a reduction in line voltage often overheat, causing damage to the motor's windings. Once the line voltage is returned to full value, the magnetic motor controller will *not* reactivate on its own. An operator must press the *start* button to re-energize the electromagnet. The necessity of doing this prevents accidents (when using machinery such as saws, presses, and conveyors). There is no way of accurately predicting when line voltage will be restored. If the machine can start up *without* the starter being reset, operators working on or near the machine could be seriously injured.

Magnetic Motor Starters. Magnetic motor starters sized in accordance with the Electrical and Electronic Manufacturers' Association of Canada (EEMAC) or the National Electrical Manufacturers' Association (NEMA) are available in eleven sizes. Each size is limited to the amount of horsepower it can be expected to handle safely. See Table 20.4. Starters that do not have a class size assigned to them are simply rated in horsepower and volts.

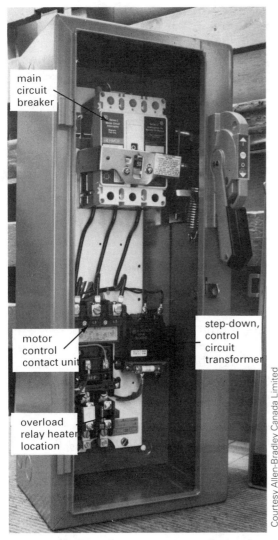

FIGURE 20.12 Typical breaker-combination starter, type 1, in a size 2 enclosure. (Door removed for photograph.)

Protection by Disconnecting Units.

Each motor must be protected by an overcurrent device such as a fuse or circuit breaker. The disconnecting switch, with or without fuses, or a circuit

breaker, must be sized in accordance with Section 28 of the Canadian Electrical Code. The disconnecting unit for a single motor must have a current rating of not less than 115% of the full-load current rating of the motor it controls. This disconnecting unit must be able to open and close the circuit safely, without exposing the operator or surrounding equipment to danger from arcing or flashing. Chapter 17 explains this danger more fully.

Operation of the Magnetic Motor Starter.

There are three basic methods for opening and closing the contacts of a magnetic motor starter. (See Fig. 20.13) When current is passed through the coil, the *magnet* attracts the *armature* and the *moveable contacts* are brought in to connection with the *stationary contacts*.

The pulsating effect of alternating current causes several problems. The armature and magnet assemblies tend to heat up as a result of the pulsating magnetic field around the main coil. To reduce this heating effect, called *hysteresis loss*, the armature and magnet are made of many thin layers of steel (*laminations*).

The *coil* loses its strength for an instant each time the alternating current falls to zero (120 times per second in a 60 Hz system). Damage results: there is a constant, high-speed attraction and release of the armature by the magnet, which can be heard as an annoying buzz or hum from the controller.

Extensive wear and heat on the magnet's *pole faces* can also be expected. Small copper rings, called *shading coils*, are embedded in the face of the magnet to counteract and reduce this undesirable effect. The pulsating magnetic field around the main coil induces (generates) a voltage and current in the

Applications of Electrical Construction

TABLE 20.4 EEMAC (NEMA) Sizes and Ratings of Magnetic Motor Starters

Size	Ampere Rating	115 V		200 V		230 V		460 V/575 V	
		HP	kW	HP	kW	HP	kW	HP	kW
00	9	³⁄₄	0.6	1½	1.1	1½	1.1	2	1.5
0	18	2	1.5	3	2.2	3	2.2	5	3.7
1	27	3	2.2	7½	5.6	7½	5.6	10	7.5
2	45	7½	5.6	10	7.5	15	11.2	25	18.7
3	90	15	11.2	25	18.7	30	22.4	50	37.3
4	135	25	18.7	40	29.8	50	37.3	100	74.6
5	270	50	37.3	75	56.0	100	74.6	200	149.2
6	540	—	—	150	112.0	200	149.2	400	298.4
7	810	—	—	—	—	300	224.0	600	447.6
8	1 215	—	—	—	—	450	335.7	900	671.4
9	2 250	—	—	—	—	800	596.8	1 600	1 193.6

NOTE: All kilowatt figures are approximate.

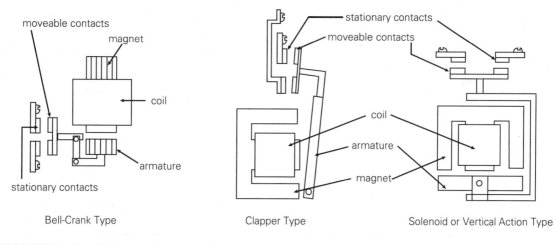

Bell-Crank Type   Clapper Type   Solenoid or Vertical Action Type

FIGURE 20.13 Three basic methods for operating magnetic motor starter contacts

shading coils. This induced current allows the shading coils to produce their own magnetic field, which helps to attract and hold the armature during the times that the main coil is weakened. (See Fig. 20.14)

A magnetic starter is made up of two electrical circuits. The *main*, or *power*, circuit, which has line terminals, main contacts, overload heaters, and motor terminals, is used to supply current to the motor. The *control* circuit, which has

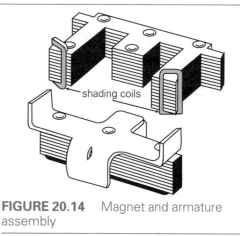

FIGURE 20.14 Magnet and armature assembly

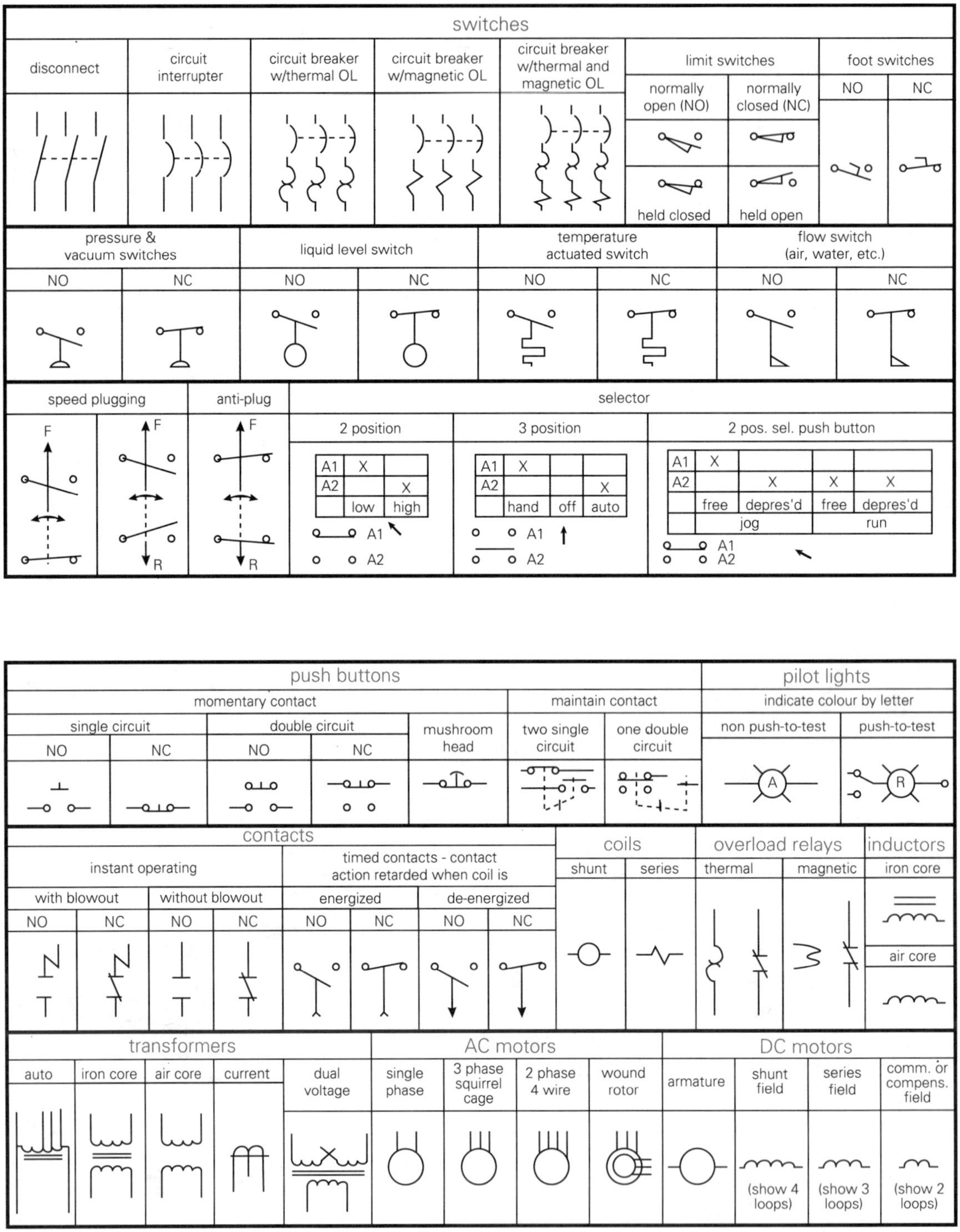

switches

disconnect	circuit interrupter	circuit breaker w/thermal OL	circuit breaker w/magnetic OL	circuit breaker w/thermal and magnetic OL	limit switches		foot switches	
					normally open (NO)	normally closed (NC)	NO	NC
					held closed	held open		

pressure & vacuum switches		liquid level switch		temperature actuated switch		flow switch (air, water, etc.)	
NO	NC	NO	NC	NO	NC	NO	NC

speed plugging	anti-plug	selector					
F / R	F / R	F / R	2 position	3 position	2 pos. sel. push button		

2 position:
A1	X	
A2		X
	low	high

o—o A1 ↖
o   o A2

3 position:
A1	X		
A2			X
	hand	off	auto

o   o A1 ↑
o   o A2

2 pos. sel. push button:
A1	X			
A2		X	X	X
	free	depres'd	free	depres'd
	jog		run	

o—o A1 ↖
o   o A2

push buttons / pilot lights

push buttons							pilot lights	
momentary contact					maintain contact		indicate colour by letter	
single circuit		double circuit		mushroom head	two single circuit	one double circuit	non push-to-test	push-to-test
NO	NC	NO	NC				A	R

contacts								coils		overload relays		inductors
instant operating				timed contacts - contact action retarded when coil is				shunt	series	thermal	magnetic	iron core
with blowout		without blowout		energized		de-energized						
NO	NC	NO	NC	NO	NC	NO	NC					air core

transformers				AC motors				DC motors			
auto	iron core	air core	current	dual voltage	single phase	3 phase squirrel cage	2 phase 4 wire	wound rotor	armature	shunt field	series field	comm. or compens. field
										(show 4 loops)	(show 3 loops)	(show 2 loops)

Based on a Square D Canada chart

FIGURE 20.15 Standard elementary diagram symbols

an electromagnet, overload contacts, auxiliary (or maintain) contacts, and an assortment of control stations, is used to activate the motor starter and ensure its safe operation.

Both of these circuits can be represented in the form of schematic wiring diagrams. Special symbols are used to represent each component in the circuit. (See Fig. 20.15)

Figure 20.16 shows a 3 phase magnetic motor starter. Figure 20.17 shows the wiring diagram for this controller. The *main power* circuit is outlined in black and the *control* circuit in red. This type of diagram, which shows each part in its correct location, is used when wiring or troubleshooting a controller.

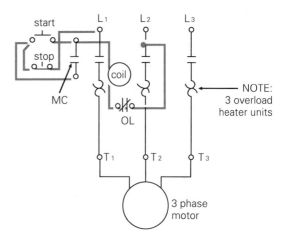

FIGURE 20.17   Wiring diagram for a 3 phase magnetic motor starter

FIGURE 20.16   A self-contained, 3 phase magnetic motor starter (with pilot light)

Courtesy Square D Canada

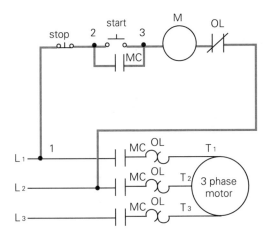

FIGURE 20.18   Schematic (elementary) diagram for a 3 phase magnetic motor starter

Figure 20.18 shows an elementary (schematic) diagram of the same controller. Although the parts are not in their true locations, this type of diagram gives an easily understood picture of the circuit and is most helpful when troubleshooting more complicated controller types. When a complex control system is to be developed, electricians often make a schematic diagram of the control circuit only. Figure 20.19 shows a variety of these diagrams.

Holding, or Maintain, Contacts

Magnetic motor starters must be able to open the power circuit to a motor whenever an overload relay indicates a dangerous condition, an under-voltage condition develops, or any form of control switch indicates that operation of the motor should be stopped. To do this, most control stations use a set of *start* contacts that remain closed only as long as the operator keeps a finger on the start button. The moment the button is released, a spring forces the contacts

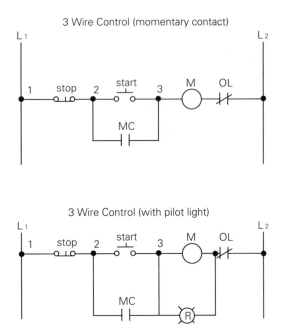

3 Wire Control (momentary contact)

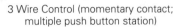

3 Wire Control (with pilot light)

NOTE: Pilot light can be wired in parallel with starter coil to show when starter energized and motor running.

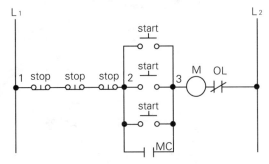

3 Wire Control (momentary contact; multiple push button station)

NOTE: Where a motor must be started and stopped from more than one location, any number of start and stop push buttons may be wired together. Also, it is possible to use only one start/stop station in combination with several stop buttons at different locations for use in emergencies.

FIGURE 20.19   Schematic diagrams for motor control circuits

Applications of Electrical Construction

open again. This de-energizes the coil in the starter and allows the motor to stop.

A set of holding, or *maintain*, contacts are either mounted on the same contact carrier bar as the main power circuit contacts or are activated by the contact carrier bar. Therefore, the maintain contacts open and close together with the main power circuit contacts. They are electrically connected *in parallel* with the start button contacts and keep the coil energized once the start button has been released. Maintain contacts are represented by a *normally open contact (NO)* symbol in both schematic and wiring diagrams. Often, the letters *MC* will appear beside the contacts to show that they are the maintain contacts. (See Figs. 20.17 and 20.18)

Control Stations

Figure 20.16 shows a self-contained motor controller with the start/stop control station mounted on the front of the unit. When there is to be *multiple-point* control, a *remote push-button* station is used. Other types of control stations are available, such as *limit* switches to control the up and down movement of an electric door, *float* switches to control the level of liquid in a tank, and *dual-action* pump control switches. (See Figs. 20.20, 20.21, 20.22 and 20.23)

Sometimes it is necessary to by-pass the automatic control device and operate the controller manually to fill a tank or start a pump. A push button station with a selector switch is used with the controller to provide both *manual* and *automatic* control functions. (See Fig. 20.24)

Occasionally, an operator wishes to control the length of time a motor operates or the length of time before a second control circuit is energized. A

FIGURE 20.20 A remote push button start/ stop control station

timing controller is used for this purpose. (See Fig. 20.25) To calibrate the time cycle, an *adjusting screw* at the top of the unit regulates air (or oil) flow in and out of a *compression chamber*. This type of control device is useful in reduced voltage starting units for large motors.

As mentioned earlier, regulations require three overload relays (one per phase) on a 3 phase motor controller. (See Fig. 20.26, page 354.) The reason why is that a 3 phase motor will continue to operate on any two of its three phases if one should fail. It will not restart on two phases, but if one phase is disconnected, it can handle approximately half of its rated power. A motor that is putting out *less* than its rated power will soon heat up from the overload, and its windings will be damaged if it is not disconnected from the line. An overload relay on each phase ensures that two or more phases will indicate overload conditions and remove the motor from the line.

Overload relays on a magnetic starter function much like those on manual motor starters. The usual thermal element, solder pot, and ratchet wheel are used. The spring-loaded mechanical section of the relay, however, does not open the power circuit of the controller directly. Instead, when the *ratchet wheel* is allowed to rotate, a set of *normally closed contacts (NC)* are forced open by the *mechanical section*. These contacts are connected *in series* with the control circuit. Once they have been opened, the magnetic holding *coil* is de-energized, and the *armature* falls back to open the main power circuit contacts. As in the case of the manual starter, a reset button in the controller must be pressed to reactivate the

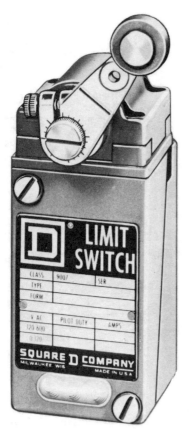

FIGURE 20.21 Limit switch

Courtesy Square D Canada

FIGURE 20.22 A float-operated switch

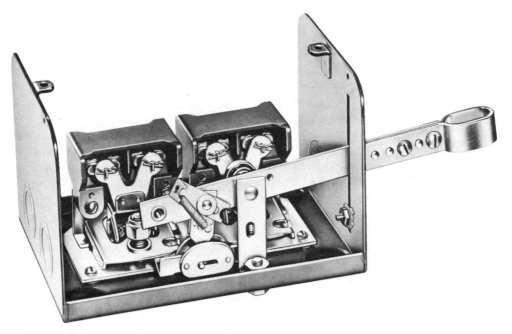

FIGURE 20.23   A dual-action pump control switch

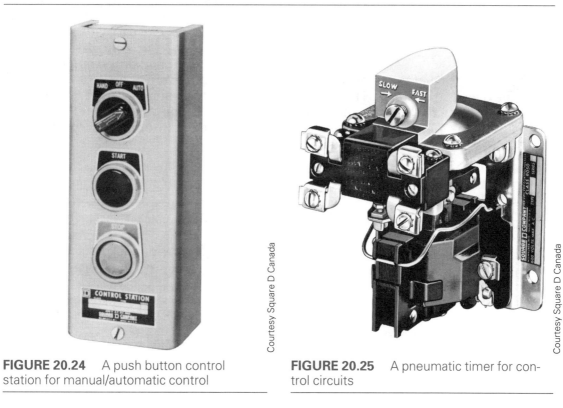

FIGURE 20.24   A push button control station for manual/automatic control

FIGURE 20.25   A pneumatic timer for control circuits

Jogging Circuits

Work must be positioned carefully on metal-cutting or forming machines before operation. Large machines often have motor-driven tables or material platforms. These motors are activated several times in quick succession to position the material accurately. This capability is called *jogging*. The main power contacts must open and de-energize the motor as soon as the operator releases the jog button.

Two basic types of circuits are used for jogging: the *push button* jog and the *selector* jog.

Push Button Jog. This jog uses a standard start/stop control circuit with a 4 terminal (double circuit) push button for the jog operation. (See Fig. 20.27) The *upper contacts* of the jog button open the regular current path, while the *bottom contacts* close a by-pass circuit around the maintain contacts. With the normal current path interrupted by the jog button, the maintain contacts cannot keep the coil energized.

There is one danger with this type of circuit. An operator in a hurry may release the jog button by allowing a fin-

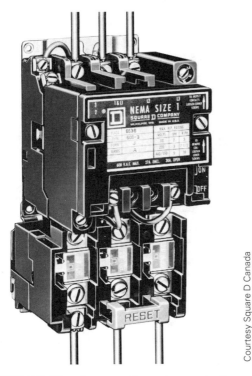

FIGURE 20.26 A magnetic motor controller with 3 overload relays

Courtesy Square D Canada

mechanical section of the relay and close the overload contacts. These contacts are shown on schematic and wiring diagrams by the *normally closed contact* symbol (*NC*) with the letters *OL* beside it.

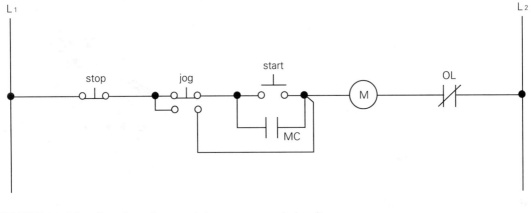

FIGURE 20.27 Start/stop/jog push-button control circuit

ger to slide off. The spring in the jog button might close the upper contacts quickly, before the armature and power contacts have had a chance to open the power circuit. If this happens, the maintain contacts will still be closed and the coil will remain energized, keeping the motor operating. To stop the motor, the stop button *must* be pressed. Although this situation does not happen often, when it does, the operator can be injured if the motor fails to stop when the jog button is released.

Selector Jog.　A safer form of jogging circuit is the selector jog. (See Fig. 20.28)

The selector switch jog circuit eliminates the danger of an operator releasing the jog button too quickly. This circuit places a single-pole selector switch *in series* with the maintain contacts. When the switch is closed (*run position*), the circuit operates normally. When the switch is open (*jog position*), the maintain contacts cannot keep the coil circuit energized. The power circuit opens as soon as the operator releases the start button.

As well, there are more specialized jogging circuits that make use of a control relay. (See Fig. 20.29)

If a motor circuit is to be jogged

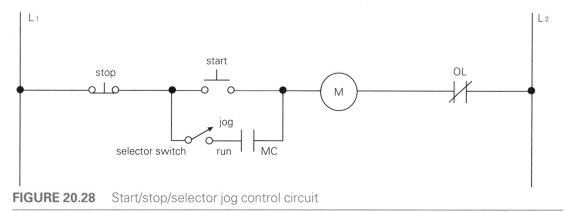

FIGURE 20.28　Start/stop/selector jog control circuit

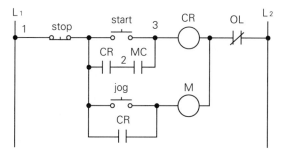

NOTE: Pressing the start button energizes the control relay (CR), which in turn energizes the starter coil. The normally open starter interlock and relay contact then form a holding circuit around the start button. Pressing the jog button energizes the starter coil independently of the relay. No holding circuit forms, and so jogging can be obtained.

FIGURE 20.29　Jog circuit with control relay

continuously (more than five times per minute), controllers must be *de-rated*, that is, either motors smaller than the nameplate rating of the controller must be used or a larger controller must be installed. The constant exposure to high inrush currents (motor starting current) will soon damage the contacts of the controller if jogging operations are carried on continuously.

Reduced Voltage Control

Many 3 phase motors operate on 480 V and 600 V (CSA standard C-235-83, preferred voltage levels for AC systems). Voltage at this level can cause painful injury to an operator or service mechanic if contact is made with live equipment. It is not uncommon to control the motor-starting equipment, at this voltage level, with a reduced voltage control system. Such a system makes the control circuit safer for installers and service personnel to work on and simplifies the actual equipment in the control circuit: less danger of a flashover or arc exists. There are two basic methods of connecting this type of reduced voltage control circuit. Figures 20.30 and 20.31 illustrate these two methods.

Figure 20.30 makes use of a *control circuit transformer* to reduce the line voltage down to a safer level (120 V). With this method, both motor and control voltage can be cut off by the same disconnect switch used to supply the motor circuit.

Figure 20.31 uses a separate source of supply for the control voltage. This system is useful when direct current (battery) or other separate source is needed to control the motor starter. In some work areas, windows or natural light sources are not available. If the

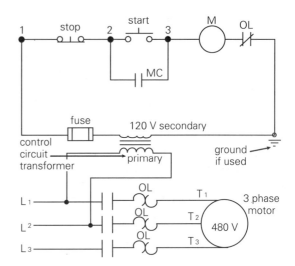

FIGURE 20.30 A reduced voltage control circuit using a source common to the motor and control circuit

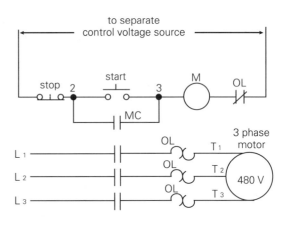

FIGURE 20.31 A reduced voltage control circuit using a separate voltage source

lighting system should fail, machine operators are left in a darkened room with all the machines running, a potentially dangerous situation. By connecting the control circuit source to the lighting panel in such an area, a loss of lighting voltage will also cancel operation of the machines. Operators will be able to move about in the darkened area more safely.

While only 480 V and 600 V were mentioned at the start of this section, supply voltages of many levels can be controlled by a reduced voltage system. The higher the supply voltage, the more likely a control circuit using reduced voltage will be required.

Reversing Controllers

Interchanging any two of the leads to a 3 phase motor will cause it to run in the *reverse* direction. Single-phase motors tend to be more complicated and often need connection changes within the motor itself. For this reason, only 3 phase controllers will be discussed in this section.

A 3 phase reversing starter consists of *two contactors* enclosed in the same cabinet. Only one set of overload relays is used, however, since both forward and reverse control circuits are interconnected.

When reversing a motor, it is vital that both contactors *not* be energized at the same time. Activating both contactors

would cause a short circuit since two of the line conductors are reversed on one contactor. Preventing a short circuit from this cause is called *interlocking*. Mechanical and electrical interlocking systems are available in most reversing controllers.

The *mechanical* interlock uses a system of levers to prevent the armature of the reversing contactor from engaging when the forward contactor's armature is in operation. Figure 20.32 shows a schematic diagram of the control circuit for this type of unit.

The *electrical* interlock makes use of two double-contact (4 terminal) push buttons. When the forward button is pressed, the upper contacts open the reverse coil circuit. Even if the reverse coil is energized, the control circuit will be broken. The power circuit is opened as soon as the forward button is pressed. (See Fig. 20.33)

The *electrical interlock* circuit allows the motor to be reversed simply by pressing the reverse button. The *mechanical interlock* system requires that the stop button be pressed. When it

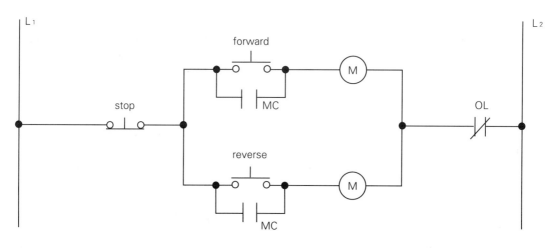

FIGURE 20.32 Mechanical interlock forward/reverse/stop control circuit

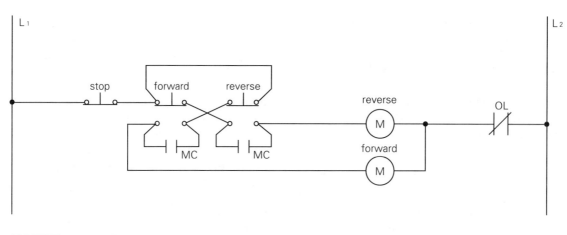

FIGURE 20.33 Electrical interlock forward/reverse/stop control circuit

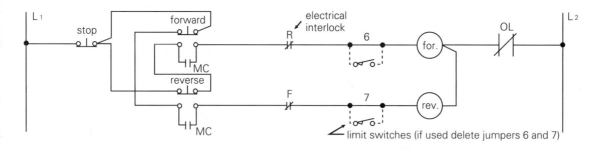

FIGURE 20.34 Schematic diagram showing a reversing control circuit with limit switches and separate electrical interlock contacts

is, the forward contactor will be disengaged before the reversing contactor can be brought into service.

Figure 20.34 shows a schematic diagram for a reversing control circuit utilizing limit switches and electrical interlock contacts. Operation of either the forward or reverse contactors will open the matching interlocking contact. This makes it unnecessary to press the stop button before changing the motor's direction of rotation.

Safety Note: Take care when reversing large motors. The sudden jar of direct reversal can damage the machine or equipment the motor is driving. High inrush currents can cause damage to both the motor and the controller if the motor is reversed without allowing enough time for the speed of the motor to decrease.

Figure 20.35 shows a complete wiring diagram for a reversing controller.

Figure 20.36 shows an alternate control circuit for such a controller schematically.

Multiple push-button stations can be used with the reversing controller. Figure 20.37 shows a schematic diagram of two forward/reverse/stop stations.

Applications of Electrical Construction

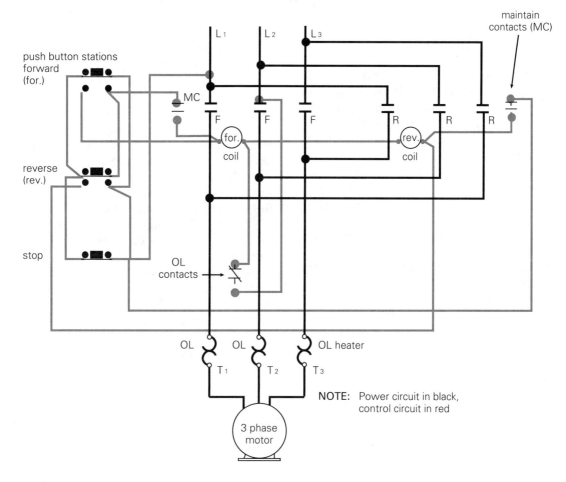

NOTE: Power circuit in black, control circuit in red

FIGURE 20.35 Complete wiring diagram of a 3 phase reversing controller

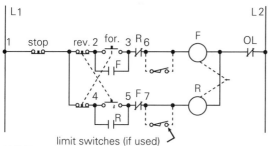

NOTE: 3 wire control of a reversing starter is possible with this forward/reverse/stop push button station. Limit switches can be added to stop the motor at a certain point in either direction. Jumpers from terminal 6 and terminal 7 to the forward and reverse coils must then be removed.

FIGURE 20.36 Alternate diagram of a reversing control circuit (with limit switches)

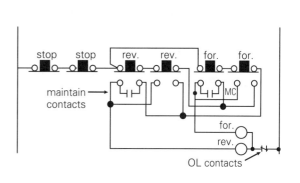

FIGURE 20.37 Schematic diagram showing multiple push button control for a reversing starter

---

Motor Control

Multi-Speed Motor Control

Some 3 phase motors, referred to as *multi-speed* motors, are designed to provide two separate speed ranges. There are two main types of multi-speed motors, the *separate winding* and *consequent pole* motors.

The separate winding motor, as the name implies, uses two or more windings which are electrically separate from each other. Each winding is capable of delivering the motor's rated horsepower (wattage) at the rated speed. The mechanical arrangement of the windings determines the number of magnetic poles per phase built into the motor, and thus the different speeds. The more poles per phase, the slower the operating rpm of the motor when that set of poles is being used. Since the windings are independent of one another, the speeds designed into the motor can be quite varied, such as 3600 rpm/600 rpm or 900 rpm/700 rpm. Figures 20.38 and

20.39 illustrate two-speed motor control circuits for use with separate winding motors.

The consequent pole motor uses a special winding which can be reconnected, using contactors, to obtain different speeds. *Two-speed* consequent pole motors always have a speed ratio of 2:1. Three types of consequent pole motors are available—*constant horsepower*, *constant torque*, and *variable torque*. The names of the three types indicate the output characteristics of the motors.

A two-speed consequent pole motor starter consists of two contactors, mechanically and electrically designed not to be activated at the same time (interlocking). One contactor has three poles, or contacts; the other has five.

For one speed, the 3 pole contactor supplies full-line voltage to the motor. For the second speed, the 5 pole contactor is activated, and the 3 pole contactor is automatically disconnected by the interlocking system. Three of the five poles provide full-line voltage to the motor through a separate set of motor leads. The remaining two poles reconnect the motor leads used by the 3 pole contactor, thereby creating a consequent pole configuration. This allows the motor to operate at a different speed.

With constant horsepower motors, the 5 pole contactor is energized on the lowest of the two speeds available. With constant and variable torque motors, the 5 pole contactor is utilized during the *high* speed operation of the motor. Wiring diagrams for the starters of these motors can be seen in Figures 20.40 and 20.41.

With all multi-speed controllers, overload relays are provided for both the high- and low-speed circuits to

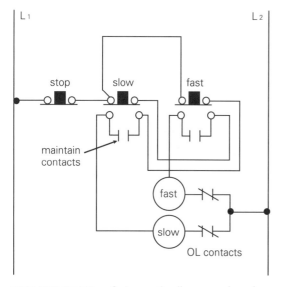

FIGURE 20.38   Schematic diagram showing a control circuit for a 2 speed (dual winding) motor

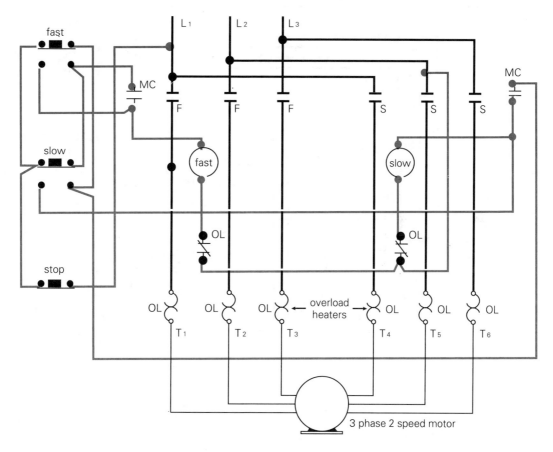

FIGURE 20.39 A dual-speed, 3 phase motor control circuit (dual winding motor)

ensure that there is adequate protection on each speed range (set of motor windings).

Reduced Voltage Starters

When motors are started with full-line voltage, high (*locked rotor*) currents and *maximum torque* can be expected. Under full-line voltage conditions, large motors often develop enough torque to damage belts, gears, and other drive line components. Starting current is often high enough to endanger the controller contacts, as well as to create a voltage disturbance on the line that can bother

other motors or electrical equipment. For these reasons, a starting system with *less* than line voltage is used.

These starters are made in Size 2 and larger. They are of two types: the *primary resistor* and the *auto transformer*.

Primary Resistor Starter.   This older starting method is not in common use today, but can be made available through many motor control manufacturers. Motor controllers of this type connect a set of resistors in series with the line which reduces line voltage for a predetermined length of time (3 s to 15 s). A *timing relay* activates the

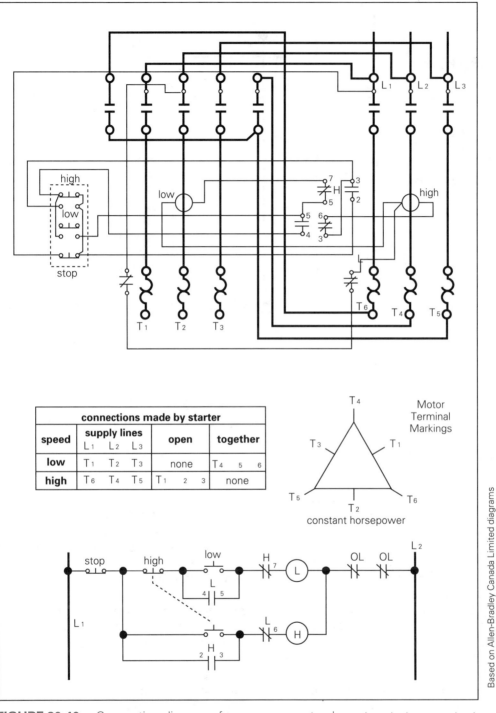

connections made by starter					
speed	supply lines			open	together
	L₁	L₂	L₃		
low	T₁	T₂	T₃	none	T₄ 5 6
high	T₆	T₄	T₅	T₁ 2 3	none

Motor Terminal Markings

constant horsepower

FIGURE 20.40 Connection diagrams for a consequent pole motor starter, constant horsepower type

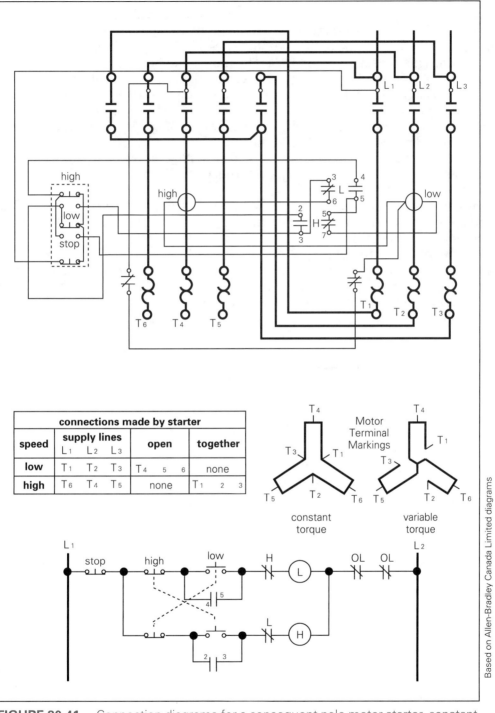

connections made by starter				
speed	supply lines L_1 L_2 L_3		open	together
low	T_1 T_2 T_3		T_4 5 6	none
high	T_6 T_4 T_5		none	T_1 2 3

Motor Terminal Markings

constant torque

variable torque

FIGURE 20.41 Connection diagrams for a consequent pole motor starter, constant torque or variable torque type

controller automatically for the second stage. (See Fig. 20.25) The motor-starting process begins with the pressing of the *start* push button. This pressing activates the line contacts coil, allowing current to flow through the primary starting resistors, and into the motor. The primary starting resistors reduce the line voltage to a level that allows the motor to start, but not draw an excessive amount of starting current.

A timing relay is activated at the same time as the line contacts coil. This relay can be preset by the operator to close its contacts at a predetermined time. When the contacts of the timing relay close, they activate the by-pass contacts coil. This in turn closes the resistor by-pass contacts, and current now flows around the primary resistors

to the motor. No voltage is lost or reduced by this path to the motor and full speed is soon reached by the motor. The motor can be stopped simply by pressing the *stop* push button.

This type of starting gives smooth acceleration to full operating speed without loss of speed during the change-over cycle.

Figure 20.42 shows a wiring diagram for this type of motor starter.

Auto Transformer Starter. The auto transformer starter, which is often called a *compensator*, uses a set of single-winding (auto transformer), step-down transformers to reduce the line voltage. As with the primary resistor starter, *partial voltage* is supplied to the motor on starting, with a *timer relay*

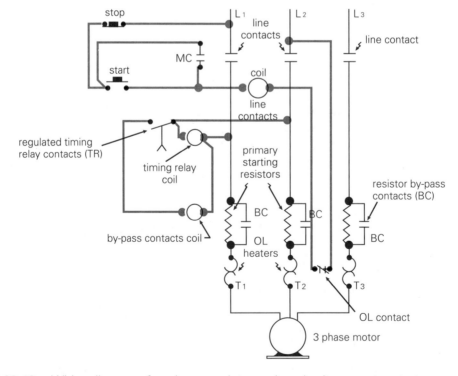

FIGURE 20.42 Wiring diagram of a primary resistor, reduced voltage motor starter

calibrated to activate the starter's second stage.

The starting sequence of this unit is set in motion by the pressing of the *start* push button. The pressing activates three electromechanical devices at the same time:

a) The timing relay is set for its predetermined time and begins its countdown. The *maintain contacts* in this controller are activated by the timing relay and keep the control circuit energized once the start button has been released.

b) The *start contacts coil* is activated

and in turn closes the *starting contacts*.

c) The *transformer contacts coil* is also energized and closes the *transformer contacts*, allowing curent to flow into the 3 phase transformer circuit (Wye connected).

As can be seen in the wiring diagram, Figure 20.43, the starting contacts receive their voltage and current from approximately midpoint on the transformer. This starting voltage is about 65% of the full-line voltage available. It can thus start the motor with a moderate amount of starting current, while preventing dangerous current surges.

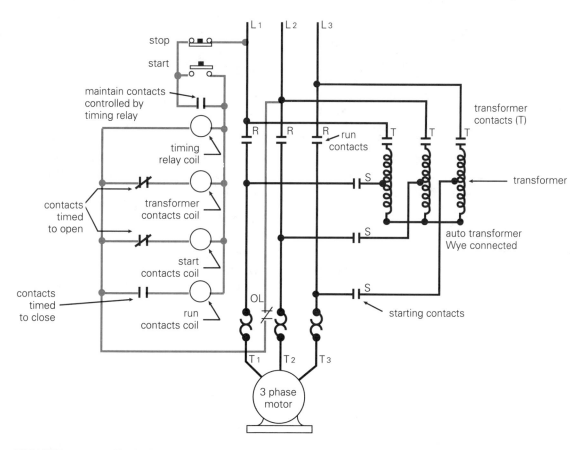

FIGURE 20.43 Typical auto transformer, reduced voltage motor starter diagram

When the timing relay reaches the end of its countdown (5 s to 15 s), it automatically opens the *starting* and *transformer* contacts by de-energizing the coils controlling those contacts. At the same time, the *run contacts coil* is activated and the *run contacts* are closed. Current is now allowed to travel straight through to the motor instead of detouring through the transformer circuit. Full-line voltage is now applied to the motor which soon reaches its full operating speed. Operation can be brought to a halt by pressing the *stop* push button.

More *starting torque* is available with this type of controller than with the primary resistor type of unit. Starting torque under reduced voltage conditions, however, is never equal to that obtained when using full-line voltage. When the commonly used 65% voltage tap is connected for start-up of the motor, the torque output is reduced to approximately 50% of the torque that would be available if full-line voltage was used. This may be quite sufficient for many applications, but high-inertia loads, such as large *ball mills*, may not be able to use the reduced voltage, starting-controllers due to the loss of torque during the starting sequence. Use of the 80% tap on the transformer will provide about 75% torque, while the 50% transformer tap will reduce torque level to approximately 30%.

Figure 20.43 shows this type of controller circuit.

To prevent the current surge normally occurring as the controller shifts from partial to full-line voltage, some manufacturers have produced a unit with a different type of control circuit.

Operating Sequence for a Reduced Voltage Starter, Closed Circuit Transition (Fig. 20.44):

a) Pressing the *start* button activates the 1S coil circuit. This 1S coil, in turn, closes the group of three 1S contacts located at the right-hand side of the auto transformers in the circuit diagram. This completes a *Wye* connection in the transformers themselves. At the same time, the 1S (normally open) contact in the 2S coil circuit is closed.

b) Closing the 2S coil circuit activates the 2S (normally open) contacts at the left-hand side of the auto transformers, feeding *line voltage* into the transformers. The 2S coil also closes the 2S contacts (normally open) in the R coil circuit and starts a pair of *timed*, TR2S contacts into their timed sequence operation.

At this point, the motor starts on 65% of the full-line voltage.

c) After 5 s to 15 s, preset by either the manufacturer or the installer, the "timed-to-open" TR2S contacts open the 1S coil circuit, thereby causing the transformers' 1S contacts to drop open and deactivate the Wye connection.

d) Simultaneously, the "timed-to-close" TR2S contacts activate the R coil circuit. This in turn closes the main R (Run) contacts in the starter, supplying full-line voltage to the motor and permitting it to reach operating rpm and torque.

e) The R coil also closes the R control circuit contact (which acts as a maintain contact) and opens the normally closed R contacts in both the 1S and 2S coil circuits. This cuts power to the transformers while they are in the *run* mode.

f) Pressing the *stop* button will open Line 1 to the entire control circuit and allow the motor to stop.

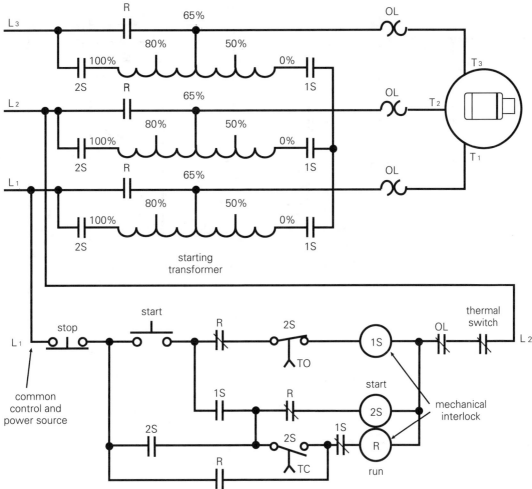

FIGURE 20.44 Closed-circuit transition, reduced voltage starter of the auto-transformer type

Wye-Delta Starter

A commonly used method of starting large refrigeration compressors and other similar motor applications where low starting torque is permissible is the Wye-Delta type of unit. The major advantage is the absence of starting resistors and/or transformers that produce heat as an undesirable by-product of their operation. A specially wound motor equipped with six leads (both ends of

each phase) must be used. The motor is started up with the internal windings connected into a Wye connection (sometimes referred to as a *star* connection), and the resulting voltage in each of the phases is reduced by $\frac{1}{\sqrt{3}}$ or 57.7% of the line voltage. Starting current and torque are reduced to approximately one-third of the full-voltage values, enough to move and start up a lightly loaded machine. (See Fig. 20.45)

Timer circuits within the controller

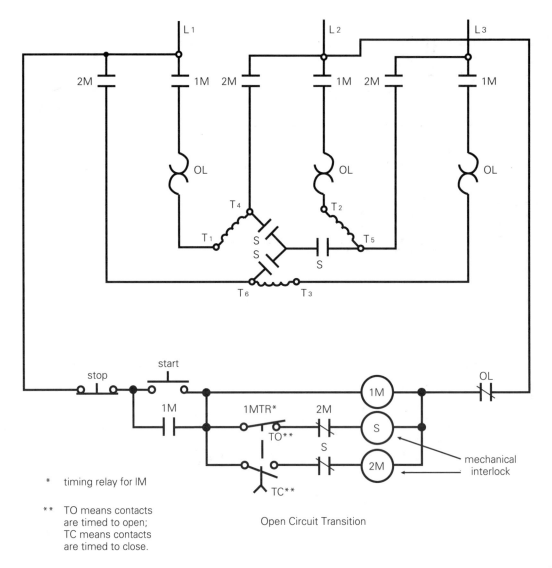

FIGURE 20.45   A Wye-Delta, reduced voltage motor starter using open circuit transition

switch the motor windings from the Wye/star configuration to the delta connection. This supplies full-line voltage to the motor, allowing it to come up to full-rated speed. During operation in the delta configuration, the contacts supplying current to the motor and the overload relays are subjected to 57.7% of the line current. As a result, the controller will carry a lower current to the motor and have a higher horsepower rating.

Wye-Delta controllers, like the auto transformer type of unit, take advantage of the closed-circuit transition principle whereby the motor can start up without any large current surges on the supply lines to the motor. Figure 20.46 illustrates this type of circuit.

Applications of Electrical Construction

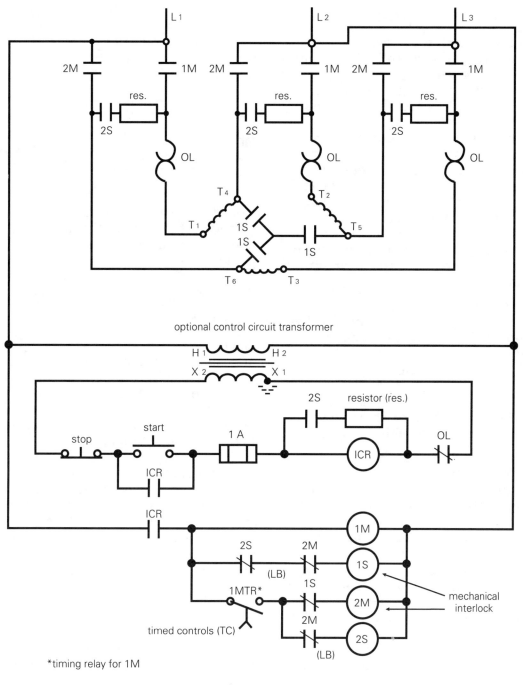

FIGURE 20.46 A Wye-Delta, reduced voltage motor starter using closed circuit transition

Solid-State Starter

Solid-state starters use microprocessor-based circuitry to control silicon control rectifiers (power SCRs), providing smooth, *stepless*, torque-controlled acceleration. This acceleration, known in the industry as *soft start*, can provide a current limit to the motor. On some motor applications, it is necessary to limit the maximum starting current, and with this type of unit, the current can be adjusted from 50% to 500% of the full load amperes. Programmed time and current levels can be selected through various switch settings (*DIP*) located on the microprocessor's printed circuit board.

Note: DIP switches, known as *dual-in-line-package* switches, frequently have moulded plastic bodies and are always small and thumbnail sized. Placed into an open or closed position by a small pointer or ballpoint pen, they are soldered into the controller's printed circuit board where they remain for the life of the circuit board. (See Fig. 20.47)

Pressing the *start* push button signals the solid-state starter to begin operating. Internal hold-in circuit latches and auxiliary contacts change state. At this point, an initial voltage is delivered to the motor windings. In the soft-start mode, this voltage continues to rise until the motor receives full-line voltage over a preprogrammed period of time and reaches full-rated rpm. Figure 20.48 illustrates a circuit for this controller.

The solid-state motor starter provides several useful, additional features. Visual indication of fault conditions, for example, stalled motor, phase loss, too high temperature, etc., can be built into the unit. An energy-saving feature for lightly loaded motors, along with con-

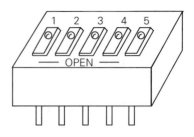

FIGURE 20.47 A dual-in-line-package switch, also known as a DIP switch

trolled stopping times (braking or extended stop time) further adds to the convenience and usefulness of these modern units.

Reversing Single-Phase Motors

Single-phase motors (split-phase type) are usually used in home workshops and light industry. They have two independent windings, called *start* and *run*.

The fine-wire *start* winding is connected to the circuit during the starting period only. It determines the direction of the motor's rotation and provides some starting torque.

The larger, heavier gauge *run* winding is connected to the line at all times that the motor is operating. It keeps the motor running.

To reverse the split-phase motor, the start winding or run winding leads must be *interchanged*. A mechanical reversing unit, called a *drum switch*, can be used to do this. (See Fig. 20.49)

Figure 20.50 shows the internal connections of this three-position switch, and Figure 20.51 shows a wiring diagram for the split-phase motor. If a drum switch is used for reversing the motor, it may be necessary to take the motor apart to gain access to the start winding leads.

Applications of Electrical Construction

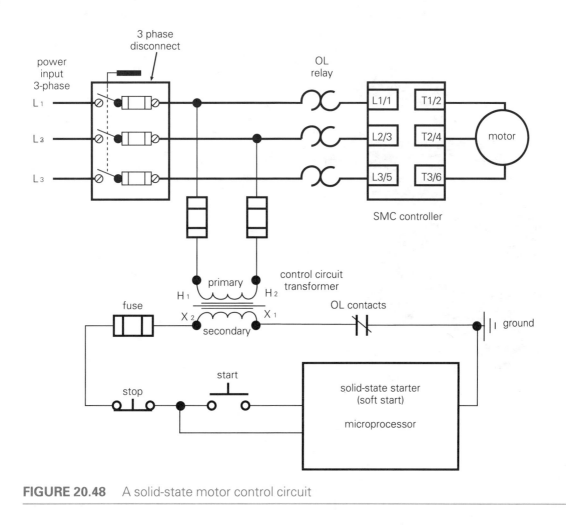

FIGURE 20.48 A solid-state motor control circuit

Direct-Current Motor Control

Direct-current motors are used far less often than alternating-current units, and they need special starting equipment.

As with large AC motors, *high starting currents* are a problem with large DC motors. The starting controllers used have a tapped *resistor* to raise the motor's speed gradually without excessive starting current.

DC motors of the compound type have two field windings, called *series* and *shunt (parallel)* fields.

To provide starting torque, the *shunt* field receives line voltage at all times. The *armature* and *series* field, however, have a *variable resistor* connected *in series*. Figure 20.52 shows a 3 point (three connections to the starter) reduced-voltage starter.

As the starting arm is moved gradually across the face of the starter, *series* field and *armature* circuit, resistance is *decreased* gradually. As a result, these

handle end					
reverse		off		forward	
1○——○2		1○ ○2		1○ ○2	
3○——○4		3○ ○4		3○ ○4	
5○——○6		5○ ○6		5○——○6	

FIGURE 20.50   Internal connections for a 3 position reversing drum switch

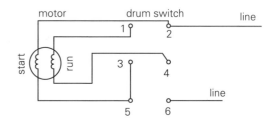

Courtesy Allen-Bradley Canada Limited

FIGURE 20.49   A 3 position—*reversing, off,* and *forward*—drum switch, with the cover removed

FIGURE 20.51   Wiring diagram showing a split-phase motor (with drum switch) reversing

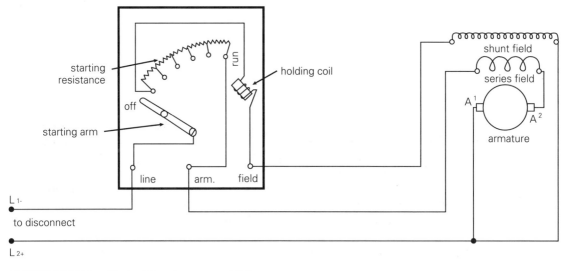

FIGURE 20.52   Typical 3 point manual DC motor starter

windings are exposed to more and more voltage, until full operating speed is reached. A holding coil keeps the starting arm in the *run* position until the operator returns it to the *off* position.

If the *shunt* field circuit is opened or damaged in any way, the holding coil, which is connected in series with the shunt field, *releases* the spring-loaded starting arm and the motor shuts off. This *no-field-release* protection is very important for DC motors. Without it, an extremely dangerous speed will be reached quickly if the shunt field is disconnected. Large DC motors can speed up to the point at which the armature windings are thrown out of their retainers (by *centrifugal force*) and the motor destroyed. The direct-current controller can be used to regulate the starting of a shunt motor, which has no series field, by placing the A2 side of the armature directly on the armature terminal of the controller.

Figure 20.53 shows the wiring for a second type of manual motor starter, the *4 point* (four connections to the starter) motor starter. These controllers provide *no-voltage-protection*, which prevents the motor from restarting by itself if a power failure has caused the holding coil to release the starting arm.

Successful starting of the motor with a manual starter depends entirely on the operator. If the starting arm is moved too quickly, strong inrush currents will damage the controller and/or the motor. An *automatic motor starter* to prevent such human error is available. The operator simply presses a start button, which in turn activates a *solenoid (electromagnet)*. The solenoid is another form of *timer relay*, which controls the speed at which the contacts of the controller close. Closing of the contacts gradually eliminates the starting resistance and the motor accelerates smoothly.

Figure 20.54 shows a wiring diagram for an automatic DC motor starter.

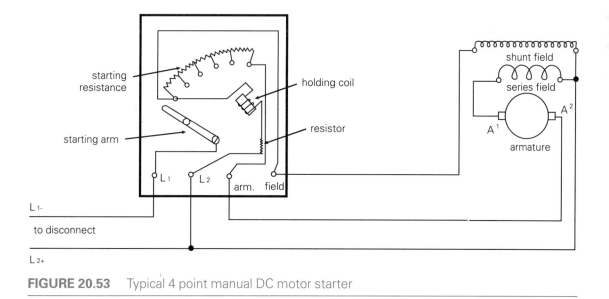

FIGURE 20.53   Typical 4 point manual DC motor starter

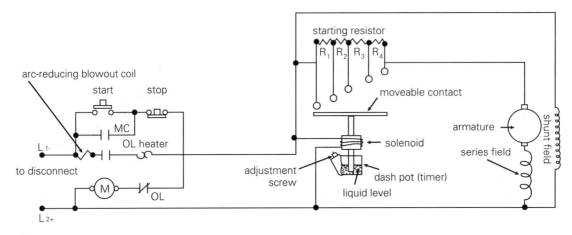

FIGURE 20.54 Wiring diagram showing an automatic DC motor starter

For Review

1. Why do motors need special control switches?
2. How does the voltage generated in a motor affect the input current?
3. Define *locked rotor current*.
4. What are the two basic methods for starting motors? What are the advantages of each method?
5. Why are motor-starting switches rated in kilowatts or horsepower?
6. According to the Canadian Electrical Code, where should control switches be located?
7. How is the fuse size for a motor determined? What effect has the fuse size on the type of disconnect switch installed?
8. Explain how motor conductor sizes are determined.
9. What are the three types of relay used to give thermal overload protection? Explain briefly how each works.

10. What two factors determine the size of the overload relay? Explain in your own words two methods for determining relay size for a motor.
11. When restarting a motor after an overload has tripped the relay, what precaution must be taken?
12. List the three main advantages of magnetic motor control.
13. List the eleven sizes of magnetic motor starters and the motors each can control safely.
14. What is the purpose of the shading coil in a starter?
15. What are the two circuits that make up a magnetic motor starter?
16. Draw a schematic diagram of the control circuit used when four start/stop stations are to be connected to the magnetic starter.
17. What is the purpose of the holding (maintain) contacts? How do they work?

18. List three types of control stations that may be used with magnetic motor starters.
19. Why is more than one overload relay required in a 3 phase starter?
20. How does an overload relay used with a magnetic motor starter differ from a relay used on a manual starter.
21. What is the purpose of the jogging circuit?
22. What is the danger when using a push button jog circuit?
23. Explain how a 3 phase motor is reversed.
24. What is the purpose of the interlock on a reversing controller?
25. What might happen if a large motor is reversed too quickly?
26. What is the difference between a dual-speed motor controller and a standard reversing controller?
27. What is the purpose of the timer relay in a reduced voltage starter?
28. Explain briefly how the auto transformer starter operates.
29. What are the advantages of using the two types of reduced voltage starters?
30. Explain briefly how a single-phase motor can be reversed.
31. What problem do both DC and AC motors have when starting?
32. Define *no-field release*.
33. Why is no-field-release protection important when using a DC motor?
34. Define *no-voltage-protection*.
35. Why is an automatic starter safer than a manual motor starter when controlling DC motors?
36. What special features does a solid-state motor starter provide that other types of starters do not?

21

Fastening Devices

As in most building construction areas, the electrical trade relies heavily on fastening devices to mount boxes and panels, to support conduits and cables, and to secure the many fittings associated with the trade. Fastening devices are available in many forms for the support of electrical equipment on wood, steel, and all types of masonry surfaces. Due to the variety of materials that the fasteners must penetrate, they are designed to accommodate either a wood- or machine-type screw. To make the best use of the fastening systems available, an understanding of wood screw and machine screw features is desirable.

Screw Fasteners

The following parts of a screw fastener are important to the installer and will be discussed: head, driving configuration, neck and/or shoulder, shank and body, thread, and point.

Head Design. The enlarged, preformed shape on one end of the screw is known as the head. Heads are shaped to meet the many requirements of the electrical and other industries. (See Fig. 21.1) The oval and flat-undercut heads are designed to allow a semi-flush or flush fit with the surface of the object being held, (e.g., switch or receptacle cover plates).

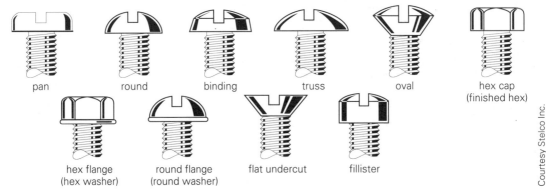

Courtesy Stelco Inc.

FIGURE 21.1 Common head designs for screw fasteners

The other types of head are for use where it is desirable to have the entire head of the screw on the surface being held (e.g., the cover on an octagon or square box). The head of any threaded fastener is the bearing surface which supports the load.

Driving Configuration. To meet the many installation problems that are encountered, a variety of tools and/or drivers are available for the screw fasteners. Figure 21.2 illustrates the various driving configurations produced in the heads of screws.

The *slotted* type is an old standard, designed for the common screwdriver found in most home or shop areas. It is produced in many sizes to meet the requirements of the various screw sizes.

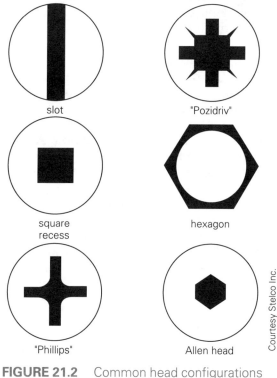

slot   "Pozidriv"

square recess   hexagon

"Phillips"   Allen head

Courtesy Stelco Inc.

FIGURE 21.2 Common head configurations for screw fasteners

The electrician's tool kit would not be complete without a set of *square recess* (commonly known as Robertson head) screwdrivers. This form of driving configuration is patented under the trade name of *Scrulox*. It is produced in six sizes ranging from No. 4 (largest) to No. 00 (smallest).

A colour-coded handle indicates the size of the driver as follows. No. 4 black is for No. 16 screws and larger. No. 3 black, which is somewhat more common, is used for No. 12 and No. 14 gauge screws. No. 2 red drivers install No. 8, No. 9, and No. 10 gauge screws. No. 1 green is for the No. 5, No. 6, and No. 7 gauge screws. No. 0 yellow installs gauge No. 3 and No. 4. The No. 00 orange-handled driver is for the tiny No. 1 and No. 2 gauge screws.

The Scrulox design has two highly desirable features. The square recess for the driver is tapered slightly and causes the screw to "cling" once it has been placed on the tip of the driver. This allows the installer to use one hand on the driver/screw combination and the other hand to support the equipment being installed. Heat treating (hardening) along with proper recess design produces an excellent driving configuration, thus reducing the frequency of "cam-out." *Cam-out* is the action between driver and recess that causes the driver to disengage from the recess in the head. Scrulox units are ideal for use with mechanical or power-operated screwdrivers.

Phillips-type screws and drivers are used extensively for appliance assembly. This type of driving configuration permits the use of air- or electric-powered driving tools, thus speeding up assembly-line procedures. Screwdrivers are available in sizes similar to those of the square recess driver, but are seldom colour coded. The *Pozidriv* unit is designed primarily for use with power-driven tools.

Hexagon-shaped heads are found on most of the larger machine screws. They require a wrench to tighten them securely without damage to the head. Socket wrenches or specially designed nutdrivers (screwdriver handle and shaft with a socket at the tip) are useful in hard-to-get-at areas.

Hexagon recess screws, frequently referred to as Allen head screws, make use of an L-shaped wrench or *key* to tighten them. They are frequently used as "grub screws" to hold motor pulleys to shafts. They are also often used in the assembly of mechanical devices or frameworks. The fastening tools, known as *Allen keys*, are available in sets of different sizes, covering the broad range of screws equipped with this type of driving configuration (See Fig. 22.14).

Torx fasteners provide another fastening alternative. They have been used in the construction of automobiles and trucks for several years. Headlight mounts, interior body trim, and moulding, to name a few specific applications, have been secured by these efficient, slip-reducing fasteners, available in a number of sizes. Manufacturers of electrical and electronic equipment have also used the Torx fastener to assemble their products. Figure 21.3 shows a Torx driver and its tip configuration.

Some manufacturers produce a screw-type fastener that can be installed or removed with either of two driver configurations. This concept of the *dual-drive* fastener has two advantages. Manufacturers of electrical equipment are able to use both power screw and nut drivers in the assembly of their products. Doing so speeds up the production process and helps to keep equipment

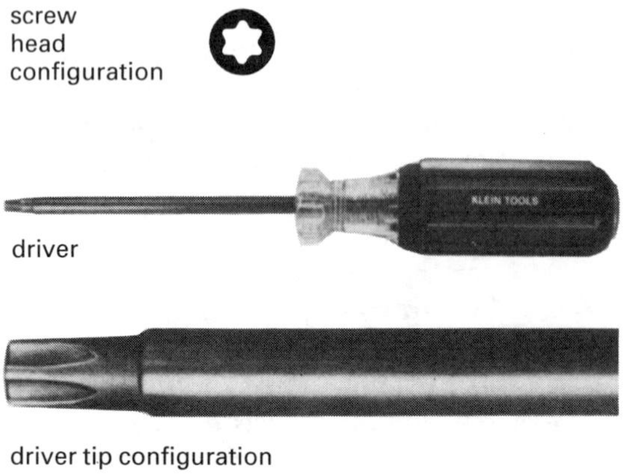

screw head configuration

driver

driver tip configuration

FIGURE 21.3 An efficient Torx screwdriver

Courtesy Klein Tools Inc.

manufacturing costs down. Secondly, installers and service personnel can now choose which of the two drivers they prefer to use when servicing this equipment. Figure 21.4 illustrates three of these dual-drive fasteners.

Neck/Shoulder. The area directly under the head of a screw is frequently given a special design treatment. Figure 21.5 illustrates several designs intended to prevent rotation of the screw while a nut is being tightened on the bolt. A screw that relies on the tightening of a nut to secure it is more likely to have this feature than one with some form of driver configuration in the head. Such a unit is ideal for use in areas where access to both ends of the screw is impossible or (for security reasons) to prevent removal of the screw from the outside (e.g., on the hinges of a supply box).

Shank and Body. The length of screw from under the head to the tip is known as the *shank*. Any unthreaded portion of the shank is called the *body* of the screw. Long screw fasteners will often have

combination hex and slot driver configuration

combination square recess and slot driver configuration

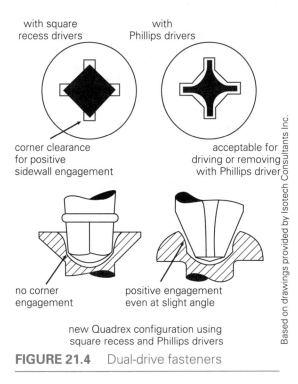

with square recess drivers

with Phillips drivers

corner clearance for positive sidewall engagement

acceptable for driving or removing with Phillips driver

no corner engagement

positive engagement even at slight angle

new Quadrex configuration using square recess and Phillips drivers

FIGURE 21.4 Dual-drive fasteners

Based on drawings provided by Isotech Consultants Inc.

thread only at the tip area, leaving a long body. Shorter screws are frequently threaded over the entire length of the shank, leaving no body at all. When only the necessary amount of thread is formed on the shank, screw production time and costs are reduced and the upper portion of the screw is stronger.

oval shoulder

round shoulder

fin neck

square (carriage) neck

Courtesy Stelco Inc.

FIGURE 21.5 Common neck and shoulder designs

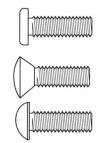

Courtesy Stelco Inc.

FIGURE 21.6 Small diameter (Gauge No.) machine screws

Thread Types.   Many thread types are available, each designed to fasten securely in a particular building material. Some of the more common types are as follows.

Machine screws. These bolt-and-nut units were designed primarily to join metal to a variety of other materials. They are produced in many thicknesses and thread pitches, depending on the amount of support strength and/or compression between surfaces required. The smaller diameter machine screws (having a gauge number) are available with a variety of head shapes and designs as seen in Figure 21.6. Larger

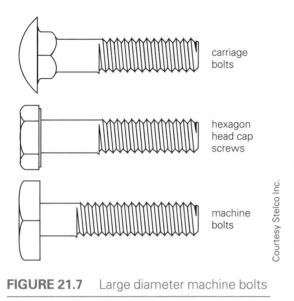

carriage bolts

hexagon head cap screws

machine bolts

Courtesy Stelco Inc.

FIGURE 21.7   Large diameter machine bolts

diameter units, frequently called bolts, are produced in three basic head shapes as seen in Figure 21.7.

Most machine screw and/or bolt sizes are produced with either coarse or fine threads. The coarse-thread bolt installs faster, since the nut advances along the bolt (*thread pitch*) a greater distance for each complete turn. Fine-thread units require more turns of the nut to tighten them, but excellent compression is obtained between the surfaces joined. Table 21.1 compares common coarse- and fine-thread screw sizes with their metric equivalents. Instead of producing metric screws that exactly match their imperial measure (inches) counterparts, manufacturers have established a new set of popular metric sizes. They have thus avoided producing

TABLE 21.1   Comparing Common Coarse- and Fine-Thread Machine Screw Sizes

Figures in this table are based on *Unified* screw thread sizes as established by the American National Standards Institute. The initials, tpi, signify threads per inch.

Unified Screw Threads				Metric Screw Threads
Coarse		**Fine**		
no.	**tpi**	**no.**	**tpi**	**Screw Thread**
2	56	2	64	M2.2 × 0.45
3	48	3	56	M2.5 × 0.45
4	40	4	48	M3 × 0.5
5	40	5	44	
6	32	6	40	M3.5 × 0.6
8	32	8	36	M4 × 0.7
10	24	10	32	M4.5 × 0.75
12	24	12	28	M5 × 0.8
diam.	**tpi**	**diam.**	**tpi**	
1/4	20	1/4	28	M6 × 1
5/16	18	5/16	24	M7 × 1
3/8	16	3/8	24	M8 × 1.25
7/16	14	7/16	20	M10 × 1.5
1/2	13	1/2	20	M12 × 1.75
9/16	12	9/16	18	M14 × 2
5/8	11	5/8	18	M16 × 2

Example: To convert a 10-24 machine screw to metric, select M5 × 0.8.
Caution: Never mismatch metric screws with imperial (inches) nuts or tapped holes.

Applications of Electrical Construction

screws in awkward dimensions (decimals). The new screw sizes have simplified metric dimensions that are reasonably close to the inch units they replace. Table 21.2 offers a comparison between the old and new fastener dimensions.

For proper installation, machine screws require a clearance hole to be drilled or a pre-threaded hole in the materials being joined. The internal threads of the hole or nut being used to secure the assembly must match the external threads of the bolt or machine screw to prevent damage to either part. Machine screw nuts are shown in Figure 21.8.

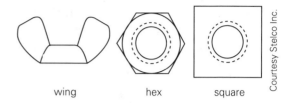

wing   hex   square

Courtesy Stelco Inc.

FIGURE 21.8   Machine screw nuts

TABLE 21.2 Comparison Guide for Popular Metric and Imperial Fastener Screw Sizes			
Diameters The range of diameters listed in metric standards is from M1.6 to M100. Wherever possible designers are asked to use the stock and preferred sizes listed here.		**Lengths** The metric lengths given are the preferred lengths, and these should be used wherever possible. Some of the shorter lengths will not be available in larger diameters. If lengths over 200 mm are required, then increments of 20 mm should be used, and for lengths over 300 mm, increments of 25 mm should be used.	
Metric Diameter	**Imperial Diameter (Gauge/Inches)**	**Metric Length (mm)**	**Imperial Length (Inches approx.)**
M2	#2	10	3/8
M2.5	#3	12	1/2
M3	#5	16	5/8
M3.5	#6	20	3/4
M4	#8	25	1
M5	3/16	30	1 1/4
M6	1/4	35	1 3/8
M8	5/16	40	1 1/2
M10	3/8	45	1 3/4
M12	7/16	50	2
	1/2	55	2 1/4
M14	9/16	60	
M16	5/8	65	2 1/2
M20	3/4	70	2 3/4
M24	1	75	3
M30	1 1/8	80	3 1/4
M36	1 1/4	90	3 1/2
	1 3/8	100	4
M42	1 1/2	110	4 1/2
M48	1 3/4	120	4 3/4
	2	130	5
M56	2 1/4	140	5 1/2
M64	2 1/2	150	6
M72	2 3/4	160	6 1/4
M80	3	170	6 1/2
M90	3 1/2	180	7
M100	4	190	7 1/2
		200	8

Self-tapping screws. Self-tapping screws are made of low-carbon, heat-treated (hardened) steel and are available in five basic types.

(1) *Thread-forming* screws, as seen in Figure 21.9, reshape the material in the pilot hole and do not remove any of the surrounding material. They provide an excellent fit and fast assembly when joining metal to metal.

(2) *Thread-cutting* screws are designed with cutting edges and chip cavities to remove material as they are being installed. They are used in thick, brittle, or granular materials where thread-forming screws are not suitable. (See Fig. 21.10)

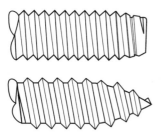

FIGURE 21.9   Thread-forming screws

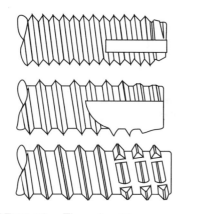

FIGURE 21.10   Thread-cutting screws

(3) *High-performance thread-forming* screws, shown in Figure 21.11, engage approximately 30% more efficiently than ordinary self-tapping screws and 60% more efficiently than machine screws in tapped holes. Figure 21.12 illustrates how the tight-fitting thread-forming screw fits the material much more securely than the tapped hole and machine screw with its thread clearance. Thread stripping and screw breakage are virtually eliminated by the well-designed projections (spaced 120° apart at the tip) which form an accurate, mating thread in the screw-supporting material. These screws are ideal for assembly-line production.

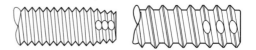

FIGURE 21.11   High-performance thread-forming screws

(4) *High-performance thread-cutting* screws are particularly suited for use with brittle materials. Chips are quickly removed, reducing stripping or cracking hazards. These screws permit quick, easy assembly and can be seen in Figure 21.13.

(5) *Metallic-drive* screws are designed for permanent fastenings. They are pressure driven by punch press or hammer into the holding material and are well suited for attaching nameplates and covers, for example. Figure 21.14 illustrates this type of screw.

Wood screws. Wood screws are used in many different areas of the construction field and are available in four basic thread configurations.

(1) *Single lead* (thread) tapered wood screws have been in use for many

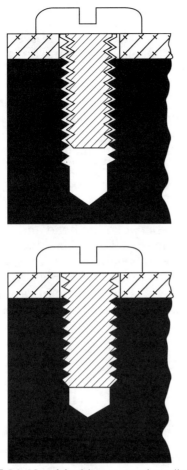

FIGURE 21.12　Machine screw installation (top) and thread-forming screw installation (bottom)

FIGURE 21.13　High-performance thread-cutting screws

FIGURE 21.14　A metallic-drive screw

years. (See Fig. 21.15) The tapered neck and shank design, however, caused frequent splitting problems when installed in wood. The amount of torque required to install the screw increases greatly as it penetrates deeper into the wood. As a direct result of this increase in torque, more stress is placed both on the screwdriver and on the installer.

FIGURE 21.15　A tapered thread wood screw

(2) A more recent development is the *double lead* (thread) and fast-spiral-thread screw. Produced under different trade names *(Kwixin* and *Twin-fast)*, this type of screw has a neck and body thickness that is less than the thread diameter. Splitting is virtually eliminated and driver/installer stress reduced significantly due to this design. In addition, the sides of the double-threaded shank are parallel, providing greater holding power and faster penetration.

Figure 21.16 offers a comparison between single- and double-lead wood screws. The sharp point on the double lead screw reduces the need for pre-

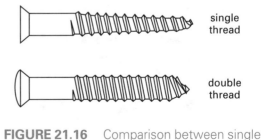

single thread

double thread

FIGURE 21.16　Comparison between single and double thread screws

drilling and permits easier starting of the screw.

(3) Fastening to particle board, plastic, and similar materials often presents a powdering or stripping problem in the material being fastened to. A specially designed screw, with wider thread spacing, sharper and deeper threads, and a well-designed self-starting point, is available. One manufacturer has named this product the *Lo-root screw*. The screw provides approximately 70% greater thread engagement with the material. Figure 21.17 illustrates this thread type and the eight-point driver configuration that makes this screw convenient for use with power screwdrivers.

(4) A wood screw for use in hard and kiln-dried woods is available. It has a special augered flute in the tip that cuts and clears the wood fibres during installation, eliminating the need for pre-drilled holes and speeding up installation considerably. Figure 21.18 illustrates this type of screw and its driver configuration. It is a fast-starting screw, suitable for use with power drivers, and is produced by one manufacturer under the trade name *Candril screw*.

Length of Screws. The length of a wood screw is determined by the distance between the head and tip of the screw. (See Fig. 21.19) Machine and sheet metal (self-tapping) screws are sized in a like manner. Screws of each thread type are produced in various lengths to meet construction needs and are available in both imperial and metric measurement.

Sharp, deep thread with minimum body diameter. Bites, holds.

Courtesy Robertson-Whitehouse Inc.

FIGURE 21.17 Lo-root screw and driver configuration

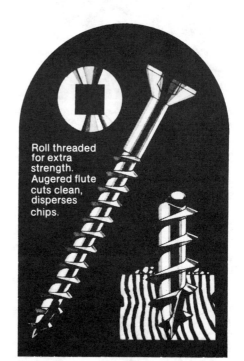

Roll threaded for extra strength. Augered flute cuts clean, disperses chips.

Courtesy Robertson-Whitehouse Inc.

FIGURE 21.18 An augered-flute tip screw and driver configuration

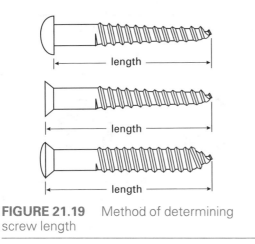

Courtesy Stelco Inc.

FIGURE 21.19   Method of determining screw length

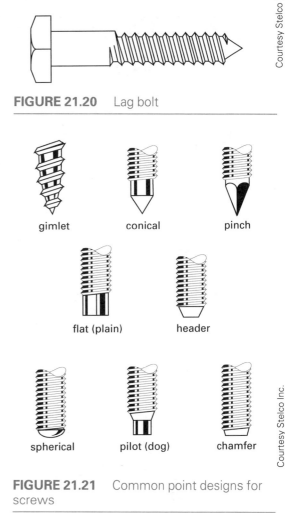

FIGURE 21.20   Lag bolt

Courtesy Stelco Inc.

gimlet   conical   pinch

flat (plain)   header

spherical   pilot (dog)   chamfer

FIGURE 21.21   Common point designs for screws

Screw Diameter.   The thickness, or diameter, of a screw is indicated by a gauge number ranging from 0 to 24. The higher the gauge number, the thicker the screw. Large, heavy-duty screw-thread units are available as illustrated in Figure 21.20. These *lag bolts*, as they are called, find extensive use in expansion-type masonry fasteners and are available in standard machine bolt diameters and lengths.

Point Design.   The tip or point of a screw is designed to perform a variety of operations. Figure 21.21 illustrates a selection of point designs. *Gimlet* and *pinch* points are for the penetration of material. *Conical*, *pilot* (dog), and *flat* (plain) points are to assist in the alignment of the parts being secured. *Spherical, header,* and *chamfer* points ease the insertion and starting of the screw thread.

Screw Construction Materials.   Many types of metal and alloy are used to produce the many types of screws available. Some of the more common metals used are as follows.

Aluminum is used to produce a screw suitable for use with other aluminum products where chemical (galvanic) action between screw and material could take place. If other than aluminum screws were to be used, severe damage could result to the material. Aluminum screws have less strength than steel and brass screws, and care must be taken during installation not to twist them off.

Brass is used a great deal where corrosion from the elements (rain, snow, air, etc.) could be a problem. It is easily

TABLE 21.3 Markings and Mechanical Properties of Hex Head Cap Screws

Strength Grades

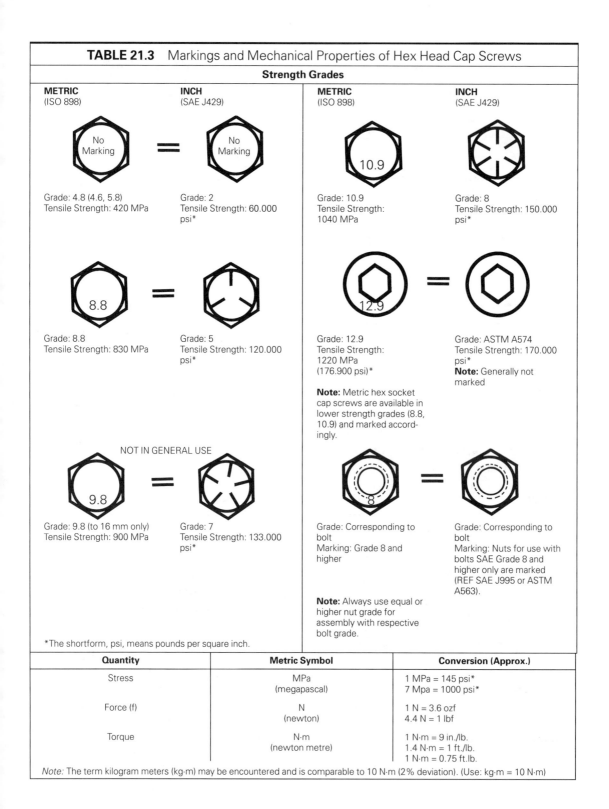

METRIC
(ISO 898)

INCH
(SAE J429)

Grade: 4.8 (4.6, 5.8)
Tensile Strength: 420 MPa

Grade: 2
Tensile Strength: 60.000 psi*

Grade: 8.8
Tensile Strength: 830 MPa

Grade: 5
Tensile Strength: 120.000 psi*

NOT IN GENERAL USE

Grade: 9.8 (to 16 mm only)
Tensile Strength: 900 MPa

Grade: 7
Tensile Strength: 133.000 psi*

METRIC
(ISO 898)

INCH
(SAE J429)

Grade: 10.9
Tensile Strength:
1040 MPa

Grade: 8
Tensile Strength: 150.000 psi*

Grade: 12.9
Tensile Strength:
1220 MPa
(176.900 psi)*

Grade: ASTM A574
Tensile Strength: 170.000 psi*
Note: Generally not marked

Note: Metric hex socket cap screws are available in lower strength grades (8.8, 10.9) and marked accordingly.

Grade: Corresponding to bolt
Marking: Grade 8 and higher

Grade: Corresponding to bolt
Marking: Nuts for use with bolts SAE Grade 8 and higher only are marked (REF SAE J995 or ASTM A563).

Note: Always use equal or higher nut grade for assembly with respective bolt grade.

*The shortform, psi, means pounds per square inch.

Quantity	Metric Symbol	Conversion (Approx.)
Stress	MPa (megapascal)	1 MPa = 145 psi* 7 Mpa = 1000 psi*
Force (f)	N (newton)	1 N = 3.6 ozf 4.4 N = 1 lbf
Torque	N·m (newton metre)	1 N·m = 9 in./lb. 1.4 N·m = 1 ft./lb. 1 N·m = 0.75 ft.lb.

Note: The term kilogram meters (kg·m) may be encountered and is comparable to 10 N·m (2% deviation). (Use: kg·m = 10 N·m)

plated (chrome, etc.) by the manufacturer for further corrosion resistance and improved appearance. Silicone bronze is similar to brass in strength and corrosion resistance.

Monel, a nickel alloy, and stainless steel are used where strength and extra corrosion resistance are required. Steel is by far the most common metal used in the manufacture of screws. It is strong and easily worked, and can be plated for resistance to corrosion.

Zinc or cadmium plating is applied over the steel to protect it from corrosion. Cadmium is the better of the two coatings.

Bolt Strength. Steel hex-head machine bolts, or *cap* screws as they are often called, are produced in a number of different tensile strengths, or SAE (Society of Automotive Engineers) grades. The grades range from 1 through 8 and the user is advised of the potential holding power of each grade of bolt. A simple pattern of ridges on the head forms a grade identification mark, indicating the SAE rating. Table 21.3 illustrates typical markings, along with size, load, strength, and hardness ratings.

Grade 1 and grade 2 bolts can be used for the simple mounting or fastening of equipment where strain and load are not severe. The mid- and high-grade units are used for heavy loads where stress and strain are greater, thus preventing bolt or thread failure. As the grade number increases, the carbon steel of the bolt is subjected to a more intensive heat/quench (hardening) process.

Masonry Fasteners

Due to the extensive use of masonry materials (concrete, brick, etc.) in both old and new buildings, a greater variety of fastening devices has been developed for use by the construction trades. Much of the electrical equipment installed must be fastened to or supported on masonry surfaces, with one or more of the following fastening systems.

Screw Anchors. Screw anchors are available in jute fibre, lead, nylon, and plastic. This common fastening device is inserted into a pre-drilled hole in the masonry surface. A screw is then used to complete the fastening system. Figure 21.22 illustrates lead and jute screw anchors suitable for use with wood screws. Figure 21.23 shows a well-engineered plastic screw anchor which can be used effectively with either a wood or sheet metal screw. Whatever anchor is used should be as long as the threaded shank of the screw (approximately ⅔ of screw length). Screw anchors are also made in a variety

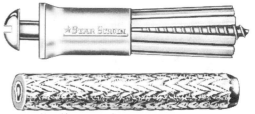

Courtesy Star Fasteners Ltd.

FIGURE 21.22 Screw anchors—lead (top) and jute fibre (bottom)

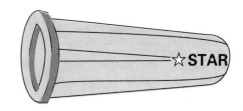

FIGURE 21.23 A well-engineered plastic screw anchor for use in masonry materials

of widths, appropriate to the gauge diameters of screws. For example, the newer plastic screw anchors can accommodate No. 6 to No. 16 gauge screws.

Figure 21.24 shows a handy plastic anchor kit with a number of anchors, matching screws and a drill bit.

Figure 21.25 matches up appropriate anchor and screw sizes.

Figures 21.26 and 21.27 illustrate fibre and lead anchor installations.

For a sturdier installation, a metal alloy expansion *(lag)* shield is available. (See Fig. 21.28) Such a unit has been designed to accept lag bolts and is useful where a larger support device is required for heavy loads.

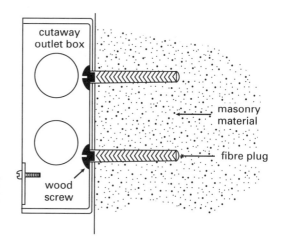

FIGURE 21.26   Installation of a fibre screw anchor

FIGURE 21.24   A versatile plastic anchor kit with anchors, screws and a masonry drill bit

Length of Anchor (inches)	Screw Size	Size of Drill (inches)
3/4	6-8	3/16
7/8	8-10	3/16
1	10-12	1/4
1-3/8	14-16	5/16

FIGURE 21.25   Plastic anchors are produced in a range of sizes so that they can be used with the most commonly used wood or sheet metal screws. (*Note*: Specifications appear in imperial measure only because the equipment is sold in that measuring system.)

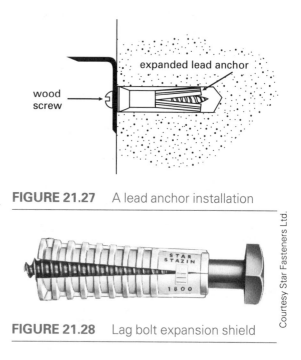

FIGURE 21.27   A lead anchor installation

FIGURE 21.28   Lag bolt expansion shield

Courtesy Star Fasteners Ltd.

Expansion Shields. Expansion shields are primarily designed for use with machine or lag bolts. These zinc alloy units are produced in various lengths and diameters, with internal threads in common machine screw/bolt sizes. A hole of the correct diameter, as indicated on the expansion shield, must be drilled into the masonry for the full depth of the shield. The shield is then inserted and the machine or lag bolt threaded into position. As the bolt is tightened, the threaded section of the shield is drawn towards the mounting surface, expanding the body of the shield against the sides of the pre-drilled hole. Since the shield can exert great pressure on the sides of the hole as it expands, care must be taken to use it in solid or firm masonry material. If not, cracking of the masonry and release of the shield can result. Figure 21.29 illustrates two types of these shields.

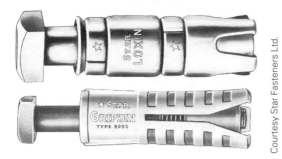

FIGURE 21.29 Zinc alloy expansion shields

Courtesy Star Fasteners Ltd.

Self-Drilling Shield. A hardened steel expansion shield has been produced. It is capable of drilling its own hole. (See Fig. 21.30) The body of the shield has a break-off, tapered end which is held in a power hammering device during installation. Once the hardened teeth at the tip of the shield have drilled the hole, a tapered steel plug is inserted into

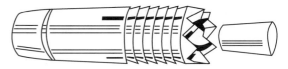

FIGURE 21.30 A hardened steel, self-drilling, expansion shield and plug

the tip. The shield assembly is then placed back in the masonry hole. Further pressure from the power hammer drives the tapered steel plug up into the shield, expanding the cutting tips and widening the bottom of the hole. This type of shield has extraordinary holding power due to the increased hole and shield diameter at the base of the hole. The tapered knock-off neck of the shield can be removed easily once the shield has been fully installed. A bolt of higher SAE grade can be used with these shields, since the internal threads of the shield are also formed of hardened steel. Figures 21.31 and 21.32 illustrate the use of expansion and self-drilling shields.

Lead Sleeve Anchor. A simplified form of expansion shield is illustrated in Figure 21.33. These units are placed in a pre-drilled hole and set into their final

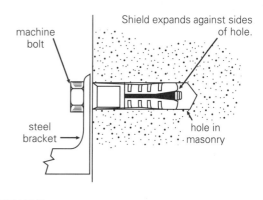

machine bolt

Shield expands against sides of hole.

steel bracket

hole in masonry

FIGURE 21.31 A zinc alloy expansion shield installation

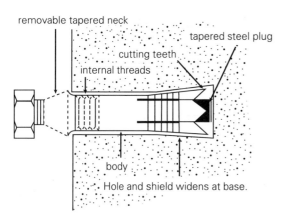

removable tapered neck

tapered steel plug

cutting teeth

internal threads

body

Hole and shield widens at base.

FIGURE 21.32 A hardened steel, self-drilling, expansion shield installation

FIGURE 21.33 Lead sleeve anchors

FIGURE 21.34 Lead sleeve setting tool

expanded position by a setting tool. (See Fig. 21.34) Blows from a hammer provide the necessary force to expand the lead anchor into irregularities in the sides of the hole. Figure 21.35 illustrates this process. An internally threaded, zinc-alloy cone assists in the expansion and accepts standard machine screws/bolts in a variety of sizes. Sleeve anchors do not provide as much holding power as other types of fasteners, but are widely accepted for use in positioning machinery on a concrete floor. The unit illustrated in Figure 21.35 uses a machine bolt instead of a threaded cone. The bolt is left protruding from the floor; the machine is then put into position and a standard

FIGURE 21.35 Installation of a lead sleeve anchor and bolt

nut placed on the threaded end of the bolt to tighten it in place.

Drive-in Anchors. Drive-in anchors are basic expansion shields that are secured in masonry material by driven nails or pins. A hole must be pre-drilled in the mounting surface; an anchor is then placed *through* the device being supported and inserted into the masonry hole. A series of blows from a hammer will drive the nail or pin into the shield, thus expanding the back portion of the shield against the sides of the hole. The shield itself supports the load with this type of fastener, since it passes through the device being supported. One example of this type of unit is illustrated in Figure 21.36. These units are normally produced in smaller diameters (5 mm to 13 mm for light-duty applications.

FIGURE 21.36   Drive-in anchor

Drilling Devices

Hammer-Driven Units.   In the past, holes in masonry surfaces were made with the use of a hand-held, hammer-driven, masonry drill. This manual drill is still used by installers, especially in areas where electric or air power is not readily available. It comes in two parts as shown in Figures 21.37A and B. The drill holder has a rubber grip assembly to lessen the hammering vibrations during drilling and to protect the user's hand. The drill bits are formed of high-quality carbon steel. They are hardened, tempered, and sharpened at the tip to produce a relatively quick hole

in concrete, brick, or stone surfaces. The shank is tapered at the end to fit easily into the drill holder. It is recommended that the drill unit be rotated by the operator during the drilling process to prevent binding and eventual breakage of the drill bit in the hole. Drill bits are produced in sizes up to 15 mm diameter for use with the smaller expansion type shields.

Larger holes can be produced with a four-point drill, available in diameters up to 40 mm. Four-point drills are also made in lengths of 380 mm and 460 mm where a deep hole must be made, or access to the work surface is awkward. Figure 21.38 illustrates this type of drilling device. This hardened-steel drill is for use on all masonry surfaces and should be rotated during use to prevent binding and/or jamming in the hole.

Power-Driven Units.   Due to the popularity and availability of good quality electric drills, a masonry drill bit

FIGURE 21.37A   A manual, rubber grip, masonry drill holder

FIGURE 21.37B   Drill bit for a masonry drill holder

FIGURE 21.38   A four-point masonry drill

Courtesy Star Fasteners Ltd.

has been produced for use with these tools. Air-powered drills can also use this type of bit for work on masonry surfaces. The shape of the bit closely resembles that of a standard steel drill bit. The cutting tip, however, is made from a small piece of carbide brazed (welded) into position. The carbide is an extremely hard metal, capable of cutting clean, fast, and accurate holes in most masonry surfaces. The hardness of the carbide tip makes it necessary to re-sharpen such a bit on a grinding wheel specially compounded for the purpose. Attempts to sharpen the bits on a standard grinding wheel usually result in the wearing down or forming of grooves in the grinding wheel itself. The bits are produced in a variety of sizes, and samples of these are shown in Figure 21.39. The carbide bit is excellent for use with most of the smaller fastening devices. Figure 21.40 illustrates the use of carbide bits.

Larger holes can be produced in masonry surfaces with a "multi" carbide-tipped, spiral flute drill. This unit has a hollow core, allowing clearance for chips and removal of larger pieces of the masonry material. Figure 21.41 illustrates this form of drill bit. Drill bits are available in assorted diameters and lengths to meet the needs of most large-hole installations. Due to the cost of these drill units, care should be taken during the drilling not to damage or remove the carbide cutting tips. Removal of the drill bit from the hole several times during the drilling operation will aid in chip clearance and prolong the life span of the bit.

Hollow-Wall Fasteners

Many fastening operations are performed in residential or other buildings where plaster materials are applied to the wall and ceiling areas. These surfaces are generally too thin and low in density to accommodate other than the small screw-type anchors. There is,

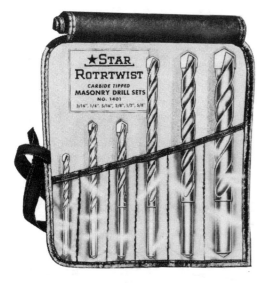

FIGURE 21.39 Carbide-tipped, power-driven masonry drill bits

Courtesy Star Fasteners Ltd.

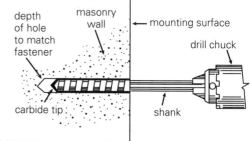

depth of hole to match fastener

masonry wall

mounting surface

drill chuck

carbide tip

shank

FIGURE 21.40 Drilling a hole in masonry surfaces

FIGURE 21.41 A "multi" carbide-tipped masonry drill bit

Courtesy Star Fasteners Ltd.

Applications of Electrical Construction

however a series of mechanical fasteners designed to make use of the space behind these plaster or similar surfaces. Figure 21.42 illustrates a spring-wing, toggle-bolt fastener. The tempered steel wings are installed on the standard thread machine screw after it has been passed through the device being mounted. The wings are then inserted into a pre-drilled hole in the mounting surface. Once clear of the back of the hole, the wings spring to an open position. Tightening of the machine screw draws the wings up against the inside surface of the mounting area, thus securing the mounted device. Figure 21.43 illustrates a toggle-bolt installation. This type of fastener cannot be reused, since it is virtually impossible to remove the open, spring-wing portion from the space behind the wall. These units are available in a number of sizes and types for the mounting of equipment on hollow-wall surfaces.

Where it is necessary to remove and replace equipment on the mounting surface, a type of fastener that remains in place is used. Figure 21.44 illustrates this type of fastener. The fastener unit fits into a pre-drilled hole in the surface, expands, and grips the back of the

mounting surface when the machine screw is tightened. The machine screw can be removed and the complete fastener will remain in position for reuse. These units are produced in several sizes for use where the load to be supported is light. Figure 21.45 illustrates the installation of this fastening device.

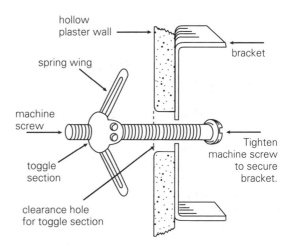

FIGURE 21.43 A spring-wing toggle bolt installation

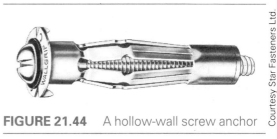

FIGURE 21.44 A hollow-wall screw anchor

FIGURE 21.42 A spring-wing toggle bolt

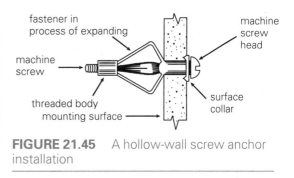

FIGURE 21.45 A hollow-wall screw anchor installation

Powder-Actuated Fasteners

The drilling of holes in masonry or steel surfaces is a tedious and often time-consuming operation. Labour versus fastener costs must be considered when choosing a fastening system. Powder-actuated fastening systems speed up fastening time and greatly reduce the physical strain on the installer. These systems release much energy at the time of firing, making it necessary to fully train competent operators. Manufacturers of powder-actuated tools and accessories are most willing to train and advise installers in the use of their equipment. In accordance with the CSA standard for powder-actuated tools (CSA, CAN 3-Z166-M85), manufacturers of these tools *must* train tool users. After training, the potential tool user is tested through a written examination and actual use of appropriate fasteners in the tool. When she or he has successfully completed the test, an operator's licence is issued. (See Fig. 21.46) Powder-actuated tools are divided into two main categories: low- and high-velocity systems.

Low-Velocity Equipment. Low-velocity tools operate at 90 m/s or less and are suitable for fastening to most masonry or steel surfaces. They are the safest form of powder-actuated tool available, partly because both piston and fastener must be accelerated by the powder charge. If at any time the fastener should pass through the receiving surface, its speed and energy will not be sufficient to endanger human life. These tools have a somewhat higher recoil than the high-velocity tools but are quieter in operation. Due to the low velocity of the fastener, the tool does not have a controlled fire angle. This feature is most necessary, however, on the

NOT TRANSFERABLE

⒜ *Ramset*°
FASTENING SYSTEMS

QUALIFIED OPERATOR OF
EXPLOSIVE ACTUATED TOOLS

This Certifies that . . . .

. . . .

has received training in the operation of the Ramset explosive actuated tool specified herein, has passed the examination, and is deemed competent to operate such a tool.

. . . . **NO.**
Date of Issue

. . . .
Manufacturer's Certified Agent

IN ACCORDANCE WITH
CSA STANDARD Z166

FIGURE 21.46   A licence for operating explosive-actuated tools

high-velocity tools. A levelling device is available as an option with the low-velocity tools and will provide fire angle control if desired.

The low-velocity tools are designed and manufactured in accordance with the CSA standard and must incorporate several safety features: *air-fire safety* to prevent the tool from being discharged into the air like a regular gun, a *safety trigger lock* to prevent accidental firing when the tool is not in use, and *drop-fire prevention* to prevent accidental firing if the tool is dropped on the floor. A well-designed buffer system prevents over-travel of the drive mechanism if and when improper charges are inserted. Figures 21.47 and 21.48 illustrate the mechanisms of low-velocity tools.

Operation.   The low-velocity tool discharges a powder cartridge inside the

Applications of Electrical Construction

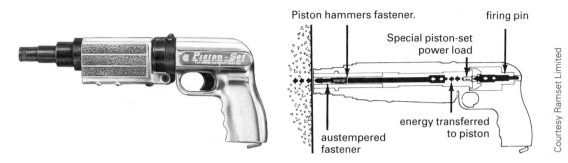

FIGURE 21.47 A low-velocity, .22 calibre, piston-set tool and mechanism used for many years in the fastening industry

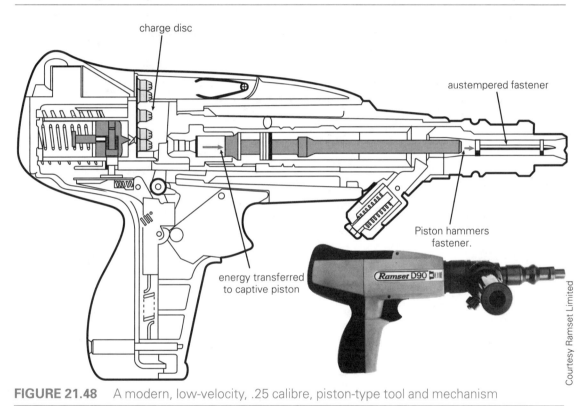

FIGURE 21.48 A modern, low-velocity, .25 calibre, piston-type tool and mechanism

breach of the tool. This energy is then transferred to a special drive piston, which in turn forces the hardened steel fastener into the mounting surface. Since the fastener is nearly in contact with the receiving surface prior to firing, it cannot reach a high speed. The energy from the piston is then used to force the fastener into the mounting surface.

Approximately 5% of the driving power in a low-velocity tool acts on the fasteners. The remaining 95% moves the piston, which is designed in such a way that it cannot leave the tool.

Powder-Charge Cartridges. The explosive used in low-velocity tools comes in the form of a brass and/or nickel rim fire cartridge. These cartridges are factory-loaded to precise power levels and are given both a number and a colour code to indicate these levels. Both cartridge and box or container are equipped with the code markings. Figure 21.49 shows some cartridges.

To speed up fastener operations, some tools have been designed to accept groups of loads assembled onto strips or discs. Figures 21.50 and 21.51 illustrate this method of load handling.

FIGURE 21.50 A multiple cartridge disc loading system

FIGURE 21.51 Alternative method of loading cartridges

FIGURE 21.49 A .22 calibre, rim fire, low-velocity cartridge (left); .22 calibre, rim fire, high-velocity cartridges (centre); a .38 calibre, centre fire, high-velocity cartridge (right)

The Canadian Standards Association has regulated that all manufacturers follow a common coding system for cartridge power levels. Cartridges supplied by a manufacturer should, however, be used only in that manufacturer's tools and not in tools from a different source. The cartridge coding system is outlined in Table 21.4.

Fasteners. Four basic types of fasteners are available. *Drive pins* are used to nail materials directly to

TABLE 21.4 Power Loads (In Accordance with the CSA Z166 Standard)

	Load No.	Case	Load Colour
Low-Velocity Tools	1	Brass	Grey
	2	Brass	Brown
	3	Brass	Green
	4	Brass	Yellow
	5	Brass	Red
	6	Brass	Purple
High-Velocity Tools	7	Nickel	Grey
	8	Nickel	Brown
	9	Nickel	Green
	10	Nickel	Yellow
	11	Nickel	Red
	12	Nickel	Purple

concrete, steel, and horizontal mortar joints. They are permanent fasteners, intended for use with equipment that does not need to be removed and

Applications of Electrical Construction

remounted. They are available in assorted lengths and diameters to fit the tool system.

Threaded studs are for use where washer-and-threaded-nut mounting is desired, e.g., to facilitate removal and remounting of equipment. They are also ideal where adjustment or repositioning of the equipment is sometimes necessary. Care should be taken not to over-tighten the nut; otherwise, the stud will be loosened in the mounting surface material. These versatile fasteners are produced in ¼ in. (6 mm) and ⅜ in. (10 mm) thread diameters, with various thread and shank lengths to suit many applications.

Eye pins find considerable use in suspended ceiling installations, hanging lighting fixtures, and attaching veneers or wire mesh to concrete. They are available in several sizes and lengths for use in both high- and low-velocity tools. Light fixtures and suspended ceilings often require the extra holding strength achieved with the high-velocity tool.

A specially designed *clip-and-fastener* is available for use with the low-velocity tool when fastening conduit or other fixtures.

Fastening Surfaces. Figure 21.52 illustrates a few combinations of materials that can be secured or supported by powder-actuated fasteners.

When a stud or pin is fired into concrete (or similar, non-brittle masonry), a compressive bond or ball is formed around the point of the fastener. This

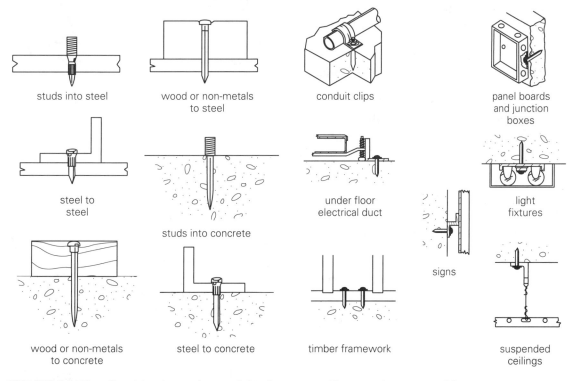

studs into steel

wood or non-metals to steel

conduit clips

panel boards and junction boxes

steel to steel

studs into concrete

under floor electrical duct

light fixtures

wood or non-metals to concrete

steel to concrete

timber framework

signs

suspended ceilings

FIGURE 21.52 Combinations of material to be secured by powder-actuated fasteners

area accounts for much of the holding power, and so the fastener should be driven into the concrete to a depth of approximately ¾ in. (20 mm) when securing to high strength concrete or at least 1¼ in. (32 mm) when fastening into low strength concrete. Maximum holding power is achieved when the strength of the concrete in area X of Figure 21.53A is greater than the bond at the point. If penetration is not sufficient, the fastener may pull out under load, removing a cone-shaped section of the concrete as indicated by the dotted lines in the figure.

Concrete requires about twenty-eight days to reach its full compressive strength. Any fastener set in "green" concrete (less than four days old) will only develop low holding power, but will improve slightly as the concrete ages. It is not recommended that fasteners be driven into any concrete that is less than four days old.

The small chips of concrete (known as *spall*) that break away from around the pin area do not lessen the holding power of the fastener. A spall guard on the fastening tool will help eliminate this, however, and improve the appearance of the finished job. (See Fig. 21.53B)

Any fastener driven into a hollow block must not be allowed to protrude through the block. The compressive bond at the point is lost completely under these conditions and little or no holding power will be the result. Fasteners should not be driven into vertical mortar seams on any brick or block wall. Mortar in these seams tends to be less in quantity and definitely lacking in compressive strength. Only horizontal mortar seams provide sufficient quantity of material and adequate compressive strength for holding power.

Care should be taken to drive several

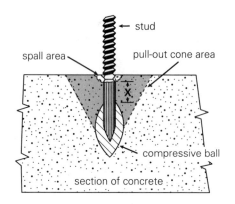

FIGURE 21.53A   A threaded stud in concrete

Courtesy Ramset Limited

FIGURE 21.53B   Threaded studs driven into concrete without a spall guard (left) and with a spall guard (right)

test pins into an out-of-the-way area of the mounting surface to determine proper cartridge strength and fastener size. Use great care to ensure that the fastener does not pass completely through the receiving surface and endanger someone on the other side. Always start with the weakest charge and progress up in strength until the correct combination is determined.

The user of a powder-actuated tool should be familiar with the minimum distance recommended between fasteners,

Applications of Electrical Construction

TABLE 21.5			
Limitation of Use When Fastening to Concrete			
Shank Diameter (mm)	Spacing (mm)	Minimum Distance Edge to Fastener (mm)	Minimum Thickness (mm)
3.8	75	50	65
4.3	75	75	100
5.5	100	150	100
Limitation of Use When Fastening to Steel			
3.8	50	25	5
4.3	50	25	10
5.5	50	25	10

TABLE 21.6		
Holding Power in Average Concrete (3000 psi)		
Shank Diameter (mm)	Penetration (mm)	Pull-out Load with Safety Factor Included (kN)
3.8	25	1.2
4.3	32	1.9
5.5	38	3.4
5.5	50	3.9
Holding Power in Steel		
Shank Diameter (mm)	Steel Thickness (mm)	Pull-out Load with Safety Factor Included (kN)
3.8	5	1.6
3.8	6	2.8
3.8	8	3.4

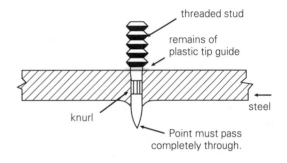

FIGURE 21.54   A threaded stud in steel

the minimum distance from the edge, and the minimum thickness of the concrete to provide safe installation and adequate holding power. Table 21.5 lists the important distances to be observed with various fasteners.

Fasteners of different shank diameters have different holding powers in concrete. Table 21.6 lists the holding power of pins or studs in average concrete.

When driving a fastener into steel, metal is displaced towards the surface of the steel by the penetrating fastener. If the steel is thinner than the depth of penetration required, a mound forms around the fastener's point of entry and around its protruding point. Owing to friction between the fastener and the steel, so much energy is converted to heat that both the surface and the fastener heat up to approximately 900°C. The fastener and the base steel then weld and fuse together. The holding power is a combination of fusion, brazing, keying action, and friction hold. (See Fig. 21.54) Once the proper amount of penetration has been achieved, the holding power is determined by steel thickness and fastener diameter. The thicker the steel and/or the larger the fastener, the more holding power is realized. For optimum holding power, fasteners should completely penetrate the steel.

TABLE 21.7		
Holding Power in Steel Using High-Velocity Fastening Tools		
Shank Diameter (mm)	**Steel Thickness (mm)**	**Pull-out Load with Safety Factor Included (kN)**
3.8	10	2.8
4.3	10	3.1
5.5	12	6.7

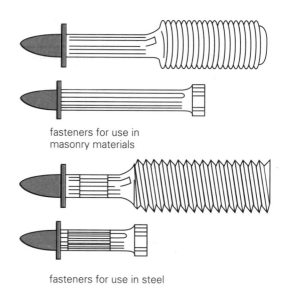

fasteners for use in masonry materials

fasteners for use in steel

FIGURE 21.55 Fasteners for masonry and steel

The holding power figures listed in Table 21.7 represent safe working loads for steel applications.

On occasion, fasteners must be driven into thick steel where complete penetration is impossible. In such a case, a reduced holding power is achieved, but it is still enough for most applications. Loads can be safely supported on these fasteners, because 50% of the fastener's potential holding power can be achieved without full penetration. When in doubt about the fastener's ability to support a load, consult the manufacturer.

A straight *spline-knurl* on the fastener's shank increases the holding power and prevents the threaded-stud type of fastener from turning while the nut is being tightened. Care should be taken not to overtighten the nut, because tremendous pressure can be exerted by the turning effort—even with a short wrench.

Fasteners for steel can be easily recognized by the knurled sections on their shanks and should not be used in masonry surfaces or materials. Figure 21.55 compares concrete and steel fasteners.

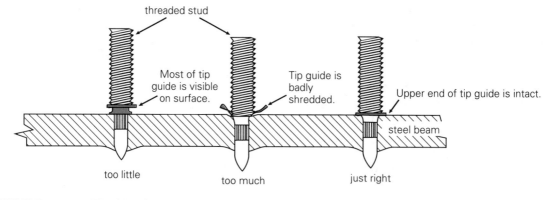

FIGURE 21.56 The plastic tip guide gives proof of proper penetration.

Many fasteners come equipped with coloured plastic tip guides and plastic washers. The plastic tip-and-washer guides the fastener while it is in the barrel of the tool and ensures a straight entry into the receiving surface. Another important function of the plastic tip is to act as a penetration guide. If most of the plastic tip is visible at the point of penetration after firing, penetration is insufficient. When no plastic is visible, too strong a charge has been used. Correct penetration will leave a small ring or flange of plastic where the fastener enters the steel. (See Fig. 21.56) Once again, start with a light charge, and increase the charge strength until proper penetration is achieved. Pins should not be driven closer than 1.3 cm from the edge of steel. Fastening closer to the edge could cause a dangerous ricochet. When steel has been welded, there is frequently an increase in the hardness of the steel surrounding the weld. For this reason, fasteners should not be driven any closer than 5 cm from a welded area.

High-Velocity Equipment. High-velocity tools (above 90 m/s) are suitable for use in denser materials where penetration is more difficult and/or increased holding power is required. The high-velocity tool is similar to a gun, in that the fastener accelerates down the length of the barrel, striking the receiving surface with considerable force. The tools are designed in such a way that they cannot be fired unless pressed firmly against the receiving surface. In this way the tool cannot be misused or accidentally discharged. The firing mechanism of the tool is also designed so that it will not function when the tool is held at an angle of 8° off the perpendicular to the

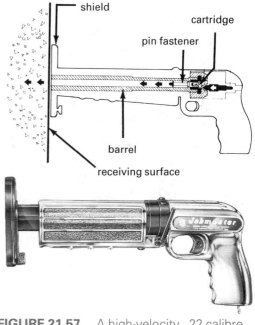

FIGURE 21.57 A high-velocity, .22 calibre tool and mechanism

receiving surface. As a direct result of the fastener's speed and striking energy, care must be taken to use manufacturer-recommended procedures, safety guards and equipment, as well as common sense.

These tools, when used by trained, competent operators, perform many otherwise tedious tasks in a fraction of the time required by other fastening methods. Danger exists only when the tool or its related equipment is used improperly. Figure 21.57 illustrates a high-velocity tool mechanism.

Powder charges used with high-velocity tools often exceed the strength of those used in low-velocity units. The powerful .38 calibre fastener, as shown in Figure 21.58, is restricted to the suspension of heavy loads. A special applications tool is available for fastening devices underwater. (See Fig. 21.59) Figures 21.60 and 21.62 show a variety of powder-actuated fastening tools.

Courtesy Ramset Limited

FIGURE 21.58 A .38 calibre high-velocity tool

Courtesy Ramset Limited

FIGURE 21.59 A high-velocity tool for underwater use

Safety Equipment. As per CSA standards, personal protective equipment should be worn by the tool operator and any helpers or observers in hazardous proximity to the operation of the tool. It should include protective headgear (hard hat) along with safety glasses or goggles. (See Fig. 21.61) Noise levels from the more powerful tools (especially in confined areas) make ear protection well worth considering. Common sense and consideration of others working nearby will greatly add to the safe use of these versatile

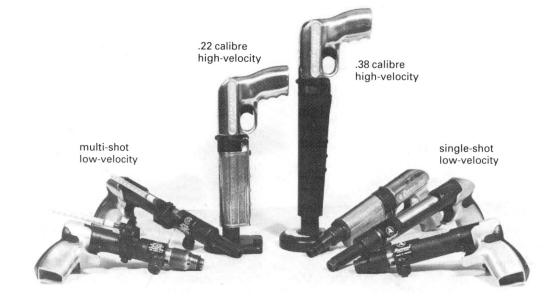

.22 calibre
high-velocity

.38 calibre
high-velocity

multi-shot
low-velocity

single-shot
low-velocity

Courtesy Ramset Limited

FIGURE 21.60 Powder-actuated fastening tools that have been chosen by many installers over the years

Applications of Electrical Construction

FIGURE 21.61 Protective headgear and safety glasses must be worn by an operator and observers in hazardous proximity to any fastening operation.

Courtesy Ramset Limited

fastening systems. Manufacturers of the tools produce a variety of special guards and fastening aids to speed up the safe operation of their equipment. The following is a list of safety recommendations well worth remembering.

Never	use a powder-actuated tool without having its operation and limitations explained to you.
Always	use the proper recommended safety shield on the tool.
Never	attempt to set a fastener through a pre-drilled hole in steel.
Always	try the weakest cartridge on the first shot. Progress to the next heaviest load only when necessary.
Never	fire a fastener into the immediate area where a previous fastener has just failed.

FIGURE 21.62 Modern powder-actuated fastening tools designed for single- and disc-loading operations

Courtesy Ramset Limited

Always	check the receiving surface to make sure it will safely accept the intended fastener.
Never	attempt to fasten into hard steel or near welds.
Always	maintain proper, safe firing distances between fasteners and the edge of mounting surfaces.
Never	attempt to modify any tool or to adapt pieces of one manufacturer's tool for use in another's.
Always	check to see whether the barrel is clear before inserting a fresh cartridge or fastener.
Never	carry fasteners or metal objects in the same pocket as the powder cartridges.
Always	wear protective safety equipment when using these tools.
Never	load a tool until you are ready to fire; never put a loaded tool away: an untrained person could fire it.
Always	keep the tool in good shape. Clean it regularly, and have it checked by a manufacturer's representative periodically.
Never	point the tool, loaded or unloaded, at yourself or another person.
Always	keep the tool against the receiving surface for at least 15 s to 20 s, if it should fail to fire when triggered.

For Review

1. List four pieces of electrical equipment that may require the use of fastening devices to assist in mounting them to masonry or similar surfaces.
2. What are the three main types of screw fasteners?
3. Name five driver configurations for screw fasteners.
4. List the different head types used for screw fasteners.
5. State three advantages of using square-recess (Robertson head) screw fasteners.
6. Where are hexagon (Allen) recess-type machine screws used?
7. What are the advantages of using self-tapping screws?
8. What is the difference between a standard single lead and a Kwixin wood screw?
9. How is the length of a wood screw determined?
10. What are the physical differences between a wood screw and a lag bolt?
11. Name three different materials used for wood-screw construction, and state one use for each.
12. Why is "bolt strength" important when using a machine bolt?
13. Name four different materials used in the construction of masonry fasteners.
14. Outline the procedure for fastening an electrical box to a masonry wall with a masonry fastener and wood screw.
15. Describe four different methods of creating a hole in a masonry wall for an expansion-type fastening device.

16. Name two types of hollow-wall fasteners and give one application of each.
17. State three advantages of a power-actuated fastening system.
18. Why is a licence necessary for operating powder-actuated fastening systems?
19. List the cartridge and charge identifications for low-velocity fastening systems.
20. Explain why a threaded stud is used to mount equipment, rather than a drive pin.
21. How does a threaded stud for steel differ from a masonry stud?
22. List three precautions to be taken when driving fasteners into steel surfaces.
23. What is the main difference between the gun mechanisms of a low-velocity and a high-velocity tool?
24. What safety equipment is needed by the operator of a powder-actuated tool?
25. What is the calibre of the powder cartridges used in high- and low-velocity tools?

Tools of the Electrical Trade

E ach trade or skill area has specialized tools designed for safety and ease of use, and the electrical trade is no exception. Over many years, the tool manufacturing industry has provided a wide variety of both hand and power tools to help both amateur and skilled professional electricians perform their tasks effectively.

Tool Quality

In today's marketplace, tools are available in many price ranges. As a general rule, higher quality tools tend to be in the higher price ranges. Their cost is often the result of extensive research by the manufacturer, higher quality materials used, more elaborate manufacturing processes, and unique or patented design features.

Many professionals have listened to part-time tool users explain why they refuse to buy the more expensive but higher quality tools. All too frequently, inexperienced part-time tool users abuse, overwork, or prematurely wear out the lower quality tools they have chosen. They have practised false economy, because now they have to purchase tools again to complete what they have started.

Consider, too, that low-quality tool material and poor design features often put great stress on the tool and the operator. Even the experienced professional will notice an increase in physical stress, frustration, and work time when using equipment not designed for the job at hand. This fact alone would suggest that quality tools are essential to both the training and continuing development of apprentices and professionals alike.

High-quality, well-designed tools have a well-balanced, easy-to-use "feel" when handled. Tools with comfortable grips designed to fit the human hand with minimal discomfort indicate care and attention to detail on the part of the manufacturer. Improperly designed tools can not only tire out the operator prematurely, but place him or her in physical danger.

Tool users vary in height, weight, and arm and hand size, so the tools they select should take their physical differences into account, as well as the requirements of the task at hand. Having proper tool size will prevent many accidents caused by exposure to a tool that is simply too powerful for a person to control safely.

The Driver Family of Hand Tools

The electrician relies heavily on a variety of tools designed to secure equipment in place and/or install conductors into terminal points or connections. Drivers in a number of different types and sizes are available. Chapter 21 of this text illustrates many of the driver configurations encountered by the installer (See Fig. 21.2)

Standard Slot Screwdriver. This versatile driver is used primarily for the installation of wood and metal screws having a slotted head. Many sizes are produced for the electrician.

An electrician's screwdriver should have a strong plastic handle capable of handling the physical stress associated with normal use. A plastic handle is also desirable because it is safe for operation on or around live equipment. Some manufacturers provide cushion grips of rubber on the handles of their screwdrivers, thereby preventing the buildup of blisters and callouses on the hand, as well.

Large diameter handles are normally indications of high-quality steel in the blade and shank. Steel quality is important because hand torque is transferred directly to the blade. (See Fig. 22.1) Blades should be specially heat treated to the proper hardness and temper to ensure that they have maximum strength and useful life span.

For both the safety of the installer and proper screw tension, the screwdriver's blade should fit the slot of the fastener. (See Fig. 22.3) This prevents damage to the fastener's slot as well as possible injury to the user's hand or surrounding equipment should the tip slip out of the slot.

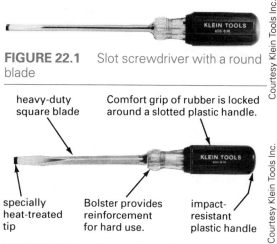

FIGURE 22.1 Slot screwdriver with a round blade

Courtesy Klein Tools Inc.

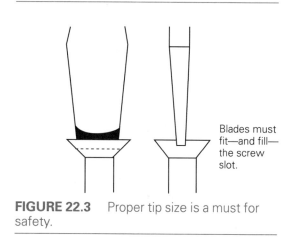

heavy-duty square blade

Comfort grip of rubber is locked around a slotted plastic handle.

specially heat-treated tip

Bolster provides reinforcement for hard use.

impact-resistant plastic handle

Courtesy Klein Tools Inc.

FIGURE 22.2 Heavy-duty slot screwdriver with a square blade

Blades must fit—and fill—the screw slot.

FIGURE 22.3 Proper tip size is a must for safety.

Blades are produced in both round and square shapes to provide the required strength for heavy-duty operations. (See Fig. 22.2)

Slot Screwdriver Maintenance. Even top quality drivers will eventually wear or chip at the corners. This condition frequently causes damage to the screw slot, needless slip-out of the screwdriver from the slot, and personal injury to the user. For the most efficient use of the driver, be sure to keep tip edges straight

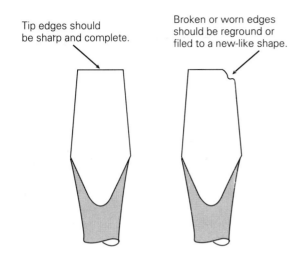

Tip edges should be sharp and complete.

Broken or worn edges should be reground or filed to a new-like shape.

FIGURE 22.4 Proper care of the tip improves driver efficiency and the personal safety of the user.

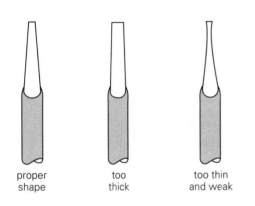

proper shape

too thick

too thin and weak

FIGURE 22.5 Proper tip thickness and shape are most important for safe and efficient driver use.

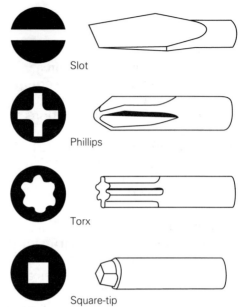

Slot

Phillips

Torx

Square-tip

Courtesy Klein Tools Inc.

FIGURE 22.6 The four most common driver tip configurations

discolouration. The tip will change to a straw colour to brown and to blue.

When reshaping the tip, cold water for the tip's rapid cooling should be kept in a container close to the grinding wheel. The tip can be dipped in the water every few seconds to help control temperature rises while grinding.

In addition to sharp, crisp tip edges, thickness is essential for safe and efficient screwdriver performance. A properly ground tip will have a gradual taper when viewed from the side. (See Fig. 22.5) Excessive tip thickness prevents the tip from fully entering the fastener slot and can result in damage to the fastener and poor torque application. Overgrinding and thinning of the tip will weaken the blade and can lead to breakage of the driver's tip, as well as damage to the fastener and/or the surrounding material.

Screwdrivers are produced in a

and crisp. (See Fig. 22.4) Reshaping the tip with a smooth hand file or a grinding wheel will achieve this.

Great care must be taken if a power grinder is used. Overheating the tip will destroy its hardness/temper ratio, allowing the blade to bend each time high torque is applied to the screwdriver. A clear indication of heat damage is tip

Applications of Electrical Construction

number of configurations (at the tip) and the more common shapes can be seen in Figure 22.6. Procedures for maintaining screwdrivers in good condition apply only to the slot.

Square-tip Screwdriver. Many installers prefer square-tip screwdrivers when installing equipment because of the unique "cling" feature of the square recess screws used with them—the screw will remain on the tip of the driver while being installed. This feature has saved many hands from injury and permitted one-handed driver operation, an advantage because the second hand is free to support the equipment being installed. The square-tip driver is commonly known as a *Robertson* driver.

Robertson drivers (See Fig. 22.7) are colour coded by some manufacturers for ease of recognition. A yellow handle indicates sizing for No. 3 and No. 4 gauge screws. The three most common colours of handles are green, for No. 5, No. 6, and No. 7 gauge screws; red, for No. 8, No. 9, and No. 10; and black, for the larger No. 12 and No. 14 gauge screws.

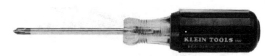

FIGURE 22.8   A Phillips screwdriver, available in tip sizes of Nos. 1, 2, 3 and 4

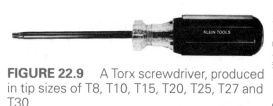

FIGURE 22.9   A Torx screwdriver, produced in tip sizes of T8, T10, T15, T20, T25, T27 and T30

matches the fastener. If not, the screwdriver will *cam-out* of the indentation in the fastener, causing possible equipment damage or personal injury. See Figures 22.8 and 22.9 for these two drivers.

Safety Note: In areas near live conductors, an insulated screwdriver blade is desirable. Using a driver with one will prevent short circuits and dangerous flashes due to accidental contact between live and grounded parts. (See Fig. 22.10)

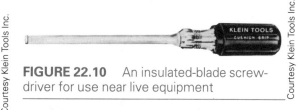

FIGURE 22.7   A square-tip or Robertson screwdriver for use with square recess screws. Robertson driver handles are colour coded to indicate tip sizes.

FIGURE 22.10   An insulated-blade screwdriver for use near live equipment

Phillips and Torx Screwdrivers.   The automotive and electrical appliance industries use two other types of screwdrivers, the Phillips and the Torx, for product assembly, and in recent years the Torx has become particularly popular. Care must be taken with a Phillips driver to ensure that the tip

Multi-Purpose Driver.   This clever type of driver stores several tip configurations in a hollowed-out handle. A magnetic tube or chuck firmly holds the steel tips in place while the driver is being used. The multi-purpose driver replaces many individual drivers and is an ideal tool for service and repair persons who are unable to carry large tool kits to the job. (See Fig. 22.11) The

FIGURE 22.11   A multi-purpose magnetic screwdriver

tips are made of high-quality steel and will withstand considerable torque from the user. However, due to its hollow handle and screw-on cap, this driver will not stand up to as much impact abuse as some of the other drivers.

Hollow-Shaft Nutdriver.   Many electrical installations require the use of a driver capable of turning a hex head bolt or nut. On occasion, the threaded bolt may be lengthy and require a driver capable of fitting over the extra length of the bolt. These unique drivers are available in both metric and imperial sizes, and may be obtained with an insulated shaft for protection near live circuits.

Nutdrivers are usually purchased in sets of approximately seven. Figures 22.12 and 22.13 illustrate this driver type.

Hex Key Driver.   Another form of driver, not usually found with a

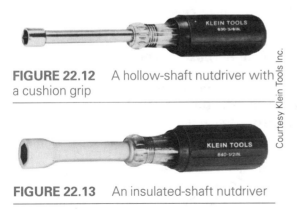

FIGURE 22.12   A hollow-shaft nutdriver with a cushion grip

FIGURE 22.13   An insulated-shaft nutdriver

FIGURE 22.14   A versatile Allen key driver set

FIGURE 22.15   A set of "T" handle, long reach Allen key drivers

screw-driver handle, is the hex key, commonly known among installers as an *Allen key*. Hex key drivers are frequently used to secure pulleys to motor shafts, as well as terminal screws in large main service panels and equipment. A convenient set of nine can be seen in Figure 22.14. These drivers are produced in both metric and imperial sizes.

Large handles, as on the "T" handle units in Figure 22.15, can be most useful when working on large service panels. Having your hands outside of the box and clear of sharp metal edges can be a genuine safety feature. "T" handle units can be purchased individually or in sets, with a large plastic grip handle available for hand comfort and extra torque when tightening.

The Plier Family of Hand Tools

The cutting and shaping of conductors when securing them to electrical equipment terminals has caused many types of pliers to be developed. Each type is unique in its design features and is produced to perform specific operations. Installers usually have a variety of types and sizes in their tool kits, their choice based on personal preference and the type of work they expect to perform.

Several manufacturers produce high-quality pliers, and many installers have a name brand they like more than the others. Pliers, after all, are an extension of the human hand, and installers want tools that will fit their hands and feel right when being used.

Side Cutting Pliers. *Side cutters*, as they are commonly known, are one of the most common types of pliers. They are so named because their cutting edges are located on one side of the plier. Side cutters grip fish tapes, twist conductors to form splices, cut and trim conductors to length, and hold nut and bolt fasteners while being tightened.

They are produced in a variety of sizes, ranging from 6¼ in. to 9¼ in. (15.9 cm to 23.5 cm). The larger models are frequently referred to as "lineman's pliers" because of their popularity with

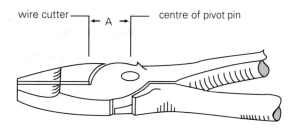

Standard side cutters have the pivot pin located approximately twice as far from the cutter (dimension A) as the high leverage models, providing less cutting pressure.

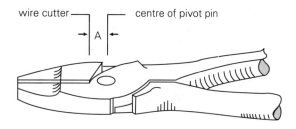

Distance "A" on high leverage models is less, providing twice the cutting pressure.

FIGURE 22.16 Comparison of standard and high leverage side cutters

the installers of pole and line equipment.

High-leverage models are available in certain larger sizes. In these pliers, the pivot point is located closer to the tool's cutting edge. This feature nearly doubles the ease of cutting large diameter conductors in the No. 3 or No. 2 AWG range of sizes. (See Fig. 22.16)

High-quality pliers are forged, proportioned, shaped, hardened and tempered so that the handles have a certain amount of *flex* or bend. This characteristic prevents excessive hand strain to the operator. (See Fig. 22.17) Modern cutting edges are specially tough and hard, with some companies using laser technology in hardening to provide the user with years of reliable service from the tool.

FIGURE 22.17 High leverage side cutting pliers with comfort-grip handles

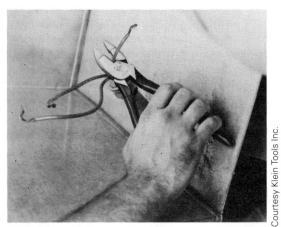

FIGURE 22.18 Choosing pliers based on the size of the user's hand ensures safe and efficient operation.

Pliers, like many other tools, should be purchased to fit the hand size of the owner, as well as to meet job requirements. Figure 22.18 illustrates a pair of high leverage side cutters well matched to the hand of the user.

Plastic comfort-grips are fitted to the handles of many pliers to provide a higher degree of user comfort. These handles can lessen hand strain but often lower the degree of cutting/gripping power the tool is able to provide. The softness and "give" of the plastic used is the cause. An analogy is that if you were to weigh yourself on a bathroom scale placed over a soft carpet, you would get a different reading than if the scale was placed on a firm surface. Plastic grips tend to enlarge the handle size of pliers, which can be a disadvantage to an owner with a smaller hand.

Plastic grips *should not be considered or treated as an insulating feature* when working with live wires. Tiny holes or cuts in the soft plastic render the tool dangerous if only the handle is used to isolate the user from a potential shock hazard on an installation or repair. Plastic grips were not designed by the manufacturer for this purpose, and were therefore not tested and approved for shock protection.

Many electricians have found that cutting two or more live wires at the same time will produce a short-circuit current at the cutting edge. Care must be taken to prevent this because the current will burn a hole right through the cutter, destroying the usefulness of a fine tool.

Diagonal Cutting Pliers. These pliers are intended for one purpose, that is, to cut wire and cable. In the hands of an experienced installer, they can do much more than trim conductors to length. They can cut small bolts, trim the sheathes of both nonmetallic and armoured cables, prepare and trim power tool cords, clean up excess strands from terminal connections, and form the loop in wires for terminal screws. They are particularly suited to work in confined areas, where the larger side cutting pliers cannot be used effectively. Over the years, skilled electricians have invented other uses for these versatile cutters, as well.

Similar to other types of pliers, diagonal cutters are available in various sizes and lengths, ranging from 4¼ in. (11 cm) to 8 in. (20 cm). (See Figs. 22.19, 22.20 and 22.21) Both regular and high-leverage models are available, using the same design features produced on the side cutters. Plastic comfort-grips are provided by some manufacturers to

Applications of Electrical Construction

FIGURE 22.19 General purpose diagonal cutter for use with electronic and small conductor circuits

FIGURE 22.20 Standard leverage general purpose diagonal cutter

FIGURE 22.21 High leverage diagonal cutters provide approximately one-third more cutting pressure than standard models.

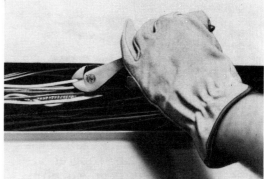

FIGURE 22.22 Use of diagonal cutting pliers in a confined cable trough

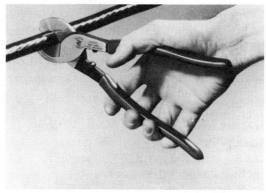

FIGURE 22.23 Hand-operated cable cutter

FIGURE 22.24 Compound action ratchet drive cable cutter

lessen hand stress while using the tool. Figure 22.22 illustrates diagonal cutter operation.

Cable Cutters. Electrical cables are frequently assembled from multiple strands of copper or aluminum wire. (See Chapter Five.) Special cable-cutting tools allow an installer to cut or trim these larger conductors effectively. Figure 22.23 illustrates a hand-operated cutter, capable of trimming soft copper cables up to No. 2/0 AWG. The strands are kept in a neat, close configuration for easy insertion into a terminal block or connector. Figure 22.24 illustrates a compound lever-action cutter which uses a ratchet assembly to provide

additional cutting force on soft copper conductors up to 350 MCM. These lightweight single-handed tools can be carried in a standard tool pouch.

Figure 22.25 shows a two-handed, 32 in. (82 mm) long-handled, shear-type cutter for copper and aluminum cables

FIGURE 22.25 Heavy-duty, two-handed cable cutter with fibreglass handles and rubber grips

up to 1000 MCM. Insulated, fibreglass handles ease the strain of operation, while providing some degree of protection to the operator from live circuits. The cutting tips can be replaced when they are too worn for effective use. This tool was not designed to cut steel cables or bolts. If it is used in this way, damage to the tool will result. The curved cutter design permits an extremely neat cut on a cable, thereby easing the task of inserting the cable into a terminal fitting.

Needle Nose Pliers. *Needle* or *long nose* pliers, as they are sometimes called, give the installer or service technician an extended reach into areas or crevices where the fingers cannot approach safely or effectively. They are used to form loops on conductors for termination, retrieve fallen or misplaced parts, hold small parts effectively for installation, and assist in the tightening of nuts and bolts. Numerous sizes are available, with or without cutters, and can be equipped by manufacturers with comfort-grip plastic handles. Figures 22.26 and 22.27 illustrate this small but versatile tool.

Many variations of these pliers are produced. Some are equipped with wire stripping notches, and some have flat, concave, or curved nose designs for special applications. Standard and long-reach designs have been produced to give installers and technicians a tool to meet any task they face.

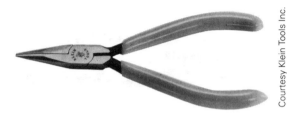

FIGURE 22.26 Small, 4¾ in. (121 mm) non-cutting long-nose pliers with comfort-grip handles

FIGURE 22.27 Heavy-duty long-nose pliers in an 8¼ in. (211 mm) length, with comfort-grip handles and wire-cutting jaws

Pump Pliers. Pump pliers are another highly useful tool in an electrician's kit or pouch. These pliers are to the electrician what a pipe wrench is to the plumber. The jaws are specially designed to grip conduit, fittings, bolts, nuts, etc., in such a way that there is an absolute minimum of slippage when torque is applied. The jaws are positioned and locked into place by a *slanted tongue* which fits into a matching groove in the opposite handle. A wide range of jaw openings are available, and the tongue and groove feature ensures that the setting cannot change under the heaviest pressure of use.

Applications of Electrical Construction

Sizes range from 6½ in. (16.5 cm) to 16 in. (40.6 cm) in length. For maximum effectiveness, the pressure of turning should be placed on the handle with the tongue, not the section with the groove. Jaw angle and tooth design will then permit superior gripping power. Applying pressure to the grooved handle will lessen the tool's grip and add to the hand strain of the user. Plastic comfort-grips are produced by a number of manufacturers. (See Fig. 22.28)

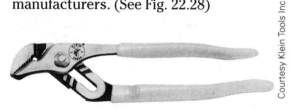

FIGURE 22.28 Heavy-duty pump pliers with comfort-grips and well-designed tongue and groove jaw adjustment

Cutting Tools

Wire and cable preparation requires the removal of a certain amount of insulation prior to termination in a connector or terminal block. Many installers prefer to use a knife for this purpose.

Knives. Figure 22.29 illustrates a versatile pocket or pouch knife with two blades. The sharpened blade can perform traditional cutting operations, while a blunted, second blade, having a slip-proof lock mechanism, can be used as a screwdriver for low-torque installations.

A curved, slitting blade is produced in both pocket and fixed-blade configurations. The curved blade pocketknife shown in Figure 22.30 can be attached to the installer's tool pouch by the handle ring. It serves as a compact part of the tool kit.

FIGURE 22.29 A two-blade electrician's knife

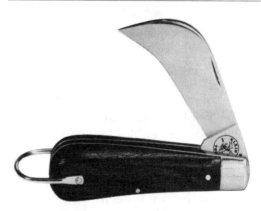

FIGURE 22.30 A curved blade, folding, cable-slitting knife

FIGURE 22.31 A heavy-duty, plastic handled, curved blade knife for line work

A larger, fixed-blade, lineman's knife can be seen in Figure 22.31 and is most suited to the difficult nature of work performed on line-work operations. The extra large handle allows the user to wear protective gloves and still maintain a good grip and control of the knife while using it.

FIGURE 22.32 A cable splicing kit consisting of a knife, scissors and pouch

Cable splicers and other installers of small conductor cables make use of a versatile scissor/knife kit. The scissors are notched on the upper part of the blade to assist in the stripping of small wires, and the two units fit snugly into a pouch designed for the purpose. (See Fig. 22.32)

Hacksaw. Conduit operations of all size require the frequent use of a metal-cutting hacksaw. The heavy-duty model shown in Figure 22.33 has a square-tube frame that holds spare blades for the installer. Blade tension is adjusted by a wing-nut unit at the base

FIGURE 22.33 A heavy-duty, square–tube frame hacksaw

of the handle: tension should be regulated to prevent undue bending of the blade while cutting.

Blades are produced in a wide range of lengths and qualities to suit the saw and task at hand. Both carbon and high-speed steel blades are available from a number of manufacturers.

The necessary number of teeth per inch of blade is determined by the thickness of the material being cut. Thin materials such as thinwall conduit and armoured cable benefit from blades having 32 teeth per inch. Rigid conduit and general cutting operations frequently require the 24 tooth per inch blade.

Triple Tapping Tool. Installers frequently have to clean up threaded holes in boxes and fittings which have become clogged or damaged during the construction process. A most useful tool for this task is pictured in Figure 22.34. Its screwdriver-like design permits one-handed operation. The most commonly used threads, No. 6-32 tpi, No. 8-32 tpi, and No. 10-32 tpi, are machined onto the blade/shaft of the unit.

FIGURE 22.34 A triple tapping tool for repairing damaged No. 6-32 tpi, No. 8-32 tpi and No. 10-32 tpi threads found in most electrical equipment

The blade, made of high-quality tool steel, can be broken easily if not held at a 90° angle to the work, or if subjected to extreme torque during its use. Blades can be replaced if necessary.

Scratch Awl. On many occasions, electricians have to make holes in metal and must first mark where the holes should be. The *awl*, with its hardened, sharpened steel point, is a tool most suitable for this task. When hand held, it is capable of scribing a clean, fine line on a metal surface.

Many installations take place on or over metal, preformed-pan ceilings, metal studs, and other lightweight sheet metal products. Sheet metal screws (see Chapter 21) are frequently used to secure boxes and cables to these surfaces. Starting holes for the screws can be made by striking the awl with a hammer and driving it into the sheet metal. A heavy-duty model is pictured in Figure 22.35.

Plastic-handled models are produced by several manufacturers for lighter duty operations.

Courtesy Klein Tools Inc.

FIGURE 22.35 Heavy-duty scratch awl

Striking Tools

Hammers form a major part of an electrician's tool kit. They are invaluable when installing equipment with nails, fastening cable with staples, or creating holes in masonry walls or surfaces. Several types have been produced to meet specific job requirements.

Claw Hammer. *Claw* hammers are most useful for work on wooden frame structures. They drive and remove nails, staples, and other fasteners with a minimum of strain on the operator. One claw hammer, shown in Figure 22.36, has been designed with the electrician in

mind. It has a non-conducting fibreglass handle—a strong, shock-absorbent shaft that will survive many years of hard work. It also has a perforated rubber handle, a feature on some claw hammers that provides the user with a sure and safe grip. Claw hammers are produced in a variety of head weights, with the most common sizes being 16 oz. (454 g) and 20 oz. (567 g).

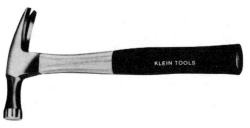

Courtesy Klein Tools Inc.

FIGURE 22.36 A straight claw, fibreglass handled, electrician's hammer with perforated rubber grip

Courtesy Klein Tools Inc.

FIGURE 22.37 A wooden handled ball peen hammer

Ball Peen Hammer. The *ball peen* or "machinist's" hammer is produced in the widest range of head weights, these being 8 oz. (227 g), 12 oz. (340 g), 16 oz. (454 g), 24 oz. (680 g) and 32 oz. (907 g). Ball peen hammers are ideal for heavy-duty striking operations such as cutting with a cold chisel, producing holes in concrete surfaces, or driving assorted fasteners into place with heavy blows. These hammers, one of which is shown in Figure 22.37, are usually equipped with strong wooden handles.

Soft-face Dead Blow Hammer.

Soft-face dead blow hammers are appropriate tools for assembling electric motors or similar equipment where the parts are made of cast iron. They have a somewhat hollowed-out head, containing hundreds of metal balls or buckshot. The soft plastic facing on the head cushions part of the blow while the shot pellets follow up with a second, softer blow—a normal hammer would just "bounce off" the casting after impact. The softer blow eliminates the bounce and prevents the casting from vibrating or "ringing." It is this absence of vibration that reduces metal stress on the casting and prevents breakage.

Figure 22.38 shows a soft-face dead blow hammer. The product is available in 32 oz. (907 g) and 48 oz. (1361 g) head weights.

FIGURE 22.39 Pocket-size, retractable, steel tape measures in both imperial and metric configurations

FIGURE 22.38 A plastic-covered dead blow hammer

Measuring Tools

Several different types of measuring tapes and rulers are available to the installer of electrical equipment. The most commonly used measuring device is the retractable tape measure. It is normally equipped with a *steel* tape, graduated in either metric or imperial units, and is produced in several lengths to meet a variety of construction needs. Replacement tapes are available and can be used to extend the life of the tool

when numbers become worn off or the steel tape damaged. Figure 22.39 illustrates this type of tape measure.

Safety Note: Great care must be taken when working near live equipment with a steel tape measure. The tape conducts electricity, so contact with live electrical parts can cause serious injury to it or the user.

Non-conducting tapes are available in two distinctive styles. Figure 22.40 shows a wooden ruler, with sections that pivot to allow it to be opened to whatever length is required. The ruler is easily "folded" back into its compact form for storage in the tool kit. It is available

FIGURE 22.40 Wooden folding ruler

Applications of Electrical Construction

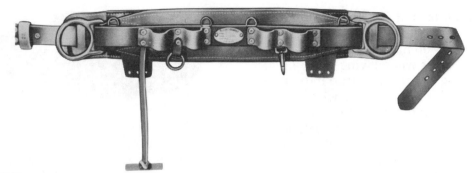

FIGURE 22.43 Body type safety belt designed to support the worker and carry basic installation tools

Courtesy Klein Tools Inc.

This lockout device can be installed through the padlock opening of switch boxes and provide spaces for up to six padlocks. In this manner, each trade can secure the power switch in an *off* position until the equipment can be safely energized or turned on.

Courtesy Klein Tools Inc.

FIGURE 22.46 A lockout device to provide multi-trade, power-off security

Courtesy Klein Tools Inc.

FIGURE 22.44 Adjustable pole strap for use with body type safety belts

Courtesy Klein Tools Inc.

FIGURE 22.45 Heavy-duty body belt with additional back support for extended work periods

Applications of Electrical Construction

in both imperial and metric measure, that is, in both 6 ft. and 2 m lengths.

Frequently an installer must measure longer distances when laying out a job or estimating lengths of material to be used. Another tape measure, with a tape of either non-conducting fabric or of steel, is available in lengths of 50 ft. and 100 ft. and in lengths of 15 m, 25 m, and 30 m. Figure 22.41 illustrates this type of measuring device.

FIGURE 22.41   Non-conducting fabric tape measure and reel

Safety Equipment

Electricians or lineworkers spend a considerable amount of installation time working from ladders, building structures, or outdoor poles. To protect them from falls and decrease their fatigue, safety belts and harnesses have been developed.

Figure 22.42 illustrates a *safety belt*, made from a tough nylon material and equipped with drop-forged tongue buckles and "D" rings. This belt is not intended to support the worker while performing a task, but to act as a fall prevention device should the worker lose his or her footing or grip during the job.

Strong, ⅝ in. (16 mm) nylon rope is recommended for use with the belt. This safety line should be secured properly

FIGURE 22.42   Non-supporting, nylon mesh safety belt, for use with sturdy safety lines

to both the belt and a structure near the worker that could withstand the stress that would be placed on both the belt and the rope if the worker slipped and fell.

Special features are required in a *body belt* meant to support the worker's weight while the job is being performed. Additional stress from body movements, pulling, tugging, twisting, and lifting must be handled by these belts. Figure 22.43 illustrates this belt type.

Figure 22.44 illustrates the adjustable type of *pole strap* frequently used when the installer or lineworker is using the belt to support his or her weight.

Figure 22.45 illustrates a far more supportive belt that can provide additional back support to the worker. This belt is intended to reduce the physical stress and fatigue of the user who may be in it for extended periods of time.

Lockout Device.   Great care must be taken when working on or around live equipment or machines that could start up without notice. To prevent injury from these sources, installers frequently place a padlock on the main power switch supplying current to the equipment being worked on. When a number of trades are required to work on one machine at the same time, each worker needs control over the start-up of the machine or equipment.

Figure 22.46 illustrates a device designed to give a worker such protection

Tool Pouches and Kits

Several manufacturers produce leather tool-carrying pouches that provide the installer with a convenient method of keeping the tools close at hand. Figures 22.47 and 22.48 illustrate two of the many designs available.

When filled with tools, these pouches are very heavy, so only those of the best quality materials and workmanship should be considered when making a purchase; lightweight, low-quality pouches will soon wear out and become both a nuisance and a work hazard. A wide, strong belt matched to the user's waist size should also be chosen for use with a pouch.

Figure 22.49 illustrates some of the tools frequently carried by installers in their pouches. Although pouches are usually purchased separately from tools, on occasion, a manufacturer or tool supplier will offer a "package deal" whereby the tools and a carrying pouch are included in the purchase price.

A high-quality tool kit, consisting of tools and pouch, costs a considerable sum of money, so care should be taken not to shorten its useful life. Exposure to wetness, extremes of heat, and corrosive chemicals can ruin this valuable work aid.

Holster-style pouches are also available. Designed for users of portable, battery-operated drills and screwdriver-type tools, these pouches help overcome

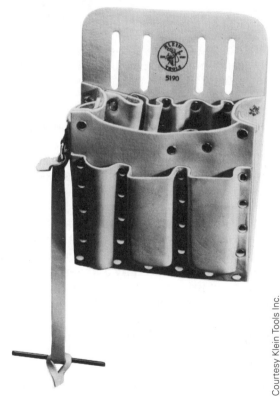

Courtesy Klein Tools Inc.

FIGURE 22.47   A three-pocket tool pouch with screwdriver support loops and tape holder

Courtesy IDI Electric (Canada) Ltd.

FIGURE 22.48   A versatile pouch with a variety of external tool sleeves, tape holder and knife-holding snap clip

Courtesy IDI Electric (Canada) Ltd.

FIGURE 22.49 Some of the many hand tools normally carried by installers or service personnel in their tool pouches

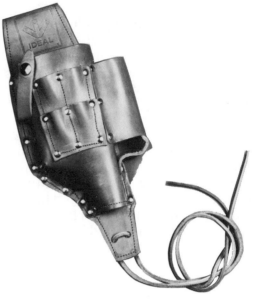

Courtesy IDI Electric (Canada) Ltd.

FIGURE 22.50 A holster-style pouch with compartments for a spare battery and bits, for use with battery-operated drill/driver tools

the problems of looking after tools when not in immediate use, particularly if someone is working from a ladder or other support platform height.

Figure 22.50 illustrates a quality holster-style pouch capable of supporting the tool, extra bits, and spare battery when needed. A leather thong has been provided to secure the pouch to the user's leg and prevent it from bouncing or flopping about when climbing ladders, etc.

Portable Power Tools

For many years, portable, electric power tools have been a major factor in both the construction and service industries. These tools have undergone immense improvement over the years, as job requirements and tool technology have

developed. New products are constantly being designed and made available to the consumer. They offer the potential owner a wide variety of styles, sizes, and features.

Quality and Cost. As with hand tools, the quality of power tools greatly affects their reliable term of use. Lower quality tools (often reflected by a lower purchase price) may appear attractive to the inexperienced user. However they seldom provide the balance, comfort of use, reliability, and power/torque required to complete a demanding task. The most expensive tool is not necessarily the best for the job, but cost does provide some indication of design forethought, materials used, and the guarantee provided by the manufacturer.

Tool life tends to be shortened considerably when an underpowered or poorly designed tool is forced to perform a task beyond its design capabilities. The need to purchase two or more replacement tools can usually be avoided if one properly designed or sized tool is chosen from the beginning. The quality tool will no doubt cost more initially, but when lost time and inconvenience are counted in, the lower cost tool is seldom cheap.

The Power Tool Motor

The heart of every portable electric power tool is the motor, and the most common type is the *universal* motor, which is a *series* motor. This hard-working electrical marvel is very similar in design to a *series direct current* motor, and it possesses most of that motor's operating characteristics. All of its internal electrical parts are connected in a single path, series circuit.

The term *universal* refers to the motor's ability to operate on either AC or DC. Universal motors are used for nearly all portable, motor-driven tools, kitchen appliances, and house maintenance equipment such as polishers and vacuums. Figure 22.51 illustrates the circuit for this motor.

Motor Parts. The two main electromagnetic parts of the motor are the *field coils* and the *armature windings*. When current is allowed to enter the motor circuit, the field coils produce a strong magnetic force. As a direct result of the single path, series circuit in the motor, the same current flows through the armature windings. These windings produce a strong magnetic force of their own. The two magnetic forces react with one another and cause the armature to

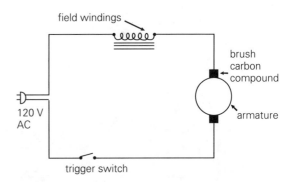

field windings

brush carbon compound

120 V AC

armature

trigger switch

FIGURE 22.51 Circuit diagram for a basic series universal motor

rotate with considerable torque, as its windings are forced away from the field poles. The more current passes through these two sets of coils/windings, the more torque is developed by the motor. Figure 22.52 illustrates the parts of this motor.

Motor Torque and Speed. The series motor will increase in torque development as more current flows through the field and armature conductors. As the armature windings rotate through the field area, a voltage, known as a *counter electromotive force (cemf)*, is induced into these windings.

This cemf opposes the applied voltage and limits input current to a level that the motor's conductors can carry safely, without overheating or burnout. Motor manufacturers carefully design the size of their coils and windings to operate with this cemf in effect.

Reducing the speed of the motor reduces the cemf produced in the armature. This reduction of cemf can be serious enough to permit excessive input current, to overheat the tool, to cause severe arcing at the brush/commutator, and to prematurely burn out the tool's motor. Conductor damage inside the motor is cumulative, and the motor will

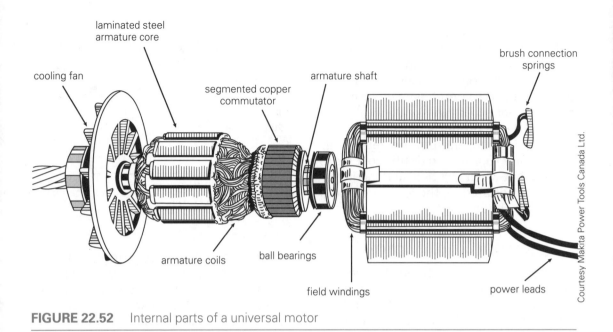

laminated steel armature core

cooling fan

segmented copper commutator

armature shaft

brush connection springs

armature coils

ball bearings

field windings

power leads

Courtesy Makita Power Tools Canada Ltd.

FIGURE 22.52 Internal parts of a universal motor

not "heal" or get better when the tool is in storage. For this reason, be sure to consider the quality and current-carrying capacity of a motor's windings when purchasing a power tool. Most power tools have a current rating stamped on their nameplate to give the prospective user some indication of the motor current recommended by the manufacturer for safe and continued use.

Most series motors have a cooling fan attached to the armature shaft to assist in the cooling of the motor and its conductors. Lower quality or light-duty tools are notorious for having conductors that are too few in number or are undersized. These conductors have a direct influence on the cost and eventual life span of the tool. Tool operators should condition themselves to listen to the motor's normal operating speed and sound and to allow the tool to perform as much as possible within its usual rpm range. Stalling the motor or overloading the tool will quickly reduce its normal operating efficiency and life span: when

speed is reduced, cooling fan efficiency drops off and the input current rises.

Motor RPM. Series motors, by nature, operate at many thousands of revolutions per minute (rpm), and their armature and field currents are one and the same. When the motor is running without a load on it, the relationship between the armature and field magnetic forces allows the armature to rotate at speeds up to 25 000 rpm. Any type of load placed on this motor reduces armature speed and increases the field and armature current. The increase in current causes the motor to develop more torque. This process will continue right up to the point where the motor stalls, and maximum torque is developed. As mentioned earlier in this chapter, care must be taken not to allow the speed reduction and current increase to cause overheating and damage to the windings.

Certain tools used in the wood and

Applications of Electrical Construction

metal industries, such as routers and die grinders, take advantage of this high-speed operation to produce clean and efficient cuts.

Most tools favoured by the electrical industry have gear drive systems within them. A gear system reduces the output rpm of a tool to a more useful level when drilling or sawing, and at the same time, increases the tool's output torque. Tremendous power/torque can be developed by these tools when equipped with the proper speed-reduction gears.

A person about to purchase a power tool should check the tool's nameplate for the output rpm rating. Some lower quality tools lack sufficient gears, thereby increasing output rpm, but reducing the output torque of the tool considerably. They may have a lower purchase price, but the short life span of such a tool and its motor soon eliminate this apparent benefit.

Motor Bearings.   High armature rpm and gear-drive torque and stress are present throughout the life of the tool. Lower quality tools frequently use a *sleeve-type* bearing, made of sintered bronze or similar metal. *Sintering* is a process whereby powdered bronze particles are compressed under great pressure into the final shape of the bearing. The compressed powder is heat treated so that it will stay in the desired bearing shape. The porous bearing is then impregnated with a lubricant such as oil. Such a bearing is considered permanently lubricated, and indeed lubrication cannot be effectively added to the sintered bearing material. Sleeve-type bearings will not stand up to continued hard use and should not be considered for tools designated as *industrial* grade.

Quality power tools *only* use a com-bination of high-speed *needle/roller and ball* bearings. These bearings can be checked and relubricated with a high-quality bearing grease throughout the life span of a tool, depending on the frequency and type of use the tool is subjected to. Needle/roller and ball bearings will raise a tool's initial purchase price but add many years of useful service to the product.

Electric Drills

Few installers lack *electric drills* which are produced in a wide variety of styles by a number of highly respected tool manufacturers. Basic differences in drills seem to be in the materials used for the construction of the outer case or housing.

Metal has been popular for many years, and is still preferred by many users for its ability to maintain proper gear/shaft/bearing alignment within the motor. Proper alignment contributes much to the life span of the tool.

In recent years, however, some manufacturers have switched to a *glass reinforced polycarbonate* material for their housings. This material can withstand years of hard use without cracking, provide the corrosion resistance that some metal-clad tools lack, and make possible the "double insulated" feature appreciated by many service personnel. *Double insulated* means that the tool's internal wiring has a primary insulation while all metal parts of the tool are electrically separated from the user by a secondary system of insulation. This type of tool does not require a grounded plug and does protect the user from electrical shock.

Figure 22.53 illustrates a drill of the double insulated type. Such a drill is produced in models having a ¼ in. (6 mm),

Courtesy Makita Power Tools Canada Ltd.

FIGURE 22.53   A ⅜ in. (10 mm) variable speed drill, light to medium duty, with a reversing switch

⅜ in. (10 mm), or ½ in. (13 mm) capacity chuck.

Many of these drills are equipped with additional features, such as variable speed control through the trigger switch and motor reversing. The motor reversing feature aids in removing some drill bits and allows the tool to be used as a screwdriver when equipped with the proper bit. (See Figs. 22.54 and 22.55 for diagrams of motor reversing circuits.)

Figure 22.56 features a more powerful drill that has been equipped with a heavy-duty motor and chuck. The drill is designed to provide extra torque for tougher jobs.

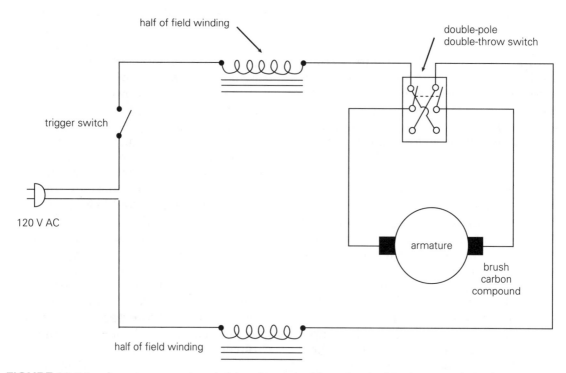

FIGURE 22.54   A motor reversing circuit using a double-pole, double-throw switch to reverse current direction through the armature windings

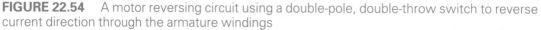

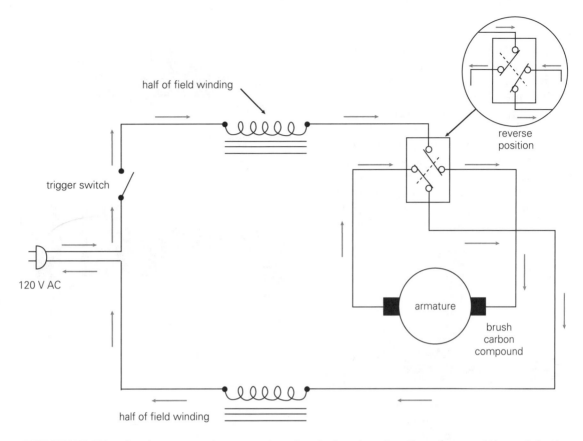

half of field winding

reverse position

trigger switch

armature

brush carbon compound

120 V AC

half of field winding

FIGURE 22.55 An alternate motor reversing circuit showing direction of current through both the field coils and armature windings

FIGURE 22.56 A heavy-duty ³⁄₈ in. (10 mm) drill with speed control and a reversing switch

Safety Note: When using these tools, be sure not to grip them carelessly. Many injuries have resulted when the operator has not taken care. The drill bit may jam, causing the tool to rotate in the opposite direction to the bit and subjecting the user's wrist to a great deal of strain. Just as an installer should use the proper size of hand tool, he or she should also match power tools to both the size of the task at hand and personal size and ability.

Speed control is accomplished by using a trigger switch with a built-in *silicone control rectifier* circuit (SCR) to regulate the actual voltage that is supplied to the field and armature windings. This trigger switch normally provides complete variable speed from zero to full speed and provides the user with a much greater degree of control for many drilling operations.

On many occasions, large holes must be drilled with *high-speed steel* drill bits or *hole saws*, and they require the power of a larger, two-handed power drill. Figure 22.57 shows this type of drill. Care must be taken by the operator to ensure firm footing, balance, and a secure grip to prevent injury during use of the tool.

FIGURE 22.58 A light-duty angle drill for use in confined spaces

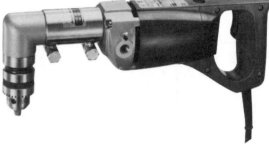

FIGURE 22.59 A ½ in. (13 mm) two speed angle drill, medium duty design, with full swivel adjustment of chuck direction

FIGURE 22.57 A two-handed power drill for heavy construction use

Special Purpose Drills. Electricians are frequently forced to drill holes in awkward or confined spaces. Several specially designed angle drills are available for these tasks and can be seen in Figures 22.58, 22.59 and 22.60.

Drilling holes in brick or cement surfaces can be time consuming and gruelling. Masonry drill bits such as those in Figure 21.39 can be put to excellent use in specially designed power drills that produce both rotary and reciprocating

FIGURE 22.60 A ½ in. (13 mm) two speed angle drill, heavy-duty model

motions. As the drill bit rotates, a built-in hammer feature causes the bit to move in and out of the cutting surface at high speed. This in-and-out action tends to break up small stones or similar hard spots in the drilling material, and quicken the task considerably.

Applications of Electrical Construction

FIGURE 22.61 A light-duty hammer drill with a hole-depth guide

FIGURE 22.62 A variable, two speed hammer drill, medium duty, with depth guide and front handle grip

Figure 22.61 shows a light-duty hammer drill equipped with a hole-depth guide to ensure uniform drilling depth.

A medium-duty hammer drill can be seen in Figure 22.62. This drill has two separate speed ranges which can be controlled from zero to full speed by the trigger switch. A reversing switch allows the tool to run in reverse so that it can remove the drill bit and residue from the finished hole. An adjustable handle is fitted to the front section of the tool to provide a better grip and safer control of the tool when in use.

Both the tool shown in Figure 22.61 and the one shown in Figure 22.62 can be used for straight drilling and hammer drilling.

Heavy-duty tools are available for the installer who must drill large holes into tough masonry surfaces. These tools are also designed to operate for long periods of time without undue wear or overheating. Figure 22.63 illustrates a heavy-duty hammer drill that is designed with a special type of chuck to accept masonry bits only. It cannot be used as a normal type of drill with a conventional chuck.

Owners of hammer drills should periodically use an air hose to blow powdered masonry particles from the inside of the tool. Eye protection should be worn during all drilling and cleaning operations.

FIGURE 22.63 A variable speed, heavy-duty rotary hammer

Battery-Operated Drivers and Drills

When battery-operated tools were introduced a number of years ago, they appeared to be little more than interesting toys. Battery-operated drivers and drills were no exception. Moderately priced driver drills with limited amounts of torque are still available for casual users; however, modern developments in motor and battery technology have enabled manufacturers to produce battery-powered drivers and drills with remarkable amounts of torque for their size.

Because these tools are *cordless*, the installer is freed from the necessity to drag along a cord or to find an electrical outlet close to the work area. He or she is also able to work in outdoor areas away from electrical outlets.

Battery-operated tools have other advantages, as well. They permit safe work in damp areas, usually sources of electrical shock hazards. Many can both drill holes and drive fasteners. Also, they use *fast-charging*, nickel-cadmium batteries, which, when cared for according to the tool manufacturer's recommended procedures and charging schedules, will provide the tool operator with years of faithful service.

Figure 22.64 shows a light-duty driver drill with a 7.2 V battery. Carrying case, battery charger, and spare batteries are options for the tool.

Figure 22.65 illustrates a heavier duty model using a 9.6 V battery. This driver drill has two separate speeds and five torque settings for driving screws. The built-in torque-drive unit operates much like a ratchet. It releases when the tool has driven in the fastener or reached the assigned torque setting. Both driver drills in Figures 22.64 and 22.65 have reverse switches for screw removal.

FIGURE 22.64   A lightweight, 7.2 V, reversible driver drill with a 3/8 in. (10 mm) chuck

FIGURE 22.65   A heavy-duty, 9.6 V, reversible driver drill with a 3/8 in. (10 mm) chuck

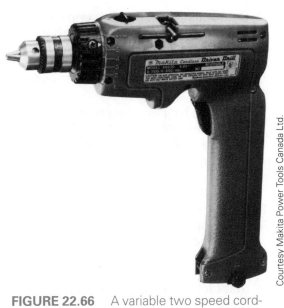

FIGURE 22.66   A variable two speed cordless driver drill with five torque settings, a reverse switch and an electric brake

Figure 22.66 illustrates a cordless, 9.6 V driver drill with two variable speeds, five torque settings and an electric brake for fast stopping when a fastener has been installed. Rpm ranges are 0 rpm to 400 rpm and 0 rpm to 1100 rpm.

Figure 22.67 illustrates a unique angle drill with a 7.2 V battery. Like its larger cousins in the power drill family, it is most suited to drilling and driving fasteners in confined spaces.

FIGURE 22.67   A cordless angle drill with a 7.2 V battery

Reciprocating Saw

When installers need to cut through wood or metal to complete their tasks, they can rely on a well-designed reciprocating saw. This saw can be fitted with metal or wood cutting blades in a variety of lengths and tooth patterns. Figure 22.68 illustrates a two-speed saw unit capable of cutting through wood, fibreboard, plaster, and ferrous metals.

FIGURE 22.68   A variable speed reciprocating saw

Disc Grinder

On certain occasions, installers deal with metal fabrications. Sharp edges, preparation of openings in panels, and removal of slag from welds become tiresome chores when approached with regular hand-operated tools. A high-speed (10 000 rpm) disc grinder, as shown in Figure 22.69, will easily remove metal from surface areas.

FIGURE 22.69   A 4 in. (100 mm) high-speed disc grinder

A disc grinder produces a great many sparks when working on ferrous metals, so care should be taken not to operate the unit near combustible materials. A removable hand grip is provided at the front end of the tool for extra safety and control when required and should be used whenever possible. A steel guard is placed at the rear of the grinding wheel to protect the operator in case the wheel should shatter and fly apart.

A disc grinder is a most useful tool when handled properly, and it can be fitted with a variety of wheel types and wire brushes for working on metal.

Power Tool Maintenance

Manufacturers normally provide a book of operating instructions and maintenance tips with their tools. By promoting proper care of their products, they are helping tool owners to gain many extra hours of tool life. The following is a list of tips and suggestions for tool care and maintenance.

1. Damage to the tool's windings is cumulative when the tool is overworked and allowed to heat up. Avoid loaning a power tool to inexperienced operators. Any damage they cause to the tool will probably not be detected until the tool stops working properly.
2. Do not cover the air circulation holes with your hands or gloves when using the tool. Air circulation helps to keep the motor windings at a safe working temperature.
3. Check the cord on those tools equipped for plug-in operation. On non-double-insulated tools, make sure the plug has all the prongs (including the ground) and is without cuts or other weaknesses that might cause a shock to the operator.

4. Do not use an extension cord that is too long or has a conductor that is too small for the tool. Voltage loss in these cords can cause damage to the tool's motor windings over a period of time.
5. Keep the tool in a carrying case or container when not in use. Doing so will prevent dampness and mechanical damage to the tool. Related parts and accessories can also be kept in the case for easy access.
6. Do not use over-size bits, sanding discs, hole saws, etc., that will overwork the power tool. The tool was designed to operate at a predetermined power/torque level. Ignoring that fact will cause early burnout.
7. Compressed air can be used to clean out grindings, chips, dust, etc., from the tool's interior. Sawdust on the windings will hold in heat and not allow the tool to cool properly. Use caution and safety glasses when performing this clean-out. Compressed air can be dangerous if the pressure is too high.
8. Fresh grease can be added to the gear case. The factory-installed lubricant will be thrown off the gears by centrifugal force after a period of time and may need freshening up. A high-quality *bearing* grease is recommended.
9. As the tool ages, the carbon brushes may wear down so much that they need replacement. These specially shaped inserts are unique to their particular tool, and should be replaced with the identical product.
10. Wipe moisture, chemicals, dust, dirt, etc., off the tool after use and coat chucks and bit holders with a dry lubricant spray. This helps to prevent rust and keep the tool in proper working order.

Choosing a Power Tool

There are some major factors to consider when choosing a tool. The following is a list of recommendations that can help you to make the right choice.

1. Decide what type of job or projects the tool will be used for. Doing so will allow you to choose a tool in the power/torque range needed to perform your task comfortably and without undue strain on the tool.
2. Try to match the tool to the size of your hand, arm strength, etc. A large high-torque tool in the hands of a person unable to control its weight and power may result in a painful accident.
3. Look at several name brands and compare their features, power, and price before making your purchase. Some manufacturers copy others in appearance and design, but leave out key features which make a considerable difference to the ease of use on the job.
4. When comparing tools of various makes and sizes, the nameplate current can be a rule-of-thumb guideline. Generally speaking, the higher the current rating, the more powerful the tool will be.
5. Check the size of the chuck on drills, tool bit or blade holders, etc., as a guideline to the quality of the tool. Manufacturers seldom apply heavy-duty, quality parts to a cheaper tool. As a rule, you get what you pay for.
6. Remember that ball and roller bearings are far superior to sleeve type bearings. They last much, much longer.
7. As a general rule, avoid "package deals" where many, low cost accessories are included with the tool. It is often better to spend your money on the tool alone and to add parts and accessory units when and if you require them. Some name brand manufacturers do, however, provide special deals that are well worth looking for. Top quality parts needed for the operation of the tool may be included.
8. When choosing the tool, inquire about the warranty or guarantee that is provided in writing by the manufacturer. Find out at the same time where repairs can be made: not all tool suppliers repair tools on the selling premises.
9. Look at the length and nature of cord and the type of plug on power tools. Rubber cords tend to be fairly flexible in cold weather; in contrast, plastic cords can be a bit stiff and awkward to use outdoors in winter months. Three prong plugs should be on all metal tools that are not double insulated and CSA approved for the intended purpose.
10. Look carefully at battery-operated tools for battery voltage, cost of spare batteries and chargers, and time required to recharge the battery. Some tools require an overnight charge, something that can be most inconvenient on an important job site project. Tools shown in this chapter are designed to charge in one hour, and can be partially recharged in minutes if only a few minutes of work are required to finish the job.

F o r   R e v i e w

1. What two features make top quality tools more desirable to the installer?
2. When choosing a screwdriver for personal use, what features should be considered?
3. a) What process of repair is used to reshape the tip of a worn, standard slot screwdriver?
   b) What care must be taken while repairing the tip of the driver?
4. Name the three most commonly used sizes of square-tip drivers. State the screw sizes they are designed to work with.
5. What special treatment is given to the long-shaft screwdrivers that are used in and around live electrical equipment?
6. Name two major applications or uses of side cutting pliers.
7. What special design feature determines if a pair of pliers is high leverage or not? What is the advantage of high leverage pliers over standard pliers?
8. What are the advantages and disadvantages of comfort-grips on the handles of various types of pliers?
9. State two specific uses of adjustable pump pliers.
10. Why are slanted tongue and groove pump pliers the safest to use?
11. What special design features are desirable in a metal-cutting hacksaw?
12. List three types of hammers used by the electrical trade and give an application of each type.
13. Why must care be taken when using steel measuring tapes near live electrical circuits and equipment?
14. What protection does the lockout device illustrated in Figure 22.46 provide when persons from a number of trades are servicing electrical equipment and their circuits?
15. What special advantages does a properly designed tool pouch have over a metal or wooden tool box?
16. What type of motor is used extensively in modern power tools, and why is that so?
17. What is the advantage of speed control on a power tool and how is speed control achieved?
18. What design features indicate that a power tool is of high quality and suitable for industrial use?
19. Name two precautions you can take to reduce the chance of electric shock when using power tools.
20. What consideration should be given to the user's physical size when selecting a portable power tool?
21. What safety equipment should be worn by the operator of a power tool such as a hammer drill?
22. List three major advantages of cordless power tools.
23. What are the two most commonly used voltages for cordless power tools?
24. Why is a motor-reverse switch desirable on electric and battery-operated drills?
25. Why is it useful for a cordless power drill to have adjustable torque settings?

26. List three electric power tools that are designed to ease the installation of electrical equipment and circuits.
27. What care must be taken with the cords and plugs of power-operated tools?
28. Why must power tools be cleaned and blown free of wood dust and other residue that has accumulated on the inside?
29. What parts of a power tool require lubrication from time to time, and what lubricant is used for the purpose?
30. Which type of bearing is most suited to long life in an industrial-grade power tool? Explain why.

Glossary of Electrical Terms

acceptable   Equipment or installation of equipment is acceptable to the authority enforcing the Canadian Electrical Code.

accessible   Not permanently closed in by the structure or finish of a building; can be removed without disturbing the structure or finish of a building.

alive   Connected to a source of voltage, or charged electrically to have a voltage different from that of the earth.

alternating current (AC)   A current which reverses its direction and magnitude in a periodic manner, rising from zero to a maximum value in one direction, falling to zero, and reversing in the opposite direction to a maximum value before falling again to zero.

ampacity   The current-carrying capacity of conductors expressed in amperes.

amperes (A)   The commonly used unit of electrical current flow in a circuit.

AWG   American Wire Gauge, used to measure solid, non-ferrous (usually copper or aluminum) conductors, providing both gauge number and diameter in thousandths of an inch.

branch circuit   That part of a circuit which extends beyond the final overcurrent device protecting the circuit or system.

CEMA   Canadian Electrical Manufacturers' Association, which regulates sizes, dimensions, and configurations of electrical devices and equipment.

CMA   Circular mil area, which indicates the nominal size of conductors larger than No. 0000 AWG.

Canadian Standards Association (CSA)   The association that approves and tests electrical devices and equipment to be sold and installed in Canada.

cycles   Cycles per second, or the number of pulsations occurring in an alternating current system within one second. Consists of one positive and one negative maximum value in an alternating current.

direct current (DC)   An electric current that flows in one direction only and is reasonably free from pulsations. Obtained in a pulsation-free form from a battery, it can also be produced by a generator or obtained electronically if alternating current is passed through a rectifier.

hertz (Hz)   The measurement of alternating current frequency indicating the number of cycles per second.

identified   *a.* A white or natural grey covering on a conductor.   *b.* A raised ridge on the surface insulation of some wiring cords.   *c.* A silver-coloured terminal screw on wiring devices.

IEC   International Electro-technical Commission. The European equivalent to CEMA and NEMA.

kilowatt (kW) A unit of power measurement to 1000 W.

kilovolt (kV) A unit of electrical pressure equivalent to 1000 V.

line A term usually referring to the input side of a switch, meter, or controlling device for a motor.

load A term usually referring to the output side of a switch, meter, or controlling device for a motor.

NEMA National Electrical Manufacturers' Association. The American organization that controls the dimensions, sizes, and configurations of electrical devices and equipment.

neutral The conductor that divides the secondary of a supply transformer into two equal sections, providing two voltages from three wires in the system. It is normally white or grey in colour.

outlet A point in the circuit at which current can be taken to supply electrical devices or equipment designed to operate on that circuit.

overcurrent device A fuse or circuit breaker designed to open a circuit automatically under predetermined overload or short-circuit conditions.

overload A flow of current in excess of normal, predetermined circuit capacity, which can cause overheating or damage to the circuit.

primary The input side of a transformer into which voltage is placed so that it may be raised or lowered in value.

PVC Polyvinylchloride. A plastic material used in the manufacture of electrical conduit and fittings.

receptacle One or more female contact devices on a common mounting bracket for the connection of plug-in equipment.

resistance The property of a conductor or electrical device that opposes the flow of current through the system. Measured in ohms.

secondary The output section or winding of a transformer from which voltage is taken after it has been raised or lowered in value.

short circuit A circuit fault caused by contact between two opposite sections of an electrical circuit. An abnormally low-resistance path is created, resulting in sudden and possibly dangerous high currents.

Underwriters' Laboratories (UL) The American equivalent to the Canadian Standards Association. The organization tests and approves electrical devices and equipment for use in the United States.

voltage (V) A unit of electrical pressure applied to a circuit.

wattage (W) A unit of electrical energy or power resulting from electrical pressure forcing a current through the circuit.

For Reference

Canadian Electrical Code, Part 1, Canadian Standards Association.

Heating and Cooling Load Calculation, Ontario Electrical League.

Intermediate Electricity, Frank J. Long, General Publishing.

Introductory Electricity, Frank J. Long, General Publishing.

Lucalox High Pressure Sodium Lamps, 8707-3, GE Lighting.

Multi-Vapour SP30 Metal Halide Lamps, 206-81245 (4/88), GE Lighting.

Typical Wiring Diagrams, publication G1-2.0, Allen-Bradley.

Index

Applications of Electrical Construction